SYSTEMS AND COMPUTER SCIENCE

PROCEEDINGS OF A CONFERENCE HELD AT THE

UNIVERSITY OF WESTERN ONTARIO

SEPTEMBER 10–11, 1965

Systems and Computer Science

EDITORS

John F. Hart

Computer Science Department
University of Western Ontario

Satoru Takasu

Research Institute for Mathematical Sciences
University of Kyoto, Kyoto, Japan

Published in Association with The University of Western Ontario by
University of Toronto Press

Preface

John F. Hart and Satoru Takasu

The ten chapters of this book are the revised and somewhat extended versions of papers given at the Conference on Systems and Computer Science, which was held at the University of Western Ontario, London, Ontario, Canada, September 10 and 11, 1965.

The primary purposes of the Conference were the promotion of research and the advancement of the teaching of computer science in Canadian universities. In more specific terms, the objective was to draw attention to certain theoretical aspects of computer and information science, such as automata, linguistics, theorem proving and general systems.

In line with these objectives, the invited lecturers were asked to bring out the nature of the different fields, the current problems, and an opinion concerning the hope for the future. In the interests of research, the speakers and topics were chosen in the expectation that there might be, within the fields themselves and globally in their interaction, a point of departure for new developments.

The situation of computer science in Canada at the time of the Conference was one in which the emphasis of scientific computing was in numerical analysis applied statistics, and operations research, while in data processing, there was extensive experience in industrial applications. It might perhaps be possible to summarize these developments by noting that Canadian development was concerned with the "applied" rather than the "pure" side of computer science. There were, however, extensions into computer science *per se*. Thus, for example, a number of universities had intensive efforts in compiler design, particularly of ALGOL, or ALGOL-like compilers. These projects had the effect of inviting researchers to look into the more abstract aspects of formal languages and liguistics, with the consequence that there was beginning to be a demand for more abstract treatment.

Another element in the Canadian scene, as elsewhere, had to do with the question of what subjects were to be included in a computer science curriculum. Many people were asking what were the subjects, after programming and numerical analysis, which made up the offering of the computer science unit. Here, also, there was a question of determining to what extent the teaching of computer science should be vocational in nature and to what extent it should be purely educational.

In considering this problem of the pure and applied in computer science the editors do not wish to come down in favour of one side or the other but to achieve a kind of balance. We wish to be able to advance the one without disparaging the other or, in other words, to train systems analysts as well as information scientists. This is perhaps an instance of the kind of parallelism with which the computer scientist is very familiar, since he must often arrange things so as to have certain processes occurring simultaneously rather than consecutively.

Nevertheless, we were concerned in this Conference with the pure side rather than the applied and the subjects in that category which we thought should be included in the computer science curriculum. We wish to emphasize that our choice of topics was neither exhaustive nor restrictive, there being certain fields which were omitted for which a real case for inclusion can be made. That our choice was a reasonably good one was evidenced by the many favourable comments we received.

In preparing this work, the editors have kept uppermost in mind that its usefulness would consist principally in pointing the way to advanced study and research. We have tried to promote the needs of the graduate student and the researcher by emphasizing general references as well as research papers. We hope that the book will serve as a stimulus to look further into the field and that it will act, to some extent, as an overall study guide.

The Theory of Finite Automata, which first appeared in 1954, is a rather new branch of science. It has been extensively studied by electrical engineers, logicians, mathematicians, and linguists and has been taught in various university departments. However, it seems to us that this theory is most appropriately considered as a branch of computer science because it affords a theoretical description of a computer, and, in more specific terms, because it allows us to define information processing very elegantly in terms of "internal states."

The concept of a finite automaton (sequential machine) was introduced in the early stages by Moore [11], and Mealy and Huffman [12] as a model of relay networks or simple digital computers. At about the same time, Kleene [10] invented the concept of *regular expression* to study the behaviour of nerve nets. He established the fact that a finite automaton can be considered as a mathematical model of a nerve net and he gave an explicit description in terms of regular expression of the sets of sequences which can be accepted by finite automata.

However, the algorithms developed for this purpose did not provide a method of overall optimization for the design of a computer, since an "internal state" means the content of all memory elements at certain moments of time; consequently, the number of states will be enormous even if the circuit is fairly small. The main reason for theoretical difficulties of this type was that the theory was not equipped with concepts corresponding to subcircuits of a given digital circuit.

Considerations relative to optimization have taken on a different emphasis due to the fact that the recent developments of electronics engineering have diminished the difference of prices of gate elements and memory elements, and the cost of wiring is becoming an important factor.

The lack of concepts corresponding to subcircuits and the change of objective optimization functions give us the following problems.

A. Is it possible to introduce concepts corresponding to the decomposition of a circuit into its subcircuit?
B. Is there a method of synthesis for design of a circuit (finite automaton) starting from a given set of building blocks (not as small as flip-flops) whose functions and wiring, i.e., their structure, are specified?

The answer to those questions has been studied by Hartmanis, Krohn-Rhodes, Brzozowski, Zeiger, and others. This type of study is called the structural theory of finite automata. In this Conference, Hartmanis talked about his results and about the recent study of Zeiger concerning question A, and Brzozowski reported on the recent development of the study corresponding to question B.

On the other hand, the relationship between regular expressions and finite automata has been studied by Kleene, Myhill [5], Nerode [6], Brzozowski [6] and others, and also, from the viewpoint of formal language theory, by Chomsky [3], Bar-Hillel *et al.* [2]. As a result, the regular expression can be considered to be a nice specification language for digital circuits, and there is a method of deriving the minimal state automaton from a given regular expression. However, the main problem of this study lies in the difficulty of manipulating regular expressions. During the Conference, McNaughton reported on the recent development of the study of manipulation and structure of regular expressions.

The study of infinite automata was begun by Turing in 1936, earlier than the study of finite automata [4, 16, 17]. Turing introduced an automaton (Turing machine), which has a finite automaton portion with a read-write head and a tape of infinite length, for the purpose of formalizing the concepts of algorithm and computability.

A Turing machine utilizes a finite portion of the infinite tape during its computation and there is no bound to the length of this finite portion. This fact is sometimes called "potentially infinite memory." Therefore, an infinite automaton means a finite automaton with some kind of potentially infinite memory, or an infinite aggregate of finite automata.

Because of this unlimited availability of work space, Turing machines are considered to be most powerful in computational capability among automata; any computation which can be done by a known (formalized) infinite automaton can be done (can be simulated) by a Turing machine. Furthermore, Turing demonstrated that a universal Turing machine exists which can simulate any computation of any Turing machine (and hence any infinite automaton).

Besides the intensive study of algorithm and Turing machines in mathematical logic, there are three main streams of the recent study of infinite automata.

A. The study of complexity of computation.
B. The study of infinite automata with restricted capability, such as linear bounded automata and push down storage automata.
C. The study of some specific types of infinite automata.

We shall discuss B later together with formal language theory. The topic A is one of the most interesting fields of study, but unfortunately it was not covered by our Conference; we only mention that it was begun by Yamada [18] and generalized by Hartmanis and Stearns [7, 8] and also by Ritchie [14].

At the Conference, two topics related to C were presented, one by Arbib and the other by Elgot *et al.* In his article, Arbib studied the problem of self-reproduction of automata, which was begun by Von Neumann. Von Neumann proposed two models to describe the process of self-reproduction: one was a 29 state cellular machine and the other was the incomplete thermodynamical model. Arbib discussed mainly the self-reproduction in cellular machines from the abstract point of view. It is quite interesting to observe that the study of self-reproduction in cellular machines, together with the indications of Von Neumann's unfinished thermodynamical model, give rise to the expectation that these theories may provide a logical framework for understanding embryological processes. Perhaps the development of the study of cellular machines will give us a basis for the logical analysis of the cancer explosion in a living organ.

The other paper related to division C is by Elgot, Robinson, and Rutledge, and it discusses multiple control computer models. There has long been a need for an appropriate model of a computer, which could explain the relation between "programme" and "computer." Because of the arbitrariness of the machine and the tape-content structure, Turing machine theory does not give us a satisfactory explanation of "programme and computer." The above authors intend to specify on a trial basis the machine structure within a model more explicitly so that these trials will ultimately bridge the problem of programming and the problem of machine design—both very challenging problems.

In mathematics, the interest in formal languages is associated with logic and linguistics. However, the interest of computer scientists in this field arises out of the desirability of having a method of describing higher level languages such as ALGOL. This interest is also a natural follow-up to the study and design of compilers which has occupied many computer scientists.

It is hoped that the study of formal languages will lead eventually to new theories of programming. A good idea of the scope of this field may be obtained from the monograph of Bar-Hillel [2], from Chomsky's article in the *Handbook of Mathematical Psychology* [3], and from chapter XV of Harrison's [6] book. Professor Gorn's paper is devoted to a study of the effect of extending languages by the introduction of explicitly defined new characters into their alphabets. He relates his work to problems in natural linguistics, logic, programming, and automata theory.

Over the years, the algorithmic non-heuristic study of theorem proving has interested many researchers in mathematical logic. However, as a result of the impact of the development of computers, new problems of finding computationally efficient algorithms for proof procedure have arisen. During the Conference, Robinson gave us the history of theorem proving and an exposition of the essential part of his new efficient algorithm as a combinatorial problem. By contrast, Amarel's paper is principally concerned with the heuristic approach to theorem proving. The initiation of this work was no doubt due to Samuel with his contribution to checker playing. Important milestones in this development are the invention of the languages IPL V and LISP. In presenting his review of this subject, Amarel has paid particular attention to problem specification and to considering simplest proofs to theorems of the propositional calculus by the method of natural deduction. The reader is advised to consult the work of Newell, Shaw, and Simon [13].

The subject of systems theory which is the concern of the last two chapters is very difficult to circumscribe because of its wide range of application. This wide extension is due in part to the generality expressed by the word "system." The word may be used in the context of real structures ranging through great differences in degrees of complexity and differences in the idealized form of abstract models. In addition, general systems may refer to an insight proper to the philosophy of science or to mathematical theories which are strongly dependent on the theory of relations. In some countries (Russia, for example), cybernetics refers to both the philosophical and the scientific aspects of general systems theory. Elsewhere cybernetics is reserved for those aspects of systems in which the emphasis is upon control.

The editors, without considering themselves to be adept in these matters, are aware of a need for the development of various aspects of systems theory. We realize that the computer scientist is inevitably drawn to the deeper systems concepts. Without an appreciation of systems theory, he cannot appreciate the effect of introducing into the environment a machine such as a computer, which may become a "system breaking device." Indeed, it is obvious now that many failures in the early application of computers to accounting were due to the lack of adequate systems concepts.

In their papers, Mesarovič and Windeknecht introduced some new directions in the study of abstract systems. Two aspects of this theory which were not dealt with explicitly are those having to do with control and internal states. These topics are treated to some extent by Zadeh and Desoer [19] in connection with engineering systems.

We have found the *Yearbook of the Society for General Systems Research* to be a good source for general background and valuable insights. As a basis for introductory courses in general systems suitable for students in computer science, we recommend Ashby's book, *An Introduction to Cybernetics* [1]. Looking to the future, we anticipate with interest the advancement of the theory in the articles to be published in the recently announced *Mathematical Systems Theory* (Springer-Verlag, 1967).

REFERENCES

[1] Ashby, W. R., *An Introduction to Cybernetics* (New York: John Wiley & Son, 1964).

[2] Bar-Hillel, Y., *Languages and Information* (New York: John Wiley & Son, 1964).

[3] Chomsky, N., *Formal Properties of Grammars: Handbook of Mathematical Psychology*, Vol. 2 (New York: John Wiley & Son, 1963), 323–418.

[4] Davis, M., *Computability & Unsolvability* (New York: McGraw-Hill, 1958).

[5] Ginsburg, S., *An Introduction to Mathematical Machine Theory* (Reading, Mass.: Addison-Wesley, 1962).

[6] Harrison, M. A., *Introduction to Switching and Automata Theory* (New York: McGraw-Hill, 1965).

[7] Hartmanis, R., Lewis, P. M., and Stearns, R. E., "Classifications of Computations by Time and Memory Requirements," *Proc. of IFID Congress* (Washington, D.C.: Spartan Books, 1965).

[8] Hartmanis, J., and Stearns, R. E., "On the Computational Complexity of Algorithms," *Trans. American Mathematical Society*, 117(5) (May, 1965), 285–306.

[9] Hermes, H., *Aufzählbarkeit, Entscheidbarkeit, Berechenbarkeit* (Berlin: Springer-Verlag, 1961).

[10] Kleene, S. C., "Representation of Events in Nerve Nets and Finite Automata," in *Automata Studies* (Princeton, N.J.: Princeton University Press, 1956).

[11] Moore, E. F., "Gedanken Experiments on Sequential Machines," in *Automata Studies* (Princeton, N.J.: Princeton University Press, 1956).

[12] ——— ed., *Sequential Machines: Selected Papers* (Reading, Mass.: Addison-Wesley, 1964).

[13] Newell, A., Shaw, J. C., and Simon, H. A., "Report on a General Problem-Solving Program," in *Information Processing*, Proc. International Conference on Information Processing, UNESCO (Paris: June, 1959), 256–64.

[14] Ritchie, R. W., "Classes of Predictably Computable Functions," *Trans. American Mathematical Society*, 106 (1963), 139–73.

[15] Smullyan, R. M., *Theory of Formal Systems* (Princeton, N.J.: Princeton University Press, 1961).

[16] Trakhtenbrot, B. A., *Algorithms and Automatic Computing Machines* (Boston: D. C. Heath and Company, 1963).

[17] Turing, A. M., "On Computable Numbers, with an Application to the Entscheidungsproblem," *Proc. of the London Mathematical Society*, 2, 42 (1936–7), 230–65.

[18] Yamada, H., "Counting by a Class of Growing Automata" (Philadelphia: Doctoral Thesis, University of Pennsylvania, 1960).

[19] Zadeh, L. A., and Desoer, C. A., *Linear System Theory: The State Space Approach* (New York: McGraw-Hill, 1963).

Acknowledgment

THE EDITORS are pleased to acknowledge financial support for the Conference from the National Research Council of Canada. They would also like to thank the University of Western Ontario for making available a grant to assist with the publication. The Information Services and Publications Department of the University has been particularly helpful in assisting with publishing details.

Among those who made contributions to the book, we would like to acknowledge a special debt to Professor R. B. Banerji of Case Institute for his overall support and criticism. Thanks are also due to Professor R. R. Korfhage of Purdue University for his criticism of Professor Gorn's paper and to Professor G. Ernst of Case Institute for his assistance in connection with Dr. Amarel's contribution.

Finally, we thank the Programme Committee of the Computer Science Association, consisting of Dr. S. D. Baxter, National Research Council, Professor W. Graham, University of Waterloo, Professor L. Robichaud, l'université Laval, and Professor N. Shklov, University of Saskatchewan.

Contents

SYSTEMS AND COMPUTER SCIENCE

On the Structure of Finite Automata

J. Hartmanis*

INTRODUCTION

The purpose of this paper is to give an elementary and unified exposition of some of the recent results about the structure of finite automata. The principal problem is how a finite automaton can be realized by the interconnection of several smaller automata. Or, more generally, how can a finite automaton be realized by several simpler automata, how these simpler automata have to be interconnected, and how are they related to the automaton being realized?

These problems are quite similar to ones pursued in modern algebra where the problem is to replace a given algebraic system by several simpler systems of the same type. A good example is provided by groups and their rich structure theory. Because of this similarity between the basic problems in the structure theories of automata and algebra, automata theory has borrowed heavily from algebra and has utilized directly several techniques and results. On the other hand, the structure theory for automata shows many interesting differences with previously studied algebraic structures which gives this work its own characteristic flavour.

Though in this paper we are concerned entirely with the problems of realizing a given automaton from several simpler automata, there are other very interesting problems in the structure theory of automata which are not of this type. These are the problems related to how the basic memory and logical units can be interconnected to realize an automaton. In these problems we are not just breaking down an automaton into smaller automata, but we are interested further in how the automaton or the component automata can be broken down into more elementary building-blocks from which physical realizations of these devices can be constructed. We want to know how the topology of the interconnections between these atomic building-blocks affect the computational characteristics of the automaton and *vice versa*. An interesting example of work in this area is provided by the recent result of A. Friedman, which shows that every finite automaton can be realized by an automaton with only one binary feedback loop. Problems of this type do not have direct counterparts in algebra but form a very characteristic

*Department of Computer Science, Cornell University.

part of automata theory. Possibly because they do not have algebraic counterparts the research in this area has not progressed as rapidly as in the area of breaking down an automaton into simpler component automata. At the same time, I believe that there is a unifying concept which is helpful in the study and exposition of both types of structure problems. This is the concept of "ignorance" or "information flow" in automata. This concept has been given precise algebraic formulation which preserves much of its intuitive appeal, and it has already found a number of interesting applications in automata theory [1].

ELEMENTARY DECOMPOSITIONS

We now turn to the problem of realizing an automaton from simpler component automata, and we use the concept of "ignorance" to yield a unified approach to these problems. This approach also shows how the generality of our concept of ignorance is related to the generality of our structural results.

A *finite automaton* is a quintuplet

$$A = (S, I, O, \delta, \lambda),$$

where S, I, and O are the finite non-empty sets of states, inputs, and outputs, respectively;

$$\delta : S \times I \to S$$

is the next state function, and

$$\lambda : S \times I \to O$$

is the output function.

For sake of brevity we omit in most of our discussion the output function and concentrate on the state behaviour of the machine.

An automaton A' is a *homomorphic image* of A if and only if there exist three mappings

$$h_1 : S \to S',$$
$$h_2 : I \to I',$$
$$h_3 : O \to O',$$

such that

$$h_1[\delta(s, x)] = \delta'[h_1(s), h_2(x)],$$

and

$$h_3[\lambda(s, x)] = \lambda'[h_1(s), h_2(x)].$$

If the three mappings are one-to-one, then they define an *isomorphism* between A and A'.

Again for the sake of simplicity we omit the output function and assume that

$$I = I' \text{ and } h_2 = \text{identity function.}$$

To gain some insight in the decomposition problem we start with a simple concept of "A realizes A'," and then derive a corresponding ignorance concept and decomposition results. Both these concepts are generalized later in this paper to obtain far more general decomposition results.

DEFINITION 1. An automaton A *realizes* A' if and only if A' is isomorphic to (a sub-automaton of) A.

DEFINITION 2. The *parallel connection* of the automata

$$A_1 = (S_1, I, \delta_1) \text{ and } A_2 = (S_2, I, \delta_2)$$

is the automaton

$$A = (S_1 \times S_2, I, \delta),$$

with the next state function

$$\delta[(s_1, s_2), x] = (\delta_1(s_1, x), \delta_2(s_2, x)).$$

We say that an automaton A has a *parallel decomposition into* A_1 *and* A_2, if and only if A is realized by the parallel connection of A_1 and A_2.

To illustrate the realization and decomposition concepts consider automaton A of Fig. 1 and the parallel connection of automata A_1 and A_2 of Fig. 2. It is easily seen that the two automata are isomorphic and the mapping h_1 is given by

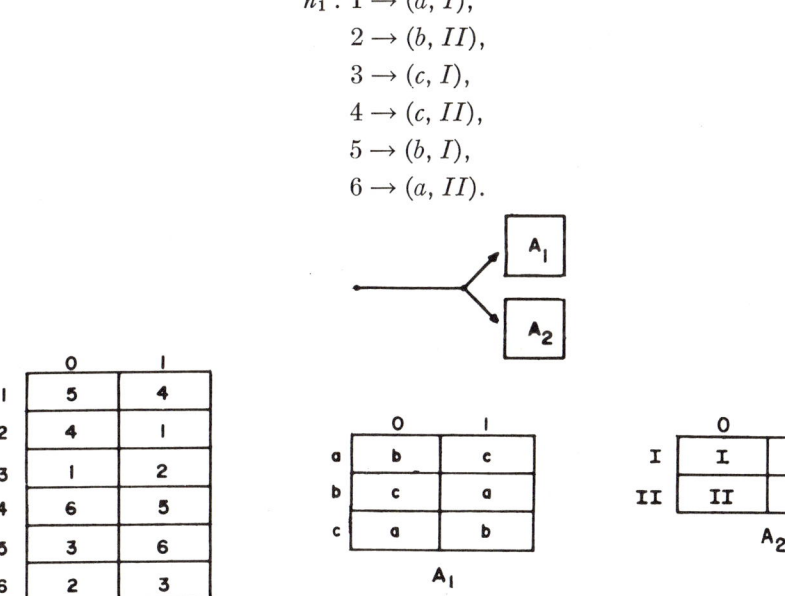

$$\begin{aligned} h_1 : 1 &\to (a, I), \\ 2 &\to (b, II), \\ 3 &\to (c, I), \\ 4 &\to (c, II), \\ 5 &\to (b, I), \\ 6 &\to (a, II). \end{aligned}$$

	0	1
1	5	4
2	4	1
3	1	2
4	6	5
5	3	6
6	2	3

Automaton A

FIGURE 1

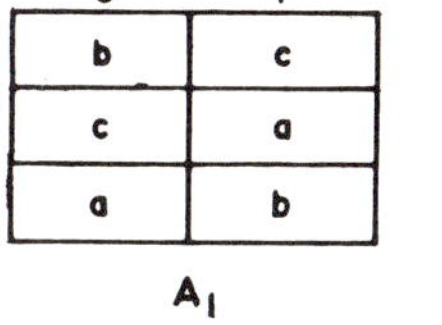

	0	1
a	b	c
b	c	a
c	a	b

A_1

	0	1
I	I	II
II	II	I

A_2

Parallel connection of automaton A_1 and A_2

FIGURE 2

Thus, in the parallel connection of A_1 and A_2, the state pair uniquely determines a state of A. On the other hand, each of the component automata separately can only have partial information about the state of A. For example, the state of A_2 can only determine whether the present state of A is 1, 3, 5 or whether it is one of the states 2, 4, 6. Thus A_2 just keeps track of whether A is in a state designated by an odd or even number.

Therefore, the ignorance which automaton A_2 has about the state of A can be described by the partition

$$\pi_2 = \{\overline{1, 3, 5}; \overline{2, 4, 6}\}.$$

Similarly, the ignorance of A_1 about the state of A is described by the partition

$$\pi_1 = \{\overline{1, 6}; \overline{2, 5}; \overline{3, 4}\}.$$

Furthermore, we see that this ignorance is such that it does not increase when the automaton A operates. Equivalently, we can say that the blocks of π_1 and π_2 are again mapped into blocks of these partitions as the automaton A operates. It will be seen, after we make these concepts precise, that they are sufficient to describe all parallel and serial realizations of finite automata in the sense of Definition 1.

DEFINITION 3. A partition π on the set of states of A has the substitution property (S.P.) if and only if for every input x and block B of $\pi(B \in \pi)$ there is a block B' in π such that

$$\delta(B, x) = \{s | \delta(t, x) = s, t \text{ in } B\} \subseteq B'.$$

It is easily verified that whenever an automaton A is realized by a parallel connection of the automata A_1 and A_2 then, just as in the previous example, the information contained in A_1 and A_2 about the state of A is characterized by two partitions with S.P., π_1 and π_2.

To see this we merely observe that if we have a realization then we have an isomorphism

$$h_1 : S \rightarrow S_1 \times S_2,$$

and we define the blocks of π_1 to be

$$B_i = \{s | h_1(s) = (p_i, q), q \text{ in } S_2\},$$

and, similarly, the blocks of π_2 are given by

$$C_i = \{s | h_1(s) = (p, q_i), p \text{ in } S_1\}.$$

The equations defining a homomorphism can now be used to verify that π_1 and π_2 have the substitution property.

Finally, it can be seen that π_1 and π_2 are such that they do not both identify the same pair of states. We write

$$\pi_1 \cdot \pi_2 = 0.$$

From these observations we can derive necessary and sufficient conditions that an automaton can be realized (in the sense of Definition 1) from two smaller component automata.

THEOREM 1. *An automaton A can be realized by a parallel connection of two smaller automata if and only if there exist two non-trivial partitions π_1 and π_2 on the set of states of A with the substitution property such that*

$$\pi_1 \cdot \pi_2 = 0.$$

If we make the natural definition of a serial connection of two automata A_1 and A_2, then we can again, utilizing a simple argument about ignorance defined in terms of partitions, derive necessary and sufficient conditions that A has a serial decomposition.

THEOREM 2. *An automaton A can be realized by the serial connection of two smaller automata if and only if there exists a non-trivial S.P. partition π on the set of states of A.*

These structure results are very similar to the results which were previously obtained in modern and universal algebra where homomorphisms and congruence relations (S.P. partitions) play a dominant role. Though these concepts are sufficient for most of the structure theory of algebra they are not sufficient for automata theory.

The reason for this, once seen, is quite simple. Algebra uses basically the same realization concept as we have done so far. This is the realization concept formulated in terms of isomorphic subsystems. This works well in algebra where the main interest is concentrated on the elements of the algebra and relations between these elements. In automata theory these roughly correspond to the states of the automaton. It turns out though that for automata the states are only auxiliary concepts which were used to define the computation performed by the automaton. We could use many other automata with more states to perform the same computation. Some of these equivalent automata (which have the same input-output behaviour) may be decomposable where others may not.

For example, automaton B of Fig. 3 has only trivial S.P. partitions and thus does not have a decomposition. On the other hand, the equivalent automaton B' of Fig. 4 has the S.P. partition

$$\pi = \{\overline{1, 2, 3}; \overline{3', 4}\},$$

and thus it has a realization by a serial connection of a two-state and a three-state automaton. Since both of these component automata have fewer states than B, this looks like a reasonable realization of B, but it is not given or admitted by our previous results.

It is actually not very hard to construct examples of automata for which a larger equivalent automaton is easier to build (same number of binary memory elements and simpler combinational logic). Thus, even from a very practical point of view, our concept of realization and the corresponding concept of ignorance are too narrow [1].

	0	1	
1	4	3	0
2	3	2	0
3	3	1	0
4	2	3	1

Automaton B

FIGURE 3

	0	1
1	4	3
2	3'	2
3	3'	1
3'	3	1
4	2	3

Automaton B'

FIGURE 4

SET SYSTEM DECOMPOSITIONS

We now proceed to generalize our concept of "A realizes A'" and to derive the corresponding ignorance concept to overcome the previously encountered difficulties and to get more general decomposition results.

DEFINITION 4. An automaton A *realizes the automaton* A' if and only if A' is a homomorphic image of (a sub-automaton of) A.

This definition is sufficiently general for the purposes of this paper, and it includes almost all the realizations which we intuitively feel should be included. Still, it omits a few cases and, at the expense of making most theorems more cumbersome, it can be generalized (see [1]).

To understand what ignorance definition should go with this definition of realization we look at our previous example. When we reduce automaton B' to B, by identifying states 3 and 3′, then the blocks of the S.P. partition

$$\pi = \{\overline{1, 2, 3}; \overline{3', 4}\}$$

are mapped onto the sets

$$\{1, 2, 3\} \text{ and } \{3, 4\}.$$

We can say that π is mapped onto the "overlapping partition"

$$\phi = \{\overline{1, 2, 3}; \overline{3, 4}\}.$$

This ϕ is such that the blocks of it are again mapped into its blocks by B, and we can build a two-state automaton which keeps track of whether B is in one of the states 1, 2, 3 or one of the states 3, 4. Thus we are led to a new concept of ignorance which we now make precise.

DEFINITION 5. A *set system* $\phi = \{S_i\}$ on S is a collection of subsets of S such that

$$\cup S_i = S,$$

and

$$S_i \subseteq S_j \text{ implies that } i = j.$$

Thus a set system on S is a collection of subsets of S whose union is S (i.e., the S_i cover S) and among which there do not exist proper containments.

DEFINITION 6. A set system ϕ has the *substitution property* with respect to an automaton A if and only if for every input x and B in ϕ there is a B' in ϕ such that

$$\delta(B, x) \subseteq B'.$$

Thus the S.P. set systems for an automaton describe this newly defined ignorance which does not spread as the automaton operates.

It should be observed that if π is an S.P. partition on the automaton A, then the image automaton (whose states are the blocks of π and which computes the block of π which contains the state of A) is uniquely determined by A and π. This, as the next example illustrates, is not the case for set systems with S.P. Consider automaton C of Fig. 5. The set system

$$\phi = \{\overline{1, 2, 3}; \overline{1, 2, 4}; \overline{1, 3, 4}; \overline{2, 3, 4}\} = \{B_4, B_3, B_2, B_1\}$$

has S.P. In the first column of C each B_i of ϕ is mapped onto a unique and different B_i of ϕ. In the second column all B_i of ϕ are mapped into

$$B_1 = \{2, 3, 4\},$$

and in the third column all B_i of ϕ are mapped into

$$B_2 = \{1, 3, 4\} \text{ and } B_1 = \{2, 3, 4\}.$$

Thus, in the image machine, we can choose for the last column of next states the states B_1 or B_2 in any order we wish. In Fig. 6 is shown one such image automaton. We will return to this example and such set systems later.

Armed with these definitions and utilizing the insight gained from the study of decompositions defined by S.P. partitions we can derive our next set of results.

	a	b	c
1	2	2	4
2	3	2	3
3	4	3	4
4	1	4	3

Automaton C

FIGURE 5

	a	b	c
B_1	B_2	B_1	B_1
B_2	B_3	B_1	B_1
B_3	B_4	B_1	B_1
B_4	B_1	B_1	B_1

An Image Automaton $C\phi$

FIGURE 6

THEOREM 3. *Let ϕ be an S.P. set system on an automaton A, such that ϕ has k blocks and the largest block of ϕ has l states. Then A can be realized by a serial connection of a k-state automaton to an l-state automaton.*

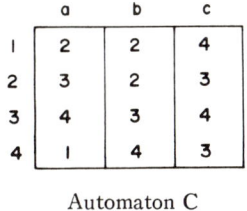

THEOREM 4. *Let* ϕ_1 *and* ϕ_2 *be S.P. set systems on* A, *such that* ϕ_1 *has* k_1 *blocks,* ϕ_2 *has* k_2 *blocks, and*

$$\phi_1 \cdot \phi_2 = 0$$

(defined in the same way as for partitions). Then A *can be realized by a parallel connection of a* k_1*-state and* k_2*-state automaton.*

Note that the two previous theorems do not give necessary and sufficient conditions for the series or parallel decompositions. This is so because any automaton with more than two states always has non-trivial S.P. set systems. For example, the set system

$$\phi = \{S_i | S_i \subseteq S \text{ and } |S_i| = |S| - 1\}$$

has S.P. for any automaton with $|S| \geqslant 3$. According to Theorem 3, therefore, this set system gives a realization of an n-state automaton by a serial connection of an n-state and an $(n - 1)$-state automata. This may not be useful for practical purposes but we will make use of this particular S.P. set system decomposition to get a very general decomposition result later.

The next result describes a necessary and sufficient condition that there exists a realization of an automaton by a serial connection of two smaller automata. The proof of this result is not hard and follows roughly the same reasoning as the proof of Theorem 2.

THEOREM 5. *An* n*-state automaton* A *can be realized by a serial connection of a* k*-state and* l*-state automata,* $k, l < n$, *if and only if there exists an S.P. set system* ϕ *with no more than* k *blocks and with no more than* l *elements in the largest block of* ϕ.

From these results we see that with the new concept of realization (Definition 4) goes the new set system concept of ignorance.

The set systems and other concepts derived from them have a wide range of applications. In particular, the idea of set system pairs, which describe how information or ignorance spreads in an automaton when it operates, is a useful concept and has found interesting applications. These ideas are discussed in more detail in [2].

A CONONICAL DECOMPOSITION

We now start the derivation of a general existence result which describes a cononical set of automata from which any automaton can be realized.

We saw before that for the set system

$$\phi = \{S_i | S_i \subseteq S \text{ and } |S_i| = |S| - 1\},$$

the automaton C had an image automaton C_ϕ which had only permutation and reset inputs. We now define a permutation-reset automaton and show that for any automaton there exists an image automaton which is a permutation-reset automaton.

DEFINITION 7. An automaton is said to be a *permutation-reset*, P-R, automaton if every input either permutes the set of states or maps all states onto a single state.

LEMMA 1. *Every automaton can be realized by a serial connection of P-R automata.*

Proof. If A has one or two states, then it is a P-R automaton. If A has more than two states let

$$\phi = \{S_i | S_i \subseteq S \text{ and } |S_i| = |S| - 1\}.$$

We know that ϕ has S.P. since every l-state block of an automaton is always mapped into some l-state block by every input. Furthermore, just like in the case of automaton C, there is an image automaton A_ϕ which is a P-R automaton. This is so because permutation inputs are again permutation inputs under ϕ and if some input is not a permutation input then a state does not appear in the input column and there is some B in ϕ which contains all the states of this column; but then B can be chosen to be the next state under this input for all states of A_ϕ. Thus, an A_ϕ is a P-R machine, and by Theorem 3 we know that A can be realized by a serial connection of A_ϕ to a machine with $|S| - 1$ states.

Now the process can be repeated, and, by induction in $|S| - 2$ steps, we arrive at a serial connection of $|S| - 1$ P-R machines which realizes A, as was to be shown.

Next we will replace every P-R machine by a serial connection of a pure permutation machine to a pure reset machine (i.e., every input either resets the state or leaves it unchanged).

LEMMA 2. *Every P-R machine can be realized by a serial connection of a permutation machine and a reset machine.*

Proof. Consider the group of permutations generated by the permutation inputs of A. Let the states of A_1 be these permutations and let the inputs of A_1 be the permutations defined by the permutation inputs of A with the identity input added if necessary. Then, for permutation input x (which performs the permutation P_x on S of A), let

$$\delta_1(P_i, x) = P_x P_i.$$

For a reset input x of A, let

$$\delta_1(P_i, x) = P_i.$$

Thus A_1 just computes what permutations are performed on the set of S of A by the permutation inputs to A, and it does not change its state when reset inputs are applied to A. The output of A_1 is its present state and input

$$\lambda_1(P_i, x) = (P_i, x).$$

The set of states of A_2 is the same as the set of states of A. The inputs to A_2 are the outputs of A_1 and are therefore of the form (P_i, x). Each input of A_2 is an identity input or a reset input. If x is a permutation input to A then

$$\delta_2[s, (P_i, x)] = s,$$

and the state of A_2 is not changed; if x resets A to state t, then

$$\delta_2[s, (P_i, x)] = P_i^{-1}(t).$$

It is now seen that if the automaton A is started in state r, and the automaton given by the serial connection of A_1 and A_2 is started in state (P_e, r), P_e = identity permutation; then any input sequence which transfers r onto s in A will transfer (P_e, r) onto (P_i, q) with the property that

$$P_i(q) = s.$$

Thus, when the present state (permutation) of A_1 is applied to the present state of A_2, we get the present state of A. Therefore A is realized by the serial connection of A_1 to A_2, as was to be shown.

Finally, we note that every partition on the set of states of a reset machine has S.P., and therefore any reset machine can be realized by a series (or parallel) connection of two-state machines. Combining this observation with the preceding lemmas we get our next result.

THEOREM 6. *Every automaton can be realized by a series-parallel connection of permutation automata and two-state automata.*

A straightforward application of elementary group theory permits us to decompose the permutation automata further. We just have to observe that the cosets of any subgroup form an S.P. partition on the group viewed as an automaton and prove, with not too much difficulty, that we can utilize a normal subgroup N to decompose the group G (viewed as an automaton) into the serial connection of two groups G/N and N. These observations yield our next result.

THEOREM 7. *Every automaton can be realized by a series-parallel connection of two-state reset automata and permutation automata whose permutations generate simple groups.*

SUMMARY

These results were obtained, up to Theorem 6, with elementary means by starting with a simple concept of realization and ignorance and then enlarging these concepts. Theorem 7 can be derived from Theorem 6 by use of elementary group theory and thus the derivation outlined here is also elementary.

It is interesting to note that in terms of the simple groups which are defined by the permutation automata we can characterize the essential and indestructible computations which an automaton has to perform. In essence, Theorem 7 shows that in conjunction with reset automata and automaton with simple groups are sufficient to realize any automaton. It can also be shown that if a simple group is "present" in the original automaton A, then it must be "present" in one of the component automata which are used to realize A. This result is unfortunately harder to obtain than Theorem 7, and more group theory has to be used [2].

HISTORY

The decompositions of automata defined by S.P. partitions were first discussed by the author [3, 4] and M. Yoeli [5, 6]. The set systems concept occurred almost simultaneously to several people including Z. Kohavi [7], M. Yoeli [8], H. P. Zeiger [10], R. E. Stearns and the author [9]. Theorems 6 and 7 were first obtained by K. Krohn and J. Rhodes [2], and the P-R automata approach to these results is due to H. P. Zeiger [10]. A detailed discussion of the structure of automata can be found in [1].

REFERENCES

[1] Hartmanis, J., and Stearns, R. E., *Algebraic Structure Theory of Sequential Machines* (Englewood Cliffs, N.J.: Prentice-Hall, 1966).

[2] Krohn, K. B., and Rhodes, J. L., "Algebraic Theory of Machines," *Proc. Symp. Math. Theory of Automata*, vol. 12 of the Microwave Research Institute Symposium Series (Brooklyn, N.Y.: Polytechnic Press, 1963).

[3] Hartmanis, J., "Symbolic Analyses of a Decomposition of Information Processing Machines," *Information and Control*, 3 (2) (June, 1960), 154–78.

[4] ——— "Loop-Free Structure of Sequential Machines," *Information and Control*, 5 (1) (March, 1962), 25–43. Reprinted in *Sequential Machines: Selected Papers*, E. F. Moore, ed. (Reading, Mass.: Addison-Wesley, 1964).

[5] Yoeli, M., "The Cascade Decomposition of Sequential Machines," *IRE Transactions on Electronic Computers*, EC-10(4) (Dec., 1961), 587–92.

[6] ——— "Cascade-Parallel Decompositions of Sequential Machines," *IRE Transactions on Electronic Computers*, EC-12(3) (June, 1963), 327–34.

[7] Kohavi, Z., "Secondary State Assignment for Sequential Machines," *IEEE Transactions on Electronic Computers*, EC-13(3) (June, 1964), 193–203.

[8] Yoeli, M., "Decomposition of Finite Automata," *Technical Report No. 10* (Haifa, Israel: Technion, March, 1963).

[9] Hartmanis, J., and Stearns, R. E., "Pair Algebras and Their Application to Automata Theory," *Information and Control*, 7(4) (Dec., 1964), 485–507.

[10] Zeiger, H. P., "Loop-Free Syntheses of Finite State Machines" (M.I.T. Ph.D. Thesis, Electrical Engineering Department, Sept., 1964).

Synthesis of Sequential Machines

J. A. Brzozowski*

INTRODUCTION

The synthesis of a sequential circuit [*A*] can be characterized as a process which begins with a statement of a problem, a given set of devices available to the designer, and a set of criteria by which the quality of the design can be evaluated. The process ends when a specific circuit, which provides the solution to the given problem, is constructed from the given devices. It is usually implied that a good design will provide, not just a solution, but the best possible solution that one can achieve with the available tools. Thus in the beginning we are given the desired *behaviour* of the circuit, and we are to produce a *structure* capable of realizing this behaviour. This realization of a behaviour by a structure is to be achieved with the given set of devices and must satisfy certain cost and reliability criteria. It is clear that a desirable theory of sequential circuits should contain a method which enables the designer to obtain the best circuit from the given data. It is also clear that no generally applicable algorithm exists at this time which is capable of solving all these problems. The purpose of this paper is to examine the techniques which are now available in order to answer the questions arising in this area.

BEHAVIOURAL DESCRIPTION

Since we are concerned with synchronous [*A*] sequential circuits, we know that the behaviour of such a circuit can be represented by a black box shown in Fig. 1,

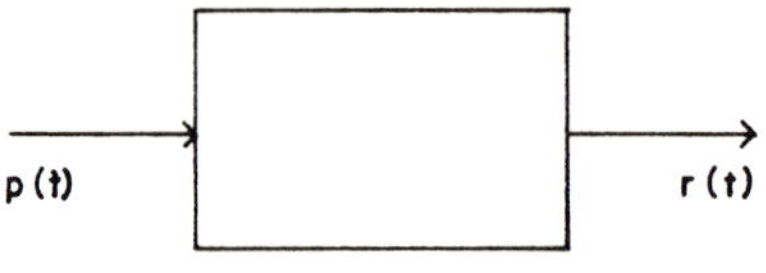

FIGURE 1. Black box representation.

*Department of Electrical Engineering, University of Ottawa.

with an input $p(t)$ and an output $r(t)$. The behaviour will be meaningful only at discrete instants of time, which we can label $t = 1, 2, \ldots$. At any such time t the input takes on one of a finite set $P = \{p_1, p_2, \ldots, p_u\}$ of values, and the output takes its value from another finite set $R = \{r_1, r_2, \ldots, r_w\}$. As time progresses, the black box produces a sequence of output values in response to an externally applied sequence of input values.

The problem to be solved by the black box is first described in words. As an example consider the following statement: Construct a circuit C which, in response to each input sequence s of 0's and 1's ($P = \{0, 1\}$), $s = p(1)p(2) \ldots p(t)$, produces an output sequence $r(1)r(2) \ldots r(t)$, which has the property that $r(i) = 1$, if $p(1)p(2) \ldots p(i)$ has an odd number of 1's, and $r(i) = 0$, otherwise. This is a rather simple problem and yet its statement is somewhat involved, because we wanted to make it precise. It is more probable that the original problem statement will be more vague, e.g., "Design a circuit which recognizes when an input sequence has an odd number of 1's." Most original problem statements are vague and incomplete, and often the person posing the problem may not be sure of all the aspects of the problem. Since it is impossible to construct a theory for vague word-descriptions of problems, we shall assume that it is possible to obtain a formal and unambiguous description of the behaviour.

The language in which the formal description [28] can be given is, of course, not unique. To illustrate this for the circuit C above, which is to determine the parity of the input sequence, we can consider the following.

(1) *Regular expression* [*B*]. C is to accept precisely those sequences which are represented by the regular expression $\rho = (0 + 10^*1)^*10^*$. Of course it is not at all obvious that this is so. However, the regular expression description is generally applicable, and in some cases easy to obtain.

(2) *Sequential function* [*C*]. C is to map input sequences into output sequences. The specific mapping for C can be given by

$$r(t) = p(1) \oplus p(2) \oplus \ldots \oplus p(t),$$

where $\oplus$ is the modulo 2 sum.

(3) *Transfer function.* This language applies only to linear sequential circuits [*D*]. The input and output sequences are considered as polynomials in D with coefficients in $\{0, 1\}$. Thus the sequence 1011 becomes $1 \oplus D^2 \oplus D^3$. In this case, the transfer function of C is $H(D) = 1/(1 \oplus D) = 1 \oplus D \oplus D^2 \oplus \ldots$. The output polynomial $r(D)$ can then be obtained by multiplying the input polynomial $p(D)$ by the transfer function $H(D)$.

Notice that we have not yet mentioned the most common formal description in use, namely, the state graph or the flow table [*A*]. This was done intentionally, because so far we have only used input–output descriptions without introducing the notion of internal state. Of course the notion of state can be derived from the descriptions (1)–(3), and every sequential circuit must have a finite number of states. Yet it appears that (1)–(3) are purely behavioural characterizations, whereas the state graph and flow table imply some structure as will be shown.

Nevertheless, we shall associate flow tables with behaviour because they are used to describe the problem.

Discussion

S. GORN: You will find that the distinction between the behavioural and structural is relative, as one man's behavioural specification is another's structural.

J. BRZOZOWSKI: I don't know how relative it should be. I would say that some models convey more structure than others.

S. GORN: For example, a state diagram will be a behavioural description for an engineer.

J. BRZOZOWSKI: That's right. I like to think about a state diagram as being somewhere in the middle, a little closer to structure than, say, a regular expression, although, of course, the structure can be made more explicit in the regular expression also.

S. GORN: Then actually, it is a comparatively relative thing to say that a model is more behavioural than structural.

J. BRZOZOWSKI: I think so. But I feel that it is sometimes worthwhile to label some aspects behavioural, others structural.

To complete our example we have:

(4) *State graph.* C must realize the graph of Fig. 2(a) (initial state $-A$).

(5) *Flow table.* C must realize the flow table of Fig. 2(b), where $q(t)$ denotes the internal state.

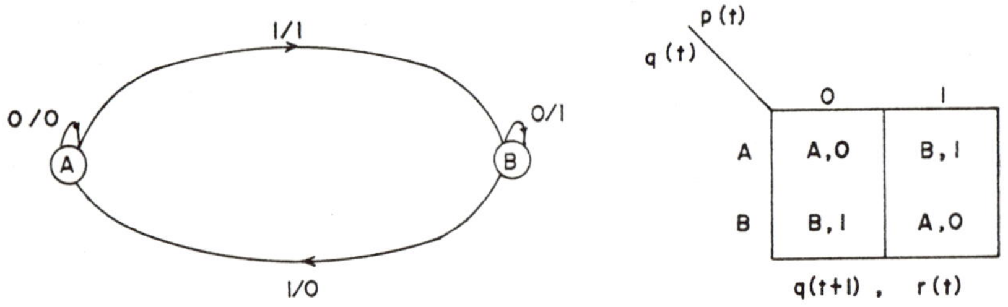

(a) (b)

FIGURE 2. (a) State graph for C. (b) Flow table for C_1.

Since (4) and (5) are by far the most commonly used, in the remainder of the paper we assume that the formal description of the behaviour is in the form of a flow table. The problem then becomes one of realizing a flow table by a sequential circuit.

It should be pointed out that it is possible to use a regular expression description and convert it directly to a circuit. However, at least with the presently available methods, the designer has very little control over the size and economy of the resulting circuit. Most of the work in economical design of sequential circuits has been done using the flow table, and it is for this reason that we shall limit the discussion to flow tables.

FROM BEHAVIOUR TO STRUCTURE

So far we have considered the following steps in the design: given the word-description of the behaviour, convert it to some formal description from which a flow table is derived. There is a natural way in which the flow table immediately suggests a structure for the realization. In the flow table, we have input alphabet $P = \{p_1, p_2, \ldots, p_u\}$, output alphabet $R = \{r_1, r_2, \ldots, r_w\}$, and a finite set of states $Q = \{q_1, q_2, \ldots, q_v\}$. Furthermore, the flow table defines the next state function σ such that

$$q(t+1) = \sigma(q(t), p(t)), \text{ for } t \geqslant 1,$$

and $q(1)$ is a specified initial state, and an output function ρ such that

$$r(t) = \rho(q(t), p(t)) \text{ for } t \geqslant 1.$$

For $t < 1$, $q(t)$ and $r(t)$ are of no interest. These relations immediately suggest the structure of Fig. 3 as a realization of the flow table. In the circuit of Fig. 3,

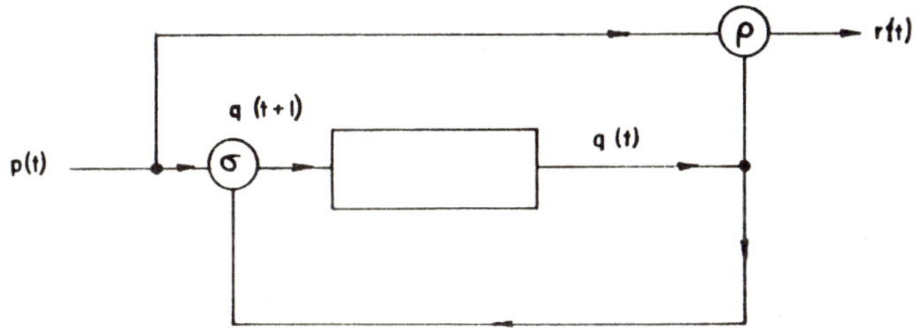

FIGURE 3. Multivalued realization of a flow table.

a single wire capable of carrying a u-valued signal represents the input $p(t)$, and $r(t)$ is similarly represented. The rectangular box is a unit delay carrying a v-valued signal. The input to the delay is the next state $q(t+1)$ and its output is the present state $q(t)$. The boxes labelled σ and ρ are combinatorial circuits capable of operating on multivalued signals.

What we have gained in the circuit of Fig. 3 is simplicity of structure. Of course the problem is that the present technology does not permit us to use multivalued devices, and most realizations use binary devices. Thus, to be more practical, we are searching for a binary sequential circuit realizing the given flow table. This can always be done by encoding each of p, q, and r into several binary signals. Then let us establish the following correspondences:

$$p\text{—}x_1, x_2, \ldots, x_n, \quad n \geqslant \log_2 u,$$
$$q\text{—}y_1, y_2, \ldots, y_s, \quad s \geqslant \log_2 v,$$
$$r\text{—}z_1, z_2, \ldots, z_m, \quad m \geqslant \log_2 w,$$
$$\sigma\text{—}f_1, f_2, \ldots, f_s,$$
$$\rho\text{—}g_1, g_2, \ldots, g_m,$$

where x_i, y_j, z_k are binary signals and f_j and g_k are Boolean functions corresponding to binary combinational switching circuits. We then obtain the general binary realization of a flow table as shown in Fig. 4, where only the details for y_j and z_k are shown.

This, then, is a general binary realization. It does not, however, represent a satisfactory solution to the design problem. For each flow table one can find an infinite number of circuits, hence one must look for a circuit which is in some sense an optimum circuit for the given problem.

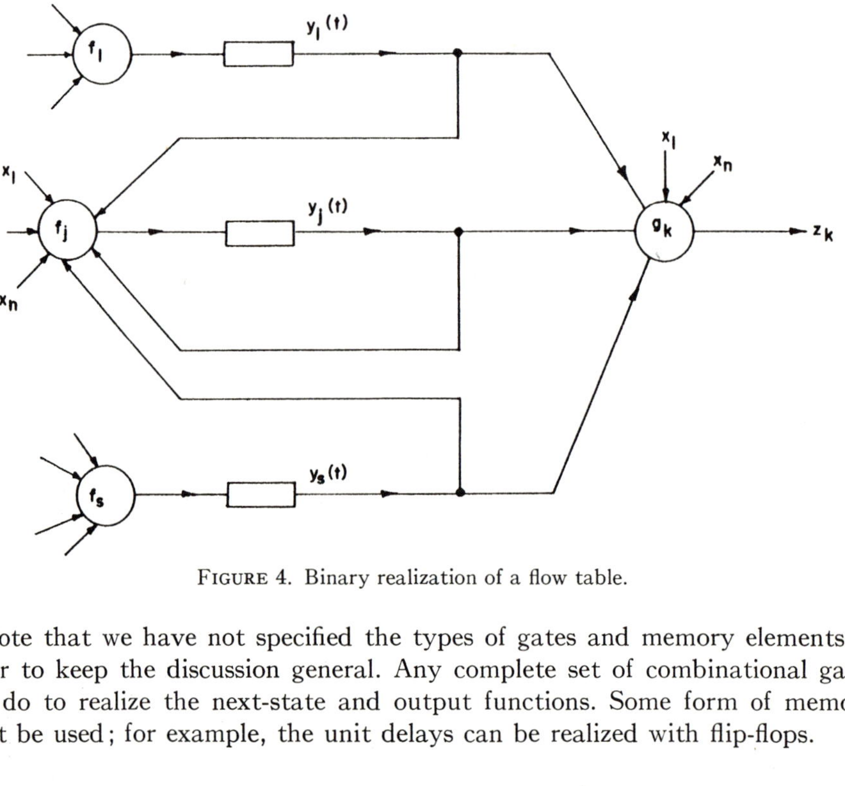

FIGURE 4. Binary realization of a flow table.

Note that we have not specified the types of gates and memory elements in order to keep the discussion general. Any complete set of combinational gates will do to realize the next-state and output functions. Some form of memory must be used; for example, the unit delays can be realized with flip-flops.

CRITERIA OF OPTIMALITY

Suppose we're given combinational gates of type $G_1, G_2, \ldots, G_t$, where the cost of a gate of type G_i is γ_i, and delays D, each with cost δ. One could then associate with each circuit C a cost function

$$\alpha_C = N\delta + \sum_{i=1}^{t} N_i \gamma_i,$$

where N is the number of delays and N_i is the number of gates of type G_i in the circuit. The formal statement of the design problem is then: Given a flow table F, find a circuit with minimum α_C realizing F. This simply stated problem is in fact amazingly complicated. Notice that among other things it contains the combinational minimization problems of switching theory. Thus we must face the fact that no algorithm exists which will produce minimum cost circuits. Furthermore, it appears safe to say that if a method is found it will be a laborious and a long process. Is the situation therefore hopeless? No; a lot can be done to shed more light on the design methods, but we must be satisfied with lesser goals.

Historically, the first result in the search for good designs was the reduction [*A*] of a flow table to an equivalent one with the minimum number of states. Using the reduced table implies using the least possible number of delays. If delays are constructed using flip-flops, which are more expensive than gates, then the idea of minimizing the number of expensive items first is certainly reasonable. The next problem is the secondary assignment [2] problem. Using the least number of secondary variables, assign them in such a way that the combinational circuitry is minimized. This usually implied using two-level diode gate circuits and minimizing them in the usual fashion. The problem still remains very difficult, even when restricted in this way. Hartmanis and Stearns, [17], [20], [39], on the other hand, took a different approach and showed that much can be gained by removing the restriction that the number of delays be minimum and by giving up the formal minimization of combinational circuits. Their approach was to look at the complexity of structure and to attempt to reduce it in the realization. This approach has been called structure theory of sequential machines. I must point out that the term structure theory is also used in another way: Dr. Hartmanis [18] is going to talk about decomposing large sequential machines into smaller ones. This is done at the level of flow tables in the sense that a given large flow table is shown to be composed in some way of smaller flow tables. Hence, we can see the structure of the flow table being exhibited in the decomposition into smaller tables. These results are certainly of significance to the design problem, but we are not going to talk about this kind of structure theory, since this is the subject of the next paper. Our structure is the structure of a circuit realizing a given flow table.

The basic idea is this: since the optimum design problem is impossible, and yet we are not satisfied with an arbitrary design, let us examine at least several different circuits realizing the flow table. If one examines several types of key structures before making the final choice, one hopes to obtain a relatively good design. This naturally leads to the question: What structures should one examine?

STRUCTURAL PROPERTIES

The diagrams of Figs. 3 and 4 suggest to us several separate items which contribute to the structure.

(1) Number of delays
(2) Complexity of next-state logic
(3) Complexity of output logic
(4) Amount of feedback

As we concentrate on one or more of these aspects we are led to a variety of structures. One can ask the following questions. Can the given flow table be constructed without feedback? If not, how much feedback is required? Can the combinational logic be linear? Can the combinational logic be eliminated altogether? Can the state behaviour be realized in shift register form, with a single combinational gate at the input of the first stage? There are a number of people asking these seemingly unrelated questions. If this work is included in the area of structure theory, the designer will have the ability to examine a variety of possible structures, and, hopefully, the best one will be among these.

At this stage we want to state more precisely what is meant by a circuit realizing a flow table. With every circuit C we can associate a flow table F_C. We say that C realizes a given flow table F if F is a reduced version of a subtable of F_C. In other words, to get from F to the circuit it may be necessary to split states in F, i.e., to expand the flow table to achieve the desired structure.

FEEDBACK [E]

The first question that one was able to answer in connection with feedback was: When can a flow table be realized without feedback? The question has been answered by several people [*F*], and the key problems in this area have been solved. There are several straightforward methods for testing whether a flow table is realizable without feedback. Furthermore, the results have been interpreted in terms of regular expressions, partitions and shift register realizations. Flow tables which can be realized without feedback are called definite.

Discussion

S. GORN: You say there are circuits without feedback. Isn't each memory unit a feedback element?

J. BRZOZOWSKI: I am considering boxes called ideal delays. If these are available then a definite table can be realized without feedback, with the possible exception of the feedback inside the delay. The delays can in fact be replaced by clocked flip-flops, in which case, each delay has a feedback loop.

J. RUTLEDGE: Why don't you build with magnetics?

J. BRZOZOWSKI: There again one has loops of current, induced voltages, etc. Perhaps this is some kind of feedback at another level?

J. RUTLEDGE: No, in a magnetic shift register the information is stored by different magnetic states.

At any rate, even if the delays are realized by flip-flops, I still think it is very worthwhile to talk about the feedback of the structure, even though my feedback may not be somebody else's feedback. If we assume that delays must be used in every realization, then our discussion of feedback simply has to do with the way in which the delays are arranged.

We are going to make a simplifying assumption [6] in order to avoid a lengthy treatment of some details. Let us suppose that the behaviour of a sequential circuit is of no interest for the first few short sequences of inputs. In other words we are willing to neglect "transient" conditions and concentrate on "steady state." This simplifies the discussion which follows of feedback-free realizations, one-feedback-loop realizations, and finite memory automata.

With the above assumption, we can realize [*F*] any definite flow table as a set of shift registers without feedback, as shown in Fig. 5. Note that no combinatorial gates are necessary for the state behaviour, but each output requires one gate, unless it is an output of one of the delays.

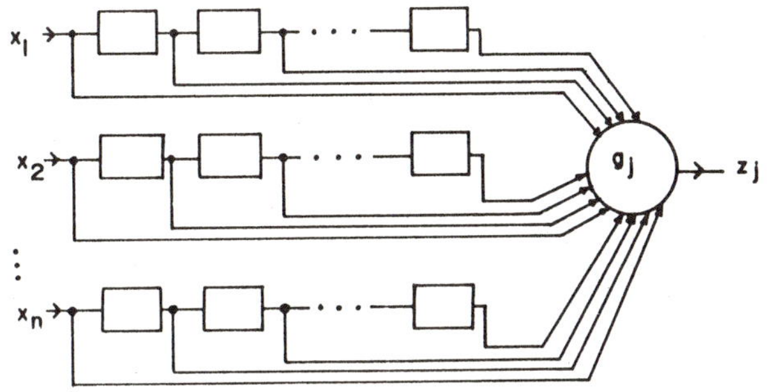

FIGURE 5. Feedback-free realization.

Suppose now that the given flow table is not definite; then some feedback is required. If each circuit is regarded as a directed graph, then the feedback index can be defined as the minimum number of wires that must be cut in order to make the resulting circuit loop-free. The question then arises, how much feedback is necessary. In Fig. 3 only one feedback loop is used, but it stores a multivalued signal for one unit of time. Can this be replaced by storing a binary signal for several units of time? This has been recently answered by Friedman [15], who showed that a single binary feedback loop is always sufficient to realize any flow table. Thus we obtain a generally applicable single-loop realization [8] as shown in Fig. 6. Any binary output can be obtained by a single combinatorial gate. Notice that, in general, one feedback-free shift register is required for each input, and one additional feedback shift register is needed.

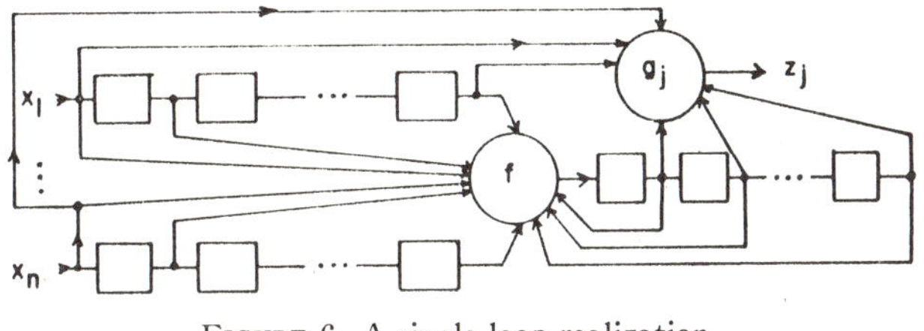

FIGURE 6. A single-loop realization.

In order to achieve a single-loop realization, it may be necessary to expand the flow table, and the circuit may use considerably more than the minimum number of delays. It may be then preferable to use more than one loop and to use fewer delays. A lot of work is still needed in this area to develop efficient techniques for finding economical realizations with a prescribed feedback index. Properties related to feedback, such as the conditions under which less feedback is better than more feedback, should be studied. Obviously the designer will make the final choice of the circuit but he should be familiar with the various forms available.

FINITE MEMORY AUTOMATA

In the general diagram of Fig. 3 we require two combinational gates. For the so-called finite memory automata [*G*] one gate is sufficient. A finite memory automaton is one in which the output can be computed from the present input and from a finite number of past values of the input and output, as shown in Fig. 7. Note from Fig. 3 that the state behaviour of any automaton has finite memory character if the state itself is regarded as the output and if the given output is removed. However, there are flow tables with outputs which are not

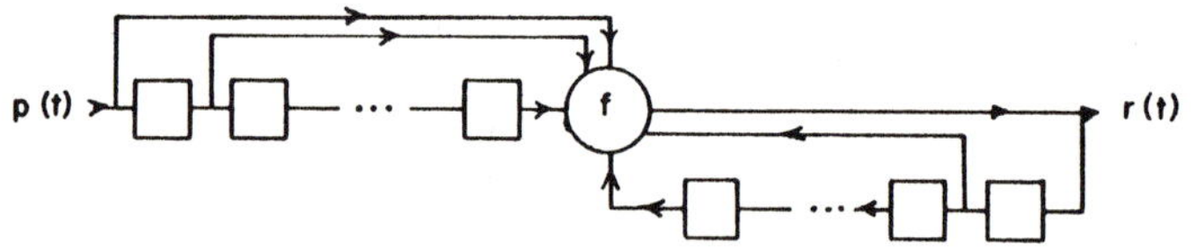

FIGURE 7. Finite memory circuit.

finite memory; hence this is not a universal form. Note also that in Fig. 7 we are allowing the output to be multivalued in general. If it is binary, then we have a single-loop realization with a single gate.

Methods for testing whether a flow table is finite memory are described in the literature [*G*].

LINEAR AUTOMATA

We define a flow table as linear if it can be realized by a circuit constructed from exclusive *OR* gates and unit delays only. Cohn [10] has recently shown that there are nonlinear flow tables; hence, once again, we are examining a subset of the set of all automata. The algebra applicable to linear automata is much nicer than in the general case.

We shall not dwell on the subject but simply point out that there are several approaches [*H*] to recognizing the linearity of a flow table, and the problem can be considered solved. Linear automata have been found very useful in connection with coding [35], and they represent an important subclass of automata.

USE OF THRESHOLD DEVICES

When one examines the combinational logic one need not limit his investigation to linearity, but may look for other properties also. For example, the combinational threshold problem (When is a Boolean function realizable with a single threshold gate?) is being studied actively. Since many combinational circuits become parts of sequential circuits, we propose the *sequential threshold problem*: "When can a flow table be realized by a circuit with a single threshold gate (and delays)?" Perhaps the threshold property may be recognized from the flow table. My guess is that the problem will be very difficult; still, it is a well-defined problem belonging to structure theory.

DEVICES WITH INHERENT DELAYS

When the speed at which one wishes to operate a sequential circuit becomes high enough, the delay associated with each gate must be taken into account. This leads to different design problems, which we feel belong together with the above-mentioned structural properties. To begin with, one encounters the question of completeness [*I*], i.e., is the given set of devices with delays sufficient to realize an arbitrary flow table. Secondly, there is the design problem itself. Although a lot is known already [*I*], algorithms for economical design are yet to be developed.

SHIFT REGISTER REALIZATIONS

If we turn our attention from combinational logic to the sequential aspects of the problem, we have the general problem of realizing a flow table with sequential modules. In particular, the problem of shift-register realizations is being studied actively [*J*]. Of course a proper interpretation must be used to make the problem meaningful. For example, every delay can be thought of as a shift register; naturally this is not what we have in mind. It appears reasonable to ask whether the flow table can be realized with a single binary feedback shift register. If not, will two be enough? etc. These problems can be easily answered for one-to-one assignments, but become much more difficult when state splitting is allowed (as it must be).

CONCLUDING REMARKS

This concludes my brief survey of the activities in the area of structure theory of sequential circuits. I will end by pointing out a few problems.

Can the description of the variety of structures be made more systematic? Is it possible to find a finite (and hopefully small) set of structural realizations which is complete in the sense that all other structures will be trivial modifications of these?

Finally, I feel that the problem of reliability has been greatly neglected. Since there are so many possible assignments for a given flow table, some will be more reliable than others, and one would like to pick one that is reasonably economical and reasonably reliable at the same time. Very little seems to be known about such problems.

Discussion

J. HARTMANIS: Are there any questions?

FLOOR: I don't understand something about the finite memory circuits . . .

J. BRZOZOWSKI: I am using Gill's [16] terminology for this. It may be a poor name. You see, the memory does not consist of a finite number of past inputs, as in the case of definite circuits. Also, the number of things that any sequential circuit can remember is of course always finite. In the parity counter, at any time the circuit remembers only one bit (even or odd). However, the entire input sequence must be looked at to determine whether the number of 1's is even or odd. Now Gill uses finite memory to describe circuits in which the output is a function of a finite number of past values of input *and* output.

FLOOR: The output is a single binary output.

J. BRZOZOWSKI: Not necessarily. If there are two binary outputs we treat them as one 4-valued output. By the way, finite memory automata are related to definite automata as follows: if the output symbol is added to the input alphabet, then the resulting automaton must be definite.

FLOOR: So this is a definite automaton in which you are allowed to take the output and feed it back as if it were an input!

J. BRZOZOWSKI: Yes. Of course the result is no longer definite.

R. BANERJI: What do you know about work on threshold functions?

J. BRZOZOWSKI: I don't know of anybody working in this field. I think it is a very difficult problem. I would probably try to solve it by first finding a shift register realization using one gate and then by testing whether the gate is a threshold gate. Now, in this approach, something is known about both problems, but it appears to be a succession of two unrelated problems. I suspect it will be very difficult to recognize threshold properties directly from the flow table.

J. HARTMANIS: It is interesting that Haring *et al.*, are working on realizing counters by using threshold gates. The paper is to be presented at the symposium in Michigan. (Author's comment inserted after the Michigan symposium: the problem there [40] is to design counters with a single threshold gate *per state variable*. Thus this is not the sequential threshold problem presented here.)

FLOOR: Returning to the problem of feedback, if you use shift registers . . .

J. BRZOZOWSKI: The circuit is feedback-free, if the cells of the register are.

S. GORN: I am suggesting that you are tossing the feedback problem to somebody else.

J. HARTMANIS: I am not so sure. You could be rolling colored balls and scanning them optically, and I don't see any feedback there.

J. RUTLEDGE: Then you probably have a feedback stabilized amplifier.

J. BRZOZOWSKI: This may actually be a good subject for another conference, by the sound of it. I think it's worthwhile to look at the feedback in the structure. The feedback complexity will affect errors. For example, without feedback, all errors disappear after a while, as the register storing past inputs is receiving new values.

M. Arbib: In automata theory we have a well-defined mathematical terminology, and within that there is no question that in a definite automaton there is no feedback.

J. Brzozowski: It all goes back to the question of whether feedback is necessary in the unit delays.

J. Hartmanis: There is no trouble. The only problem is the interpretation to the physical world.

J. Brzozowski: That's right.

REFERENCES

Note. Letters refer to several papers on related topics. The correspondence is shown below.

[A]—[30], [31]
[B]—[4], [29]
[C]—[14], [21] [36], [37]
[D]—[7], [22], [35]
[E]—[6], [8], [15], [19], [27]
[F]—[1], [3], [6], [19], [27], [34]
[G]—[6], [16], [27]
[H]—[9], [11], [13], [38]
[I]—[1], [21], [25], [26], [36]
[J]—[5], [12], [23], [24], [32], [33]

[1] Arden, D. N., "Delayed Logic and Finite State Machines," *Proc. 2nd Ann. Symp. on Switching Circuit Theory and Logical Design* (Sept., 1961), 131–51.

[2] Armstrong, D.· B., "Sequential Circuits—Part A: Structural Properties," *IEEE Trans. on Circuit Theory,* CT-11 (March, 1964), 25–7.

[3] Brzozowski, J. A., "Canonical Regular Expressions and Minimal State Graphs for Definite Events," 12, *Proc. Symp. Math. Theory of Automata* (Brooklyn, N.Y.: Polytechnic Press, April, 1962), 529–61.

[4] ——— "A Survey of Regular Expressions and their Applications," *IRE Trans. on Electronic Computers,* EC-11 (June, 1962), 324–35.

[5] ——— "Review of 'k^{th} order finite automaton'—C. L. Liu," *IEEE Trans. on Electronic Computers,* EC-13 (April, 1964), 162.

[6] ——— "An Essay on Feedback," *Technical Report No. 65-2* (Ottawa: Dept. of Elec. Engineering, University of Ottawa, March, 1965).

[7] ——— "Regular Expressions for Linear Sequential Circuits," *IEEE Trans. on Electronic Computers,* EC-14 (April, 1965), 148–56.

[8] ——— "On Single-Loop Realizations of Automata," *Proc. 6th Ann. Symp. on Switching Circuit Theory and Logical Design* (Oct., 1965), 84–93.

[9] Brzozowski, J. A., and Davis, W. A., "On the Linearity of Autonomous Sequential Machines," *IEEE Trans. on Electronic Computers,* EC-13 (Dec., 1964), 673–9.

[10] Cohn, M., "A Theorem on Linear Automata," *IEEE Trans. on Electronic Computers,* EC-13 (Feb., 1964), 52.

[11] Cohn, M., and Even, S., "Identification and Minimization of Linear Machines," *IEEE Trans. on Electronic Computers,* EC-14 (June, 1965), 367–76.

[12] Davis, W. A., "On Shift Register Realizations for Sequential Machines," *Proc. 6th Ann. Symp. on Switching Circuit Theory and Logical Design* (Oct., 1965), 71–83.

[13] Davis, W. A., and Brzozowski, J. A., "On the Linearity of Sequential Machines," *Proc. 5th Ann. Symp. on Switching Circuit Theory and Logical Design* (Oct., 1964), 197–208.

[14] Elgot, C. C., and Rutledge, J. D., "Operations on Finite Automata, Extended Summary," *Proc. 2nd Ann. Symp. on Switching Circuit Theory and Logical Design* (Sept., 1961), 129–32.

[15] Friedman, A. D., "Feedback in Synchronous Sequential Switching Circuits," *IEEE Trans. on Electronic Computers*, EC-15 (June, 1966), 364–67.

[16] Gill, A., *Introduction to the Theory of Finite State Machines* (New York: McGraw-Hill Book Co., 1962).

[17] Hartmanis, J., "On the State Assignment Problem for Sequential Machines, I," *IRE Trans. on Electronic Computers*, EC-10 (June, 1961), 157–65.

[18] ——— These proceedings.

[19] Hartmanis, J., and Stearns, R. E., "A Study of Feedback and Errors in Sequential Machines," *IRE Trans. on Electronic Computers*, EC-12 (June, 1963), 223–32.

[20] ——— "Pair Algebra and its Application to Automata Theory," *Information and Control*, 7 (Dec., 1964), 485–507.

[21] Hohn, F. E., "States of Sequential Machines whose Logical Elements Involve Delay," *Proc. 3rd Ann. Symp. on Switching Circuit Theory and Logical Design* (Sept., 1962), 81–9.

[22] Huffman, D. A., "The Synthesis of Linear Sequential Coding Networks," *Information Theory*, C. Cherry, ed. (New York: Academic Press, 1956), 77–95.

[23] Liu, C. L., "k^{th} Order Finite Automaton," *IEEE Trans. on Electronic Computers*, EC-12 (Oct., 1963), 470–5.

[24] ——— "Sequential Machine Realization Using Feedback Shift Registers," *Proc. 5th Ann. Symp. on Switching Circuit Theory and Logical Design* (Oct., 1964), 209–27.

[25] Loomis, H. H., Jr., "Completeness of Sets of Delayed-Logic Devices," *IEEE Trans. on Electronic Computers*, EC-14 (April, 1965), 150–72.

[26] Loomis, H. H., Jr., and Wyman, R. H., Jr., "On Complete Sets of Logic Primitives," *IEEE Trans. on Electronic Computers*, EC-14 (April, 1965), 173–4.

[27] McCluskey, E. J., "Reduction of Feedback Loops in Sequential Circuits and Carry Leads in Iterative Circuits," *Information and Control*, 6 (1963), 99–118.

[28] McNaughton, R., "Sequential Circuits—Part B: Behavioural Properties," *IEEE Trans. on Circuit Theory*, CT-11 (March, 1964), 27–9.

[29] ——— These proceedings.

[30] Mealy, G. H., "A Method of Synthesizing Sequential Circuits," *Bell Syst. Tech. Journal*, 34 (Sept., 1955), 1045–79.

[31] Moore, E. F., "Gedanken Experiments on Sequential Machines," *Automata Studies, Annuals of Math. Studies*, 34, C. E. Shannon and J. McCarthy, eds. (Princeton, N.J.: Princeton University Press, 1956), 129–53.

[32] Nichols, A. J., III, "Modular Synthesis of Sequential Machines," *Proc. 6th Ann. Symp. on Switching Circuit Theory and Logical Design* (Oct., 1965), 62–70.

[33] ——— III, "Minimal Shift Register Realizations of Sequential Machines," *IEEE Trans. on Electronic Computers*, EC-14 (Oct., 1965).

[34] Perles, M., Rabin, M. O., and Shamir, E., "The Theory of Definite Automata," *IEEE Trans. on Electronic Computers*, EC-12 (June, 1963), 233–43.

[35] Peterson, W. W., *Error Correcting Codes* (New York: John Wiley and Sons, 1961).

[36] Pugsley, J. H., "Sequential Functions and Linear Sequential Machines," *IEEE Trans. on Electronic Computers*, EC-14 (June, 1965), 376–82.

[37] Raney, G. N., "Sequential Functions," *Journal Assoc. Comp. Mach.*, 5 (April, 1958), 177–80.

[38] Srinivasan, C. V., "State Diagram of Linear Sequential Machines," *Journal Franklin Inst.*, 273 (May, 1962), 383–418.

[39] Stearns, R. E., and Hartmanis, J., "On the State Assignment Problem for Sequential Machines II," *IRE Trans. on Electronic Computers*, EC-10 (Dec., 1961), 593–603.

[40] Gustafson, C. H., Haring, D. R., Susskind, A. K., and Wills-Sandford, T. G., "Synthesis of Counters with Threshold Elements," *Proc. 6th Ann. Symp. on Switching Circuit Theory and Logical Design* (Oct., 1965).

Techniques for Manipulating Regular Expressions

Robert McNaughton*

A few years ago Professor Brzozowski wrote a survey paper [1] which proved to be quite helpful to those, like the readers of the present paper, who wanted a summary account of regular expressions. The present paper will be concerned with work done since that paper appeared.

To begin with, regular expressions are expressions standing for regular events (or regular languages), which are certain sets of words. A word is a string of symbols over some alphabet, denoted by Σ. Although Σ is in general any finite set, all the examples in this paper will be regular events over the binary alphabet $\Sigma = \{0, 1\}$.

Regular expressions are made up of letters of the alphabet, and signs standing for certain operators. The three operators are union, concatenation, and star (or closure). Union is the ordinary set-theoretic union, whose sign is "$\cup$." Concatenation is written as a dot and sometimes is denoted by mere juxtaposition. The concatenation of two words is obtained by writing the first word, and then the second word following it without any space. The concatenation of two sets of words tends to be thought of as a Cartesian product; however, it is not quite that. Let α and β be two events, $\alpha \cdot \beta$ or $\alpha\beta$, the concatenation of α and β is the set of all words that can be obtained by a concatenation of a word from α and a word from β in that order. For example, if $\alpha = \{0, 01, 001\}$ and $\beta = \{1, 11\}$, then $\alpha\beta = \{01, 011, 0111, 0011, 00111\}$. Note that the word 011 is obtained in two ways: as 0 concatenated with 11, and as 01 concatenated with 1. This example is enough to show that concatenation is not a Cartesian product.

Before explaining the star operator, something should be said about the null-word. First, the null-word λ and the empty set $\emptyset$ must be distinguished: for the null-word is the word of zero length, while the empty set is the set that has no words at all as members; these symbols behave quite differently in regular expressions. If the null-word is mysterious, think of a word of length 3. If you throw

*Rensselaer Polytechnic Institute. Work reported herein was supported by Project MAC, an M.I.T. research programme sponsored by the Advanced Research Projects Agency, Department of Defense, under Office of Naval Research Contract No. Nonr-4102(01). Reproduction in whole or in part is permitted for any purpose of the U.S. Government.

away one symbol, you are left with a word of length 2. Then if you throw away another symbol, you are left with a word of length 1, a single symbol. When you throw away that symbol you are left with a word of zero length, the null-word. The concept of the null-word is important because of the manner in which it concatenates with another word: thus, for any W, $\lambda W = W\lambda = W$. As a consequence, for any set α, $\lambda\alpha = \alpha\lambda = \alpha$. This behaviour of the null-word is almost its definition.

By the way, we see in the notation "$\lambda\alpha$" a tendency that may dishearten the logical purist, but is convenient in the practice of writing regular expressions. It is the notational identification of an object and its unit set. Thus "λ" in "$\lambda\alpha$" means the unit set of the null-word; in the mathematics of regular events, a confusion never results from this identification.

On the other hand, the empty set $\emptyset$ is a set-theoretic concept, meaning a set that does not have anything in it. In contrast to the fact that $\lambda\alpha = \alpha\lambda = \alpha$, for any set α, $\alpha\emptyset = \emptyset\alpha = \emptyset$. In order to see this, it is necessary to go back to a literal interpretation of the definition of the concatenation of two sets: $\alpha\emptyset$ is the set of all words obtained by concatenating a word from α and a word from $\emptyset$. But there are no words in $\emptyset$, since $\emptyset$ is the empty set. Hence $\alpha\emptyset = \emptyset$. Similarly, $\emptyset\alpha = \emptyset$.

The star operator can be defined as follows.

$$\alpha^* = \{W : (\exists n)(\exists V_1) \ldots (\exists V_n)(\forall i)\ V_i \in \alpha \,\&\, W = V_1V_2 \ldots V_n\}.$$

Here n is a non-negative integer. Since n may be 0, $\lambda \in \alpha^*$, for every α. Now the above definition is not a legitimate expression of symbolic logic because of the dots, which are not easily eliminated. Another definition is as follows: α^* is the smallest set containing λ and containing, for every word $W \in \alpha^*$ and $V \in \alpha$, the word WV. Thus star is the iteration of concatenation. It is sometimes called "closure."

The *restricted regular-expression symbolism* consists of union, concatenation (denoted by juxtaposition), star, each member of the alphabet (whatever it may be), λ and $\emptyset$. The *enlarged regular-expression symbolism* has all of this plus intersection and complementation. In this paper we shall be concerned only with the restricted language, and the term "regular expression" will refer to an expression of this language.

The language of regular expressions supresses not only the distinction between the unit set of a word and the word itself, but also the distinction between a letter of the alphabet and a word of length one. Thus "01" stands for both a word and the unit set of that word. And "0" stands for a letter of the alphabet, a word of length one, and the unit set of that word.

Some laws which have been noticed about regular expressions are the following.

(1) $\lambda\alpha = \alpha\lambda = \alpha$.
(2) $\emptyset\alpha = \alpha\emptyset = \emptyset$.
(3) $\lambda^* = \emptyset^* = \lambda$.

(4) $\alpha(\beta\alpha)^* = (\alpha\beta)^*\alpha$.
(5) $(\alpha \cup \beta)^* = (\alpha^*\beta^*)^*$.
(6) $\alpha(\beta \cup \gamma) = \alpha\beta \cup \alpha\gamma$.
(7) $(\alpha^*\beta)^* = \lambda \cup (\alpha \cup \beta)^*\beta$.
(8) $\mathrm{Reg}(\alpha, \beta) \subseteq (\alpha \cup \beta)^*$.
(9) $\alpha^* = \lambda \cup \alpha \cup \alpha^2 \cup \text{- - - -} \cup \alpha^{k-1} \cup \alpha^k\alpha^*$.

Of these, (1) and (2) have already been discussed. (3) can also be verified by going back to a literal interpretation of either of the two definitions of the star operator. (4) can be proved by a general technique that could be called "proof by reparsing." That is, consider any word $W \in \alpha(\beta\alpha)^*$.

$$W = U_0(V_1U_1)(V_2U_2) \ldots (V_nU_n),$$

where each $U_i \in \alpha$ and each $V_i \in \beta$. By reparsing the last expression (using the fact that concatenation is associative), we establish that

$$W = (U_0V_1)(U_1V_2) \ldots (U_{n-1}V_n)U_n,$$

which shows that $W \in (\alpha\beta)^*\alpha$. Since W is arbitrarily selected, this shows that $\alpha(\beta\alpha)^* \subseteq (\alpha\beta)^*\alpha$, whereupon (4) is established by symmetry.

(5) is also established easily by reparsing. (6) is too obvious to require proof. (7) is perhaps least obvious of all, and will be discussed in detail below. In (8), $\mathrm{Reg}(\alpha, \beta)$ means any restricted regular expression made up exclusively from α and β; in other words, containing no occurrence of the alphabet or letters thereof outside of α and β. The proof of (8) by reparsing is quite similar to the proof of (5). However, one must be quite careful because (8) is true only for restricted regular expressions. It is no longer true if intersection or complementation is allowed in $\mathrm{Reg}(\alpha, \beta)$. A good name for (9) is the development law, which is quite obvious from the definition of the star operator.

There are two main research problems about regular expressions. The first concerns means of proving valid equations (such as the above), while the second is the problem of finding interesting and fruitful equations to prove. The second problem will be discussed only briefly at the end of this paper. Most of the remainder of this paper will be concerned with the first problem.

Suppose we are given two regular expressions for comparison. The first question that arises is, are they equal? If not, then is one included in the other? However there is a familiar and simple way of handling inclusion as an equation: thus $\alpha \subseteq \beta$ will be true if and only if $\alpha \cup \beta = \beta$. Thus the problem of proving equations is foremost; a systematic and successful approach to this problem could be called "a calculus of regular expressions."

At this point it would be appropriate to say something explaining the connection between this paper and the purpose of the Conference as a whole. The discipline of regular expressions was originally introduced to describe automata in Kleene's paper [2], which was written in 1951 and published in 1956. For many years since, this discipline was interesting to switching theorists. Recently, however, I have discovered that there is considerable interest in this language among people who do advanced programming. It is likely that their interest in the regular-expression language is due to many different reasons; but one important reason is that the regular operators turn up frequently in the advanced study of mechanical languages. It must be said, of course, that regular events (or languages) only form a small subclass of the class of all languages that are interesting to advanced programmers. There are other classes which come closer to being candidates for the class of all programming languages. For example, there is the class of context-free languages. A less likely candidate would be the class of context-sensitive languages—a less likely candidate because it is too broad, and because most principles of formal language construction seem to be context-free principles. Indeed, one could argue that the class of all context-free languages is too broad. But, in any case, the class of regular languages is certainly too narrow, and the deficiency is made up in the direction of the class of context-free languages. Nevertheless, it is still true that the study of regular events has application to problems coming up in the investigation of the broader class of languages.

In spite of all this speculation about possible applications, I would like to suggest that our proper attitude towards the study of regular expressions (and towards the study of the theory of automata in general) should be that it is part of pure mathematics. We should attack the problems without worrying about where the application is going to be. In spite of its origin in switching theory (or, more precisely, in nerve-net theory) and present vague relation to problems of advanced programming, these connections are not altogether decisive in formulating valid research objectives. I hope these remarks suffice to explain my opinion about the applicability of regular expressions to the more practical matters in the computer sciences, and, at the same time, to provide a link between the contents of this paper and the Conference as a whole.

Now let us return to regular-expression equations. There are several methods of proof, which I shall survey. The first method is the technique of converting each regular expression to a state graph, reducing the state graphs, and testing whether the resulting state graphs are isomorphic. We thereby make use of a basic theorem from switching theory that two reduced state graphs that recognize exactly the same languages have to be isomorphic.

And so there it is, a procedure for testing an equation. This method is foolproof and works absolutely. That is, given any two regular expressions, one can perform the test mechanically and provide a yes or a no answer to the question of their equality. The only trouble is that it is not insightful; that is to say, when one actually goes through this process with a given α and β, the mechanical procedure that is used is usually not such as to provide any insight into the nature of these regular expressions. For example, α and β may be unequal but closely related to each other. But the procedure would terminate with a no answer and that would be the end of it; one would not discover any relationship. Even in cases where a yes answer results, there is usually a feeling, after the computation is over, that one lacks an understanding of why the two expressions are equal. Generally speaking, the reason that the method of test by state graph fails to produce any insight about the regular expressions is that it goes outside of the language of regular expression to test the equality.

The second method of establishing the equality of regular expressions is proof by reparsing. This method was used above to verify that $\alpha(\beta\alpha)^* = (\alpha\beta)^*\alpha$. To illustrate a more involved application it will now be used to show that $(0^*1)^* = \lambda \cup (0 \cup 1)^*1$. Note that this equation is an instance of law (7) above. It is easy to see that the proof below is adequate to establish the more general law, but the more specific instance is more easily discussed.

To show first that $(0^*1)^* \subseteq \lambda \cup (0 \cup 1)^*1$, consider an arbitrary word $W \in (0^*1)^*$. Then $W = U_1U_2 \ldots U_n$, $n \geqslant 0$. If $n = 0$ then $W = \lambda$ and then $W \in \lambda \cup (0 \cup 1)^*1$. If $n \geqslant 1$, then for each i, $1 \leqslant i \leqslant n$, $U_i \in 0^*1$. We can then write $W = (U_1U_2 \ldots U_{n-1}0 \ldots 0)1$, where the number of 0's at the end of the parenthesized part is zero or more. Clearly $U_1U_2 \ldots U_{n-1}0 \ldots 0 \in (0 \cup 1)^*$, since the latter is the set of all words of 0's and 1's. Thus $W \in (0 \cup 1)^*1$ and therefore $W \in \lambda \cup (0 \cup 1)^*1$.

To show that $\lambda \cup (0 \cup 1)^*1 \subseteq (0^*1)^*$, consider an arbitrary $W \in \lambda \cup (0 \cup 1)^*1$. If $W \in \lambda$ then $W \in (0^*1)^*$ by the definition of the star. If $W \in (0 \cup 1)^*1$ then $W = U1$, where U is any sequence of 0's and 1's. Suppose there are $n - 1$ occurrences of 1. For the sake of perspicacity, and since our primary concern is not for rigour, suppose that $n = 5$ and suppose $U = 00110100001000$. Then $W = 001101000010001$. To show that $W \in (0^*1)^*$, simply parse W so that each 1 ends a phrase, thus

$$W = (001)(1)(01)(00001)(0001).$$

That this reparsing can be accomplished in general follows from the mere fact that W ends in a 1, which concludes the proof.

A proof by reparsing such as this, although insightful, tends to be tedious and complicated, if a new proof is required for each new equation. The third and fourth methods are more systematic and practical. The third method is that of a logical system with rules of inference, in which one proves equations in the manner of a formal proof. This method has received some attention recently with interesting positive results. Arto Salomaa has written a paper [3] in which he put forth an axiom system with three rules of inference, described below, and conjectured that the system is complete (which means that all valid equations can be derived). A few months after he published this work, he was able to prove that his system was complete [11]. Furthermore, the proof of completeness for a similar system, using exactly the same rules with a slightly different set of axioms, was given by Mr. Stal Aanderaa [4]. The set of axioms Aanderaa used are similar to Salomaa's. They are simple and obviously valid.

These equations may contain variables ranging over regular sets of words, which will be the later letters of the Roman alphabet, x, y, etc., in this paper. Greek letters will be meta-language variables ranging over regular expressions, for use in describing the system.

The three rules of inference used by both Salomaa and Aanderaa are as follows.

Substitution. Substitute a regular expression for every occurrence of a variable in a given equation. Example: from $(x^*y)^* = \lambda \cup (x \cup y)^*y$ infer

$$(x^*000)^* = \lambda \cup (x \cup 000)^*000.$$

Replacement. From an equation $\alpha = \beta$, and from an equation in which α occurs as a well-formed part, the result of replacing an occurrence of α by β may be obtained. Example: from $(x^*y)^* = (x^*y)^* \cup yyy$ and $(x^*y)^{**} = (x^*y)^*$, infer $(x^*y)^* = (x^*y)^{**} \cup yyy$.

Star introduction. From $\alpha = \alpha\beta \cup \gamma$ derive $\alpha = \gamma\beta^*$, provided $\lambda \notin \beta$. (It is to be noted here that there is an easy syntactic method of noting whether or not any given regular expression contains the null-word; e.g., see page 8 of [3].) Example: take $\alpha = (0 \cup 1)^*1 \cup \lambda$, $\beta = 0^*1$, and $\gamma = \lambda$. Then from

$$(0 \cup 1)^*1 \cup \lambda = [(0 \cup 1)^*1 \cup \lambda]0^*1 \cup \lambda,$$

one can infer $(0 \cup 1)^*1 \cup \lambda = \lambda(0^*1)^*$, since in this case $\lambda \notin \beta$.

Note that this last example could be a step in the proof that

$$(0^*1)^* = (0 \cup 1)^*1 \cup \lambda$$

in an axiom system. The earlier portion of the proof would have to establish the lemma that $(0 \cup 1)^*1 \cup \lambda = [(0 \cup 1)^*1 \cup \lambda]0^*1 \cup \lambda$, and the last few steps would simply utilize the fact that $\lambda(0^*1)^* = (0^*1)^*$.

To give some idea of how simple the set of axioms can be, I shall adapt Aanderaa's axioms to the notation of this paper.

(A1) $x \cup x = x$.
(A2) $x \cup y = y \cup x$.
(A3) $(x \cup y) \cup z = x \cup (y \cup z)$.
(A4) $(xy)z = x(yz)$.
(A5) $x(y \cup z) = xy \cup xz$.
(A6) $(x \cup y)z = xz \cup yz$.
(A7) $x \cup \emptyset = x$.
(A8) $x\emptyset = \emptyset$.
(A9) $\emptyset x = \emptyset$.
(A10) $\lambda = \emptyset^*$.
(A11) $x\lambda = x$.
(A12) $x^* = \lambda \cup xx^*$.
(A13) $x^* = (\lambda \cup x)^*$.

To say that the system with these axioms and rules of substitution, replacement, and star introduction is complete, is to say that all valid equations can be derived from the axioms by means of the rules. Aanderaa's proof of completeness makes rather skilful use of Brzozowski's notion of the derivative, as found in [5]. (See [11].)

The simplicity of the axioms is not all that should be hoped for in such a system. Optimally, an axiom system should have only logical rules of inference; the axioms ought to be sufficient to yield all desired valid truths by means of such rules. But in this axiom system, the rule of star introduction expresses a great deal of mathematical content and cannot be justified on the basis of logic alone. Substitution and replacement are all right in this regard; they are justified solely by virtue of the meaning of equality.

Now Redko [6] has proved that there is no finite set of axioms yielding all valid equations when only the rules of substitution and replacement are allowed. In my opinion, an intriguing unsolved problem (although vague) is whether there exists an interesting infinite set of axioms from which all valid equations can be derived using only substitution and replacement. Such an axiom set might be more significant than the simple set listed above. The question comes up here as to what we mean by an *interesting* infinite set of axioms. Saul Gorn has suggested† that the axioms might all be instances of a finite number of schemata. This appears to me to be a reasonable suggestion, although it does not remove all of the vagueness from the question until we specify precisely what we mean by "an axiom schema."

The fourth method for proving regular-expression equations is by means of graphs. The earliest paper on the use of the kind of graphs that we shall now discuss (as opposed to the more restricted concept of state graph) is the early paper published by Chomsky and Miller in 1958 [10]. It did not discuss regular expressions, but defined the notion of "finite-state language" as a language generated by a graph with a finite set of nodes. Although several papers have appeared linking up these graphs with regular expressions, I know of no reference since their paper to the proof by graphs of regular-expression equations.

Until recently it seemed that the graphical method of proving equations had the most to offer in almost every respect. However, since then completeness proofs for the axiom systems have indicated that the axiomatic method is at least as worthy of attention. For it is so much easier to write down a sequence of regular-expression equations than to write down a sequence of graphs. My present opinion is that even if the axiomatic method is better for proving equations, the graphical approach is better in cases in which we are not certain what it is we want to prove.

To begin our discussion of graphs let us take a radically different point of view; let us think of a regular expression as a device for generating words. For example, from $[0(00)^*1 \cup 11]^*$ one generates a sequence either with a 0 or a 1. If one selects the 0 then one has the option of writing 00 any number of times and then writing 1; but if one elects to begin with a 1 one must write 1 immediately after. After that, one again has the option of writing 0 or 1, etc. Of course, by virtue of the meaning of the star, one could have settled for the null-word at the very outset. Thus the regular expression can be thought of as a generator, a sequential machine that operates in a non-deterministic manner (in the sense that it makes arbitrary choices at each juncture) to generate a word. The event of

†During the discussion that followed the oral presentation at the Conference.

such a machine would be the set of all possible words that could be generated by such a device.

Now the concept of non-deterministic machine has little practical significance, but is found very frequently in the theory of automata. Certainly not all machines considered in the theory of automata will come into existence, and probably all of the varieties of non-deterministic automata that have been considered are excellent candidates for non-existence. But the prevalence and the obvious theoretical usefulness of non-deterministic concepts cannot be denied; rather, we are left with a challenge to explain why it is that these concepts are useful.

In this case we are interested in the non-deterministic machine only as a way of describing a set, namely, the set of all words that could be generated. Thus there is no sense in building such a machine, since the regular expression is as concrete as we need. It turns out that a useful alternative to the regular expression is something equally abstract, namely, the graph. As Yamada has noted [9], the graph is a concept that arises quite naturally when one takes this point of view towards a regular expression. Thus, consider the regular expression

$$[0(00)^*1 \cup 11]^*$$

as a word generator in the manner described above. The same function could be accomplished by the graph in Fig. 1. To generate a word in such a graph one first selects a path beginning at the initial node (designated by a single, unlabelled

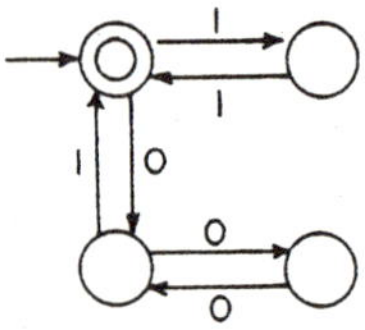

FIGURE 1

arrow pointing to it) and terminating at a terminal node (double circle). The word spelled out by this path by the labels on the branches along the way is in the event. Thus the event is simply the set of words spelled out by the set of all such paths.

Every regular expression can be converted into a graph. There is an algorithm to accomplish this which I shall not describe in detail, since, in most cases, the construction is obvious and is suggested by the relationship between the regular expression $[0(00)^*1 \cup 11]^*$ and Fig. 1. I should mention, however, that there is a pitfall: if one constructs a graph in the most obvious way from a given regular expression one may introduce sneak paths. For example, one might incorrectly render $[10 \cup 0(11)^*]^*$ as the graph of Fig. 2, which contains a sneak path, namely, the path spelling out 1011, which is not in the event represented by the regular expression. A correct rendition would be Fig. 3 which has a branch labelled λ (the null-word). The significance of such a branch is that, from any path using the branch, to determine the word spelled out, one ignores the λ. Sneak paths

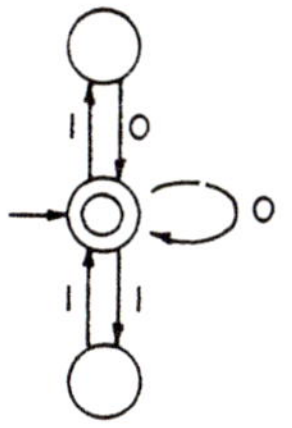

FIGURE 2

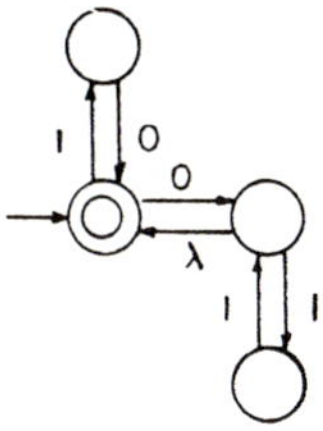

FIGURE 3

are avoided by this technique because any path using the branch must go in the direction of the arrow.

The totality of these graphs includes the well-known state graphs. But not all such graphs are state graphs; in fact, a state graph is a very special kind of graph, in which there are no branches labelled λ and, for every member of the alphabet and for every node, there is exactly one branch labelled with that letter and leaving that node. These general graphs are sometimes referred to as "nondeterministic state graphs," although I prefer to call them "transition graphs." An interesting variant on transition graphs, in which the branches may be labelled with arbitrary regular expressions and to which the techniques of signal flow graphs can be applied, has been developed by McCluskey and Brzozowski [7].

Now, although every regular expression can be transformed into a graph that has the same structure, the converse is not true. I will not define here precisely what I mean by the structure of a regular expression or graph, and hope that my point is made on the intuitive level. Thus, for example, the structure of $(0^*1)^*$ is very much different from the structure of $(0 \cup 1)^*1 \cup \lambda$, although the two regular expressions are equal. And I hope the reader knows what is meant by the statement that the state graph of Fig. 3 has the same structure as the regular expression $[0(11)^* \cup 10]^*$ or (more precisely) $[0(11)^*\lambda \cup 10]^*$. One way of explaining what this means in intuitive terms is to say that the parts of the graph correspond to the parts of the regular expression in such a way that a path through a graph will spell out the same word as a path through the corresponding parts of the regular expression. To be slightly more precise, the part of a regular expression corresponding to a node of a graph would be a point immediately to the left or to the right of one of the characters (such as 0, 1, or λ). And, for further clarification of this point, note that for each well-formed part of the regular expression $[0(11)^* \cup 10]^*$ there is a part of the graph, such that if one well-formed part of a regular expression is part of another, the same is true of the corresponding part of the graph.

Hopefully, the notion of structure of regular expressions and graphs is clear. It is important to note that although every regular expression has a graph which represents the same event and has the same, or almost the same, structure, there are some graphs whose structures are very far from the structure of any regular expression. An example of such a graph is given in Fig. 4. Such graphs are graphs in which there is no hierarchy of loops as there is in regular expression. A loop in a regular expression is always given by a star; and for every pair of stars, either one is inside the scope of the other or their scopes are completely disjoint. A loop in a graph is any directed path that ends where it begins, and otherwise does not

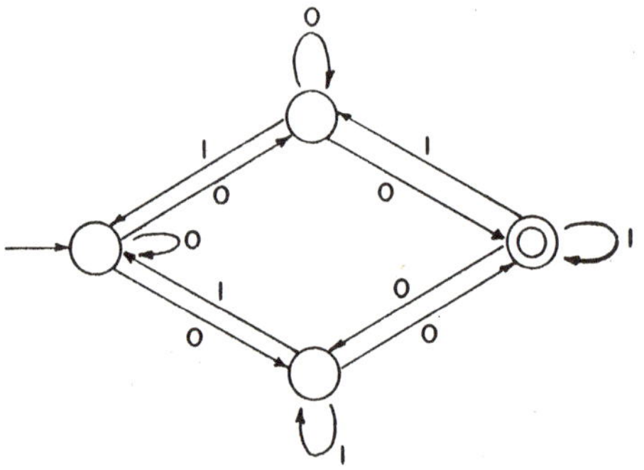

FIGURE 4

repeat any node. It should be clear that in the graph in Fig. 4, there is no way of imposing a hierarchy on the loops corresponding to stars in a regular expression. I am afraid this remark must be left as a rather vague suggestion, rather than as a precise technical proposition, as a precise account of what a hierarchy of loops might be in a graph is beyond the scope of this paper. I trust it is clear, at least, that the graph of Fig. 4 has a structure which is quite unlike the structure of any regular expression. And it is not difficult to construct many graphs of this kind.

For this reason, we can say that the set of graph structures is richer than the set of regular-expression structures. This fact contrasts with the fact that the class of events represented by graphs is exactly the same as the class of events represented by regular expressions. For it is well known that for every graph (including, e.g., that of Fig. 4) there exists a regular expression representing the same event (although it has, in general, a vastly different structure). In fact, there is an algorithm to make this conversion (see [8]).

To show how graphs can be used to prove a regular-expression equation, I will present a graph-theoretic proof of the same equation that I have proved by the other methods, namely $(0^*1)^* = (0 \cup 1)^*1 \cup \lambda$. This is done in the sequence of graphs in Fig. 5, where the graph of Fig. 5a has the structure of $(0^*1)^*$ and the graph of Fig. 5d has the structure of $\lambda \cup (0 \cup 1)^*1$. The sequence is such that each step preserves the event precisely; no words are introduced or deleted. Furthermore, each step is the result of deletion or addition of a branch or of a branch and a node. The step leading to Fig. 5b is justified in that any path using the new branch could just as well have gone via the old branches labelled 1 and λ. The addition of the branch to a new terminal node in Fig. 5c is justified by the fact that any word ending at that new node could just as well have ended at the original terminal node. And the deletion of the branch in Fig. 5d is justified by a

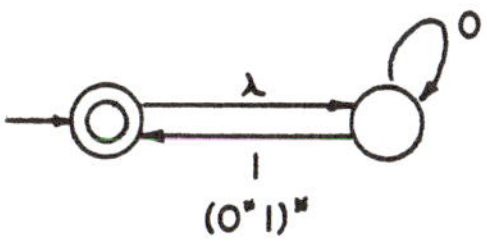

FIGURE 5a

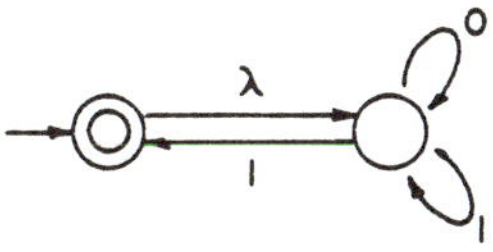

FIGURE 5b

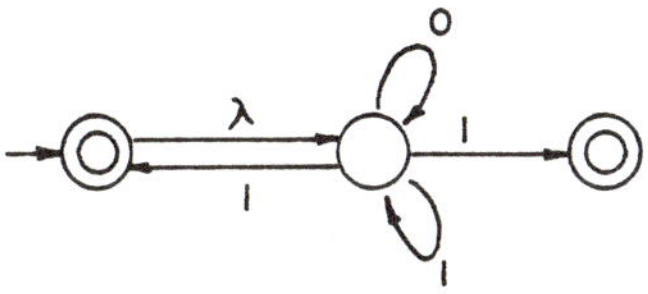

FIGURE 5c

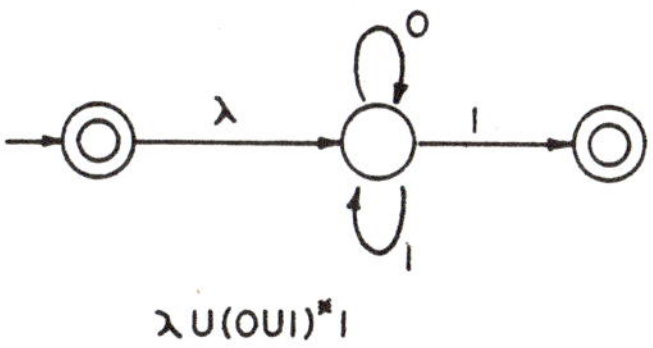

FIGURE 5d

slightly more complicated consideration: any use of that branch in a path is either terminal or non-terminal; if it is terminal then it could be replaced by the branch labelled 1 to the right-most node; if it is non-terminal then it must be followed by the branch labelled λ, in which case, both items in the path together could be replaced by the loop branch labelled 1. Note finally that λ is in the event represented by the graph of Fig. 5d by virtue of the fact that the initial node is terminal.

It turns out that, probably, the proof by graphs in Fig. 5 is not the most general kind of proof by graphs† because all of the graphs in Fig. 5 are structurally similar to regular expressions as follows: (5a) $(0^*1)^*$; (5b) $((0 \cup 1)^*1)^*$; (5c) $((0 \cup 1)^*1)^* (\lambda \cup (0 \cup 1)^*1)$; (5d) $\lambda \cup (0 \cup 1)^*1$. In general, in such derivations, we can expect graphs which are not structurally similar to any regular expressions. (Examples of such graphs occurring in a derivation are Fig. 7b and Fig. 8b.)

This will conclude my discussion of proofs by graphs. Several more illustrations of proofs (Figs. 6, 7, and 8) will be given for the enterprising reader without any explanation of the steps in each case, although each can be verified by inspection, perhaps after some reflection. Fig. 6 is a proof that

$$[00^*1 \cup 101]^* = \lambda \cup (01 \cup 0 \cup 101)^*(01 \cup 101);$$

Fig. 7 that $\{[(0 \cup 1)^*000]^*001\}^* = (0 \cup 1)^*000001(001)^* \cup (001)^*$; Fig. 8 that $[(1^*0)^*01^*]^* = \lambda \cup 0(0 \cup 1)^* \cup (0 \cup 1)^*00(0 \cup 1)^*$. (These were problems assigned to my class in "The Theory of Automata" at M.I.T., and some parts of these proofs are adapted from their homework papers.) Note that in these graphs branches are labelled with arbitrary words instead of just letters of the alphabet; this practice makes them easier to draw and easier to look at. The reader who studies Figs. 6, 7, and 8 will naturally ask the question, is there a set of formal rules for graph manipulation of this kind? As far as I know, no one has put forth a set of rules which are valid, simple formally precise, and complete. ("Complete"

†This was pointed out during the discussion following the oral presentation of this paper.

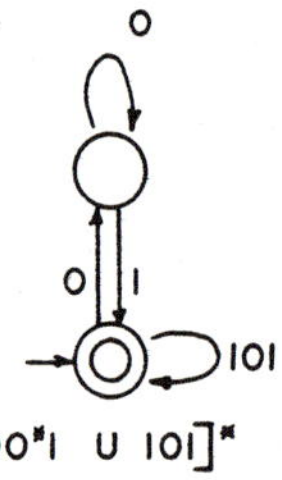

FIGURE 6a

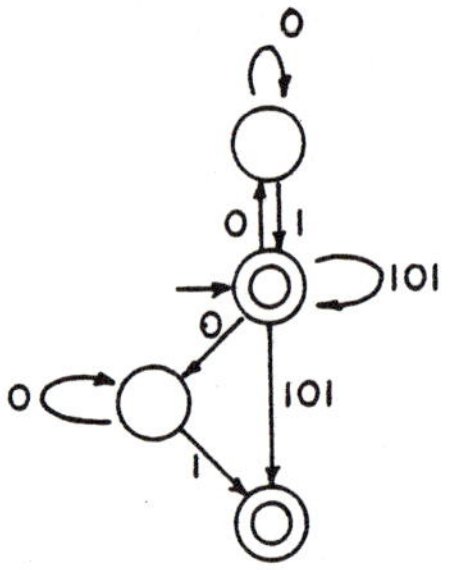

FIGURE 6b

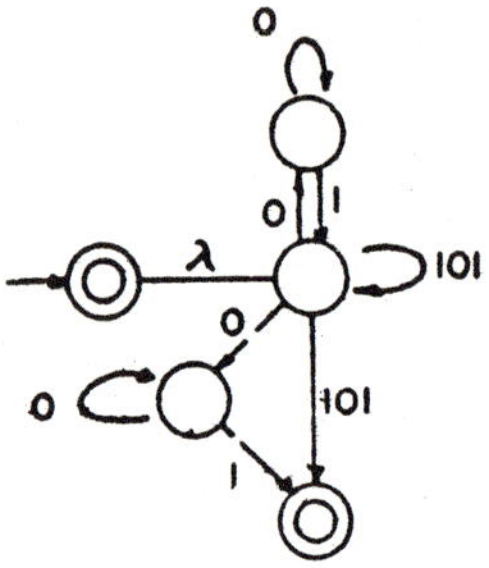

FIGURE 6c

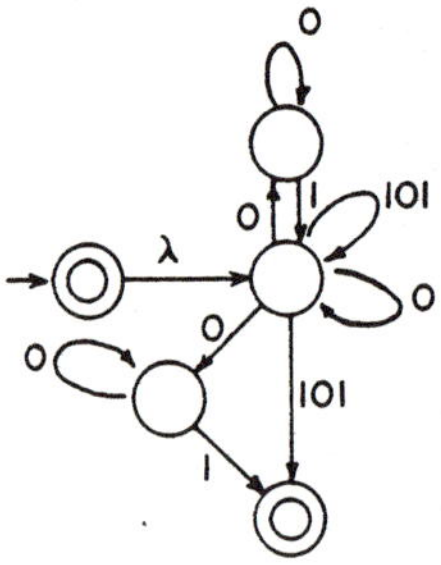

FIGURE 6d

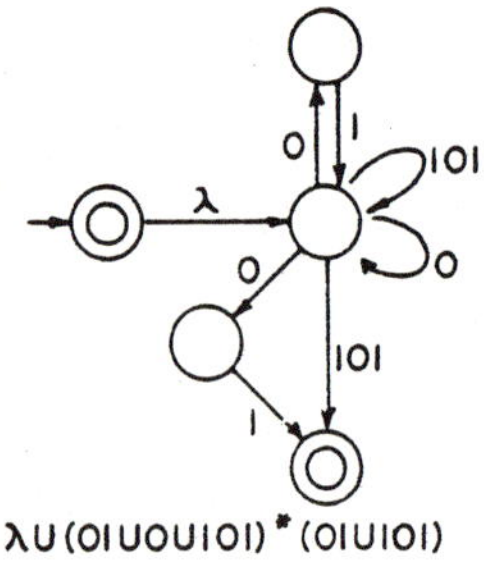

FIGURE 6e

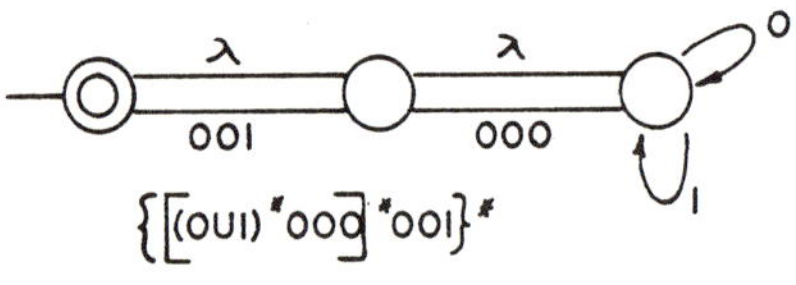

FIGURE 7a

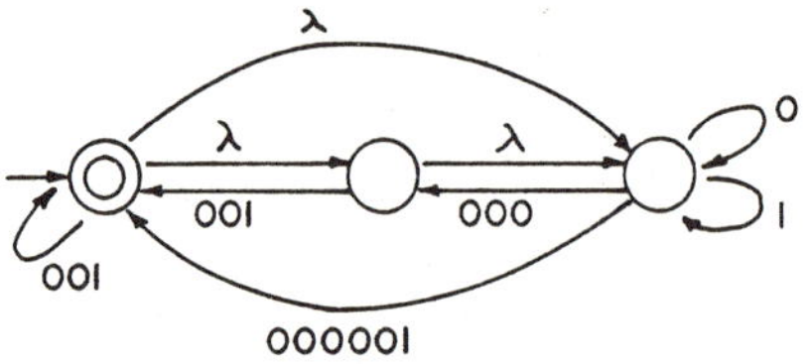

FIGURE 7b

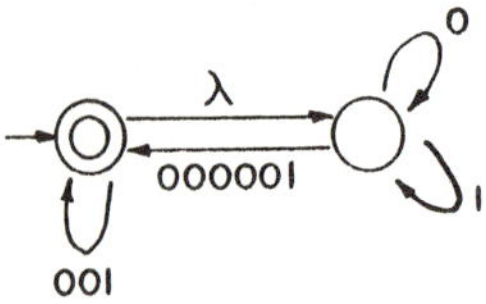

FIGURE 7c

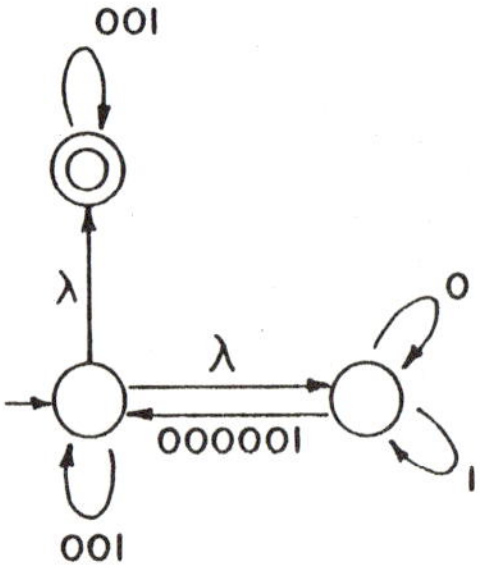

FIGURE 7d

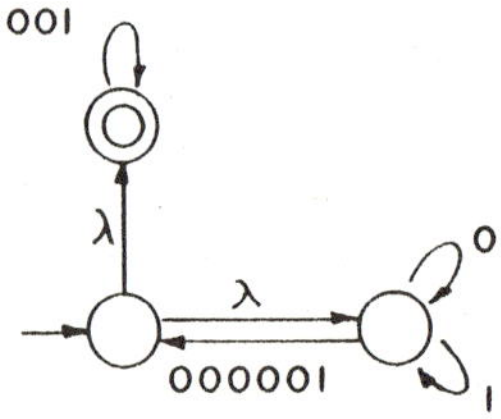

FIGURE 7e

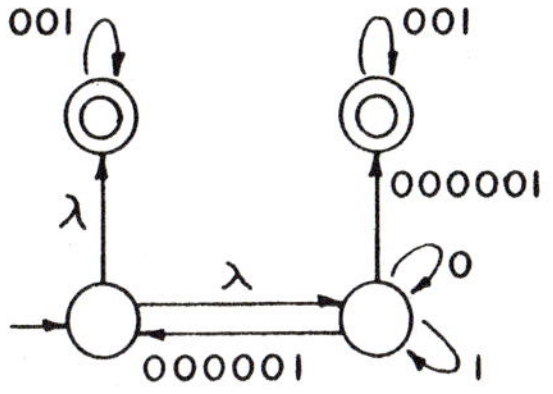

FIGURE 7f

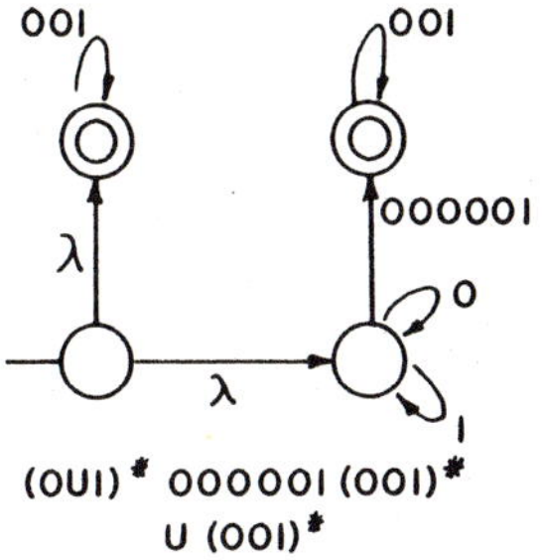

FIGURE 7g

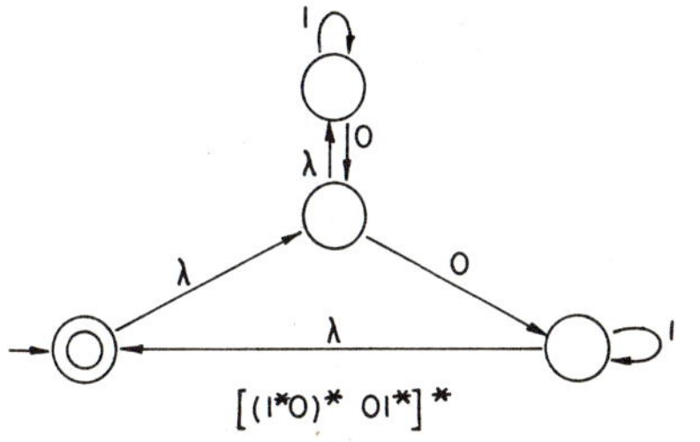

FIGURE 8a

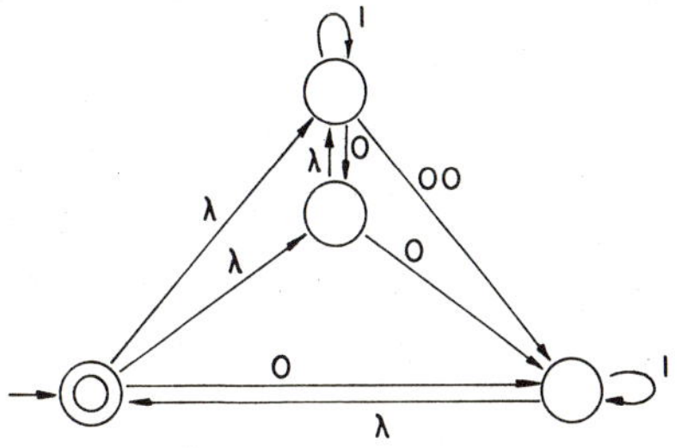

FIGURE 8b

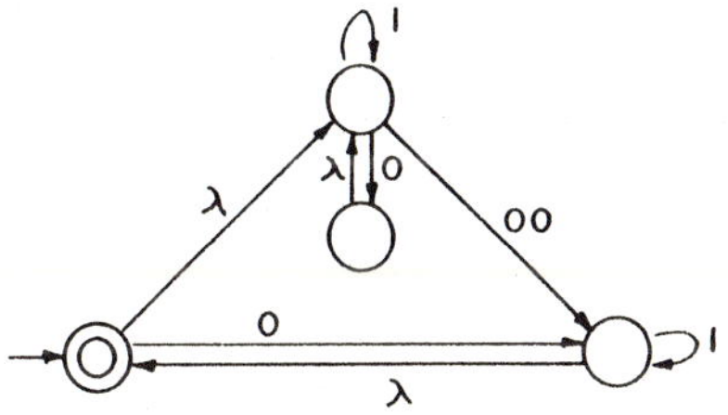

FIGURE 8c

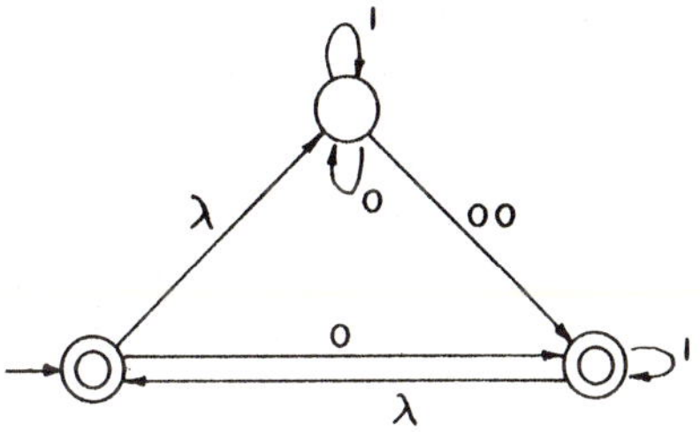

FIGURE 8d

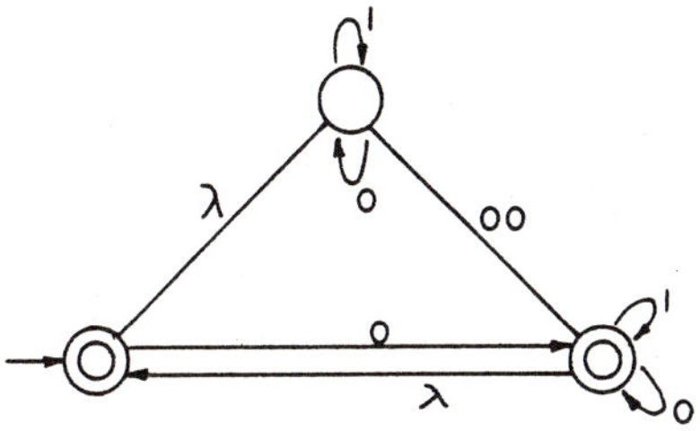

FIGURE 8e

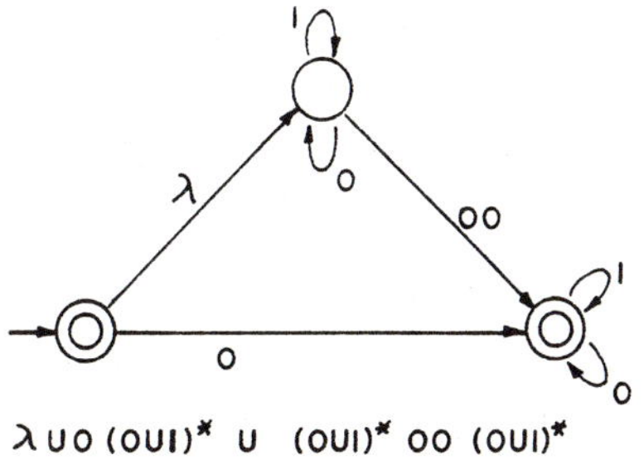

FIGURE 8f

means that for any valid regular-expression equation there is a sequence of graphs establishing the equation, where each step is according to one of the rules.) In my opinion, this open problem is worth looking into.

My final remark concerns the problem of finding interesting equations to prove. Alternatively, given a regular expression, what interesting or (in some significant sense) more simple regular expression is it equal to? A more ambitious question along these lines is that of a canonical form; as far as I know, no one has advanced a method of putting regular expressions into a canonical form, which is anything but just an arbitrary unique regular expression.

My favourite concept along these lines is the concept of star height. The star height of a regular expression is the maximum length of a sequence of stars in the expression, such that each star is in the scope of the star that follows it. Thus the star height of (0*1)*[(000*10)*000]* is 3 by virtue of the sequence consisting of the third, fourth, and fifth stars.

The star height of a regular expression is a precise explication of the loop complexity. And since it does achieve something to simplify the loop complexity of an event, it is also an achievement to reduce the star height of a regular expression. I will stop now, except to point out that this objective could also explain the choice of equations proved in Figs. 6, 7, and 8. In each case, imagine that one starts with a regular expression whose star height he wishes to reduce. He then draws a graph, reduces the loop complexity of the graph, and ends up with a graph whose corresponding regular expression has a reduced star height. Thus if we consider the problems of reducing the star height of the regular expressions (0*1)*, [00*1 ∪ 101]*, {[(0 ∪ 1)*000]*001}*, and [(1*0)*01*]*, then Figs, 5, 6, 7, and 8, respectively, are solutions showing that each is reducible to a regular expression with star height 1. (See [12].)

REFERENCES

[1] Brzozowski, J. A., "A Survey of Regular Expressions and their Applications," *IRE Trans. on Electronic Computers*, EC-11 (1962), 324–35.

[2] Kleene, S. C., "Representation of Events in Nerve Nets and Finite Automata," *Automata Studies*, C. E. Shannon and J. McCarthy, eds. (Princeton: Princeton University Press, 1956).

[3] Salomaa, A., "Axiom Systems for Regular Expressions of Finite Automata," *Turun Yliopiston Julkaisuja (Annales Universitatis Turknensis)*, series A, 75 (1964), 5–29.

[4] Aanderaa, S., "On the Algebra of Regular Expressions" (unpublished).

[5] Brzozowski, J. A., Derivatives of Regular Expressions, *Journal Assoc. Computing Machinery*, 11 (1964), 481–94.

[6] Redko, V. N., "About the Determining Set of Relations in the Algebra of Regular Events," *Ukrainian Mathematical Journal*, 16 (1964), 120–6.

[7] Brzozowski, J. A., and McCluskey, E. J., Jr., "Signal Flow Graph Techniques for Sequential Circuit State Diagrams," *IEEE Trans. on Electronic Computers*, EC-12 (1963), 67–76.

[8] Eggan, L. C., "Transition Graphs and the Star-Height of Regular Events," *Michigan Math. Journal*, 10 (1963), 385–97.

[9] Yamada, H., "Disjunctively Linear Logic Nets," *IRE Trans. on Electronic Computers*, EC-11 (1962), 623–39.
[10] Chomsky, N., and Miller, G. A., "Finite State Languages," *Information and Control*, 1 (1958), 91–112.
[11] Salomaa, A., "Two Complete Axiom Systems for the Algebra of Regular Events," *Journal Assoc. Computing Machinery*, 13 (1966), 158–69.
[12] McNaughton, R., "The Loop Complexity of Regular Events," to appear in the *Proceedings of the 1965 Symposium on Logic Compatability and Automata in Rome, New York* (ed. Frank B. Canonito).

Some Comments on Self-Reproducing Automata

Michael A. Arbib*

1. INTRODUCTION

The first answer to the question "Why study the problem of self-reproducing automata?" is the familiar "Because it's there." Von Neumann raised the problem in 1948, and at his death left a complicated and unfinished manuscript on the problem. This somewhat romantic history has given the problem a fascination which it might otherwise lack.

The second answer has perhaps been less influential in interesting people in the topic, but should, in the long run, prove to be far more important: "Because it may provide a logical framework for understanding embryological processes." The great excitement over DNA-RNA prompted many people to claim that at last we had found the "secret of life." However, the simple DNA → RNA and RNA → enzyme transductions, while important biologically, are of little interest to the experienced automata theorist. The challenging questions seem to be further up in the hierarchy: "How can a complex multi-cellular automaton grow from a single cell, given that a finite programme can be executed within each cell?" Such a question as this is non-trivial, and becomes a fit topic for automata theory, though it must be confessed that, at present, the emphasis is on ingenious programming of cellular arrays rather than on weaving a rich texture of theorems.

There are other answers, too, and we shall encounter a few of them in what follows.

It should perhaps be emphasized that the theory of automata is usually concerned with devices which transform information from an input string to an output string, changes in the automaton being regarded as incidental. In studying self-reproducing automata, the emphasis shifts to the way in which initial information serves to regulate the growth and change in structure of an automaton. It will be of some interest to see what contribution the latter approach can make to our more classical problems. The study of self-reproduction is but one chapter

*Department of Electrical Engineering, Stanford University. The research for this paper was supported in part by the Air Force Office of Scientific Research, under contract no. AF49(638)-1440. I am grateful to Karl Kornacker of M.I.T. for his contributions to sections 6 and 7.

in a thorough-going study of growing automata, and much of what follows is written with this wider aim in view.

When von Neumann presented his paper in 1948 to the Hixon symposium, he raised the following semi-paradox. If one thinks of a machine making another machine, one has the idea that machine reproduction, and I am using this word reproduction in a very broad sense, implies a degradation of complexity—a complex machine builds a simpler machine. But in biological systems, the complexity of an offspring seems much the same as that of the parent; and evolution (whether or not you take it in the narrow sense of natural selection) talks about machines (mainly meat-type machines) producing more complicated machines. This led von Neumann to the question: "Is there a machine such that you can set it loose in a pool containing all matter of components that will run around and hook the components together and eventually come up with a copy of itself—a self-reproducing machine?" This is the question on which we are going to focus our attention and which will lead us off into other questions as we go along. Now, the first thing to observe is that this problem can be made trivial, as the "domino example" of Fig. 1 shows.

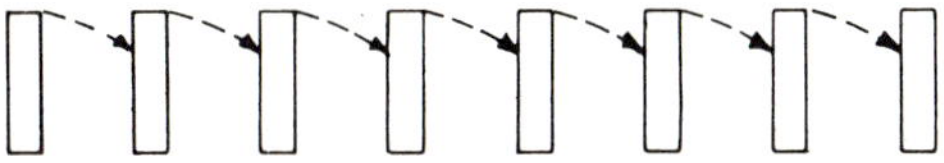

FIGURE 1. Trivial self-reproduction. A domino on edge is the basic component. We stand dominoes in a chain, as shown, and let the automaton we want to reproduce be a *falling* domino. A falling domino knocks down its neighbour, and thus "reproduces"—*falling* is propagated down the chain.

Let us return to the wonderful fact that a human can be produced from a single cell. The point here is that a single cell, bounded in size, containing a large but none the less finite number of ongoing chemical reactions, can develop into an aggregate with much greater capabilities. In the light of this fact, I would like to re-phrase the question of self-reproduction: "How, starting from one fixed kind of finite automaton as basic cell, can we design automata of arbitrary complexity with the computational ability of a universal Turing machine and which can also reproduce itself?" If the question is phrased in this way, we can avoid the domino objection.

2. BRIEF DIGRESSION ON TURING MACHINES*

A. M. Turing was one of the people who, in the thirties, was worried about evolving a precise notion of effective computation. We are all familiar with the idea of an algorithm; one has a recipe such that if one follows it you always get

*For more details, see chap. I of Arbib [1].

the right answer—if there is one. You may run on indefinitely if there isn't an answer. Turing produced a formalization as follows: Let us consider a box which has finitely many states, say q_1 up to q_n, and which operates on a tape divided lengthwise into squares, each of which can bear any one of a finite number of symbols, S_1 up to S_m. We start with a finite tape on which is printed the initial data, the machine is started in state q_1, and operates synchronously. At each time ($t = 0, 1, 2, 3, \ldots$) the symbol it scans and the state it is in determine that the machine stops *or* determine three things: the new symbol (perhaps the

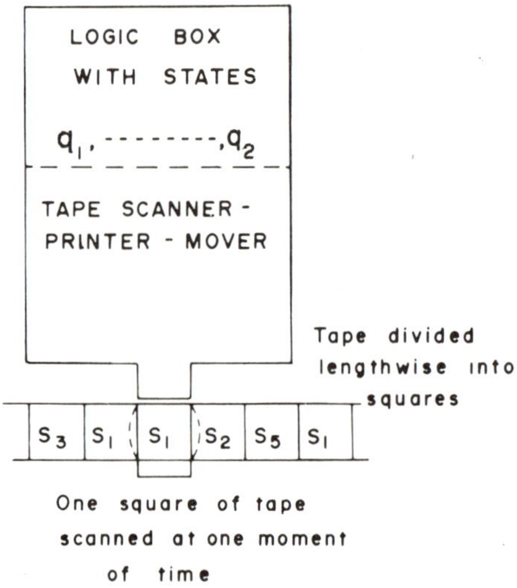

FIGURE 2. A Turing machine. (*Hartmanis*: It's upside down; the tape should be on top. *Arbib*: Ah, but it's an Australian Turing machine!)

same as the old one, or it may be a blank) it prints on the scanned square, the moving of the tape, at the most, one square, and the change of state. Then it is all ready to repeat the cycle, obeying the instruction keyed by the new state-symbol pair. If it ever comes to the end of the tape, then it will add on a new square. This is very different from the action of finite automata—here there is no bound on the amount of memory, for our Turing machine can keep adding more squares as it needs to remember more and more information. Thus, by suitable programming, Turing machines can do many things that finite automata cannot do.

Now suppose we have an algorithm for effectively producing one integer from another. In the past, such a rule has always been transcribable into a set of instructions for a Turing machine. Church, Kleene, Post, Gödel (see, for example, [24]) all gave different models of what they meant by algorithm (recipe, effective computation, mechanical computation), and, in every case, every recipe that they could handle could also be handled by a Turing machine. On the other hand, of course, it is quite clear that a Turing machine computation is in itself a recipe, an effective process. At the present time, the usual thing is to identify the intuitive notion of an effective computation with the formal notion of computation which can be carried out by a Turing machine. (Observe that one can not formally prove this equivalence between our informal notion and our formal notion without formalizing the informal notion—a vicious circle.)

For our present purposes, the important thing that Turing discovered in his paper of 1936 was that you could build a "universal" Turing machine. How does it work? Usually, given a Turing machine, Z, we use it to compute a function, f, by placing the number x coded on the tape, and letting the machine run. If and when Z stops, the result on the tape is decoded to give us $f(x)$. Every such machine Z is given by a finite list of instructions, one for each state-symbol pair. These lists can be enumerated, and thus we can designate a Turing machine, say Z_n, as nth in some appropriate list. Now Turing gave a programme for a Turing machine, U, which was universal in that if you wrote on U's tape the ordered pair* n and x then, at the end of U's computation, it would produce $f_n(x)$—precisely what Z_n would have computed with input x. If Z_n wouldn't have stopped, of course U won't stop.

To conclude this digression we must mention an important fixed-point theorem. We say a function is partial recursive (from natural numbers to natural numbers) if it is an f_n, i.e., some Z_n can compute it. Such a function is called total if it is defined for all x (i.e., if Z_n will eventually stop, no matter what x it is started with). Consider a *total* recursive function h. It induces a functional $\tilde{h}$ on the partial recursive functions

$$\tilde{h}(f_n) = f_{h(n)}.$$

The *recursion theorem* tells us that $\tilde{h}$ has a fixed point—there is an n_0 (depending on h, of course) such that

$$f_{n_0} = f_{h(n_0)}.$$

It is important to bear in mind that Z_{n_0} and $Z_{h(n_0)}$ are probably not identical Turing machines—they compute the same function, but may use entirely different programmes to do so. The machines are functionally, but not structurally the same.

3. SELF-REPRODUCTION AS TWO FIXED-POINT THEOREMS

We have seen that it is possible to programme a universal Turing machine which, given for any Turing machine Z, a suitable encoded description $e(Z)$, would simulate any computation which Z could execute. This suggested to von Neumann the existence of an automaton A which, when furnished with the

*Actually, Turing required a coded description of Z_n's programme, rather than n itself. We took the simple code here of position in the enumeration, but the result is still true for any "reasonable" choice of code.

description of any other automaton *M* (composed from some suitable collection of elementary parts), would construct a copy of *M*.

He outlined how one could use such a "universal constructor" *A* to solve the problem of self-reproduction, to build an aggregate out of our elementary parts in such a manner that "if it is put into a reservoir in which there float all these elements in large numbers, it will then begin to construct other aggregates, each of which will at the end turn out to be another automaton exactly like the original one."

However, there is a great gap between the known universal computer and the posited universal constructor. For our universal Turing machine only simulates symbolic operations—any physical realization of such a machine would use components of a nature and complexity quite different from that of the tape symbols. This point requires some emphasis. Von Neumann's paper of 1951 has left many people with the impression that there is no new logical problem here—that the existence of a self-reproducing machine is reducible to an instance of the recursion theorem of recursive function theory. This is not so. There are two fixed-point theorems to prove.

One might suspect that, given a list of elementary parts, a universal constructor for machines comprising only those parts might itself have to be constituted of a more complex variety of components. The proof that this suspicion is wrong gives us the first fixed-point theorem: *There does exist a set of components from which may be built a universal constructor for automata built from that very set of parts.*

The universal constructor *A*, supplied with a description of an automaton, will build a copy of that automaton. However, *A* is not self-reproducing: *A*, supplied with a copy of its own description, will build a copy of *A* without its own description. The passage from a universal constructor to a "self-constructor" is, in essence, a fixed-point theorem reducible to the recursion theorem.

Von Neumann gave the first construction of a self-reproducing automaton in a manuscript on "The Theory of Automata: Construction, Reproduction, Homogeneity" which was incomplete at the time of his death, and which A. W. Burks has since edited for publication by the University of Illinois Press. Von Neumann replaced the reservoir in which the organism floated by a "tessellation" of identical cells, which were finite automata with only 29 states.

The trouble is that it is very complicated to programme something with such simple components. For instance, you can't just have two wires crossing. One method of solving the cross-over problem is to code messages so that, when a message comes to a crossing, it goes both ways, but only a decoder in the desired direction will be able to "understand" the message. Another method is to introduce a large array of cells such that if you put a pulse in any corner, it will come out of the opposite corner. Thatcher [31] has since produced a polished version of von Neumann's scheme, using the same set of components. Arbib [2] has shown how to reduce the complexity of the programming immensely, but only at the cost of more complicated modules. We shall argue in section 7 that this is a cost we may be happy to pay.

4. *CT*-MACHINES AND TESSELLATIONS

CT-machines (Thatcher [31]) combine the functions of a *W*-machine (the programmed version of a Turing machine, Post [24], Wang [36], and Lee [15]) and of a construction machine which is a print-only machine with a half-plane divided into equal squares for its tape. The constructing arm operates on one square at a time, where it can print one symbol from the alphabet V_c; the arm may also be instructed to move one square up, down, left or right.

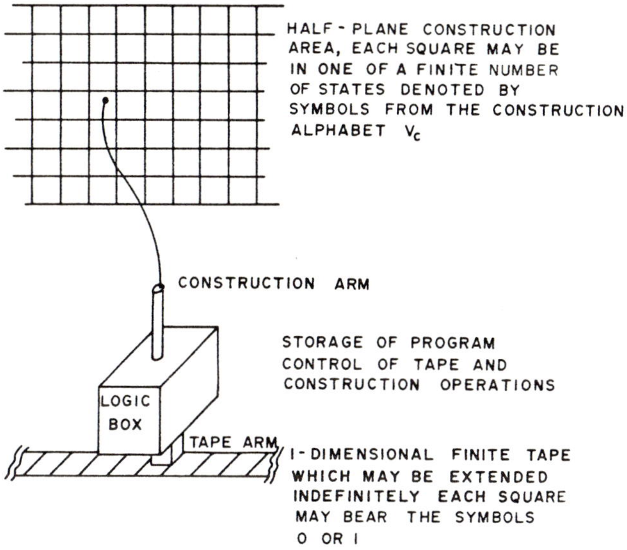

FIGURE 3. A *CT*-machine.

A *CT*-machine is programmed with a finite list of instructions—*C*-instructions for constructions, and *T*-instructions for tape computations. Our first task is to embed *CT*-machines in a plane tessellation in such a way that they can construct new *CT*-automata in the half-plane above them. Each cell in the tessellation is an identical finite automaton; cells differ only in their internal states. One state is designated quiescent.

We shall give a brief outline of the construction of Arbib [2]. One factor which drastically simplifies the construction is the use of a "welding" operation* whereby cells can be formed into aggregates which may be moved about the plane *en masse*. Thus, in distinction to the von Neumann-Thatcher model, the tape of a Turing machine is actually modeled as a one-dimensional string of cells welded together so that they can be moved left or right on command. Another

*Suggested by von Neumann in his 1948 Princeton lectures—see Burks [4]—but abandoned in his tessellation model; it was used by Myhill [19] in his outline of the design of a self-reproducing *C*-machine, as distinct from *CT*-machine.

factor is the use of programmed modules. The basic cell has the structure shown in Fig. 4. There are input and output channels, and weld positions, in each of the four directions. There is also a bit register *BR*, and 20 registers to hold the internal programme.

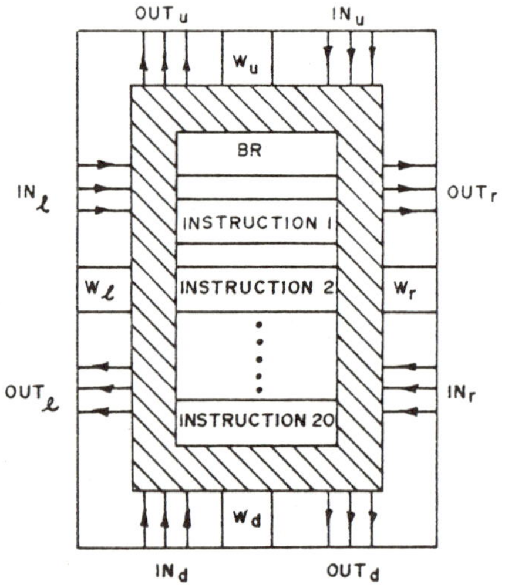

FIGURE 4. The basic module.

The hatching in Fig. 4 denotes the combinatorial circuitry which combines the inputs and the setting of the registers and welds at time t to determine the move between t and $t + 1$; new register settings and output of the module at time $t + 1$. The quiescent state is that in which all 25 registers (including the four weld registers) are set to zero. Each weld may be on (state 1) or off (state 0); and two neighbouring cells are said to be welded if either of the welds on their common boundary is in state 1.

Let W be the ancestral relation of welding on cells, i.e., $W(a, b) \Leftrightarrow a$ and b are welded neighbours or $\exists\ c$ such that $W(a, c)$ and $W(c, b)$. A collection C of cells is called *co-moving* if $a \in C \Rightarrow : b \in C \Leftrightarrow W(a, b)$.

We make the following convention: If any cell in a co-moving set C receives at time t an instruction to move in direction x (u, d, l, or r), and no cell in the set receives at time t an instruction to move in any direction other than x, then all cells of C will move one square in the direction x, unless such a move would cause C to "collide" with another co-moving set. Conventions to handle the latter case, and to ease programming, are given in detail in the original paper.

The behaviour of a module between times t and $t + 1$ will be governed by the state of its four neighbours at time t, with the sole exception of the move operations described above.

The overall plan of the machine is very simple, as is shown in Fig. 3. The programme consists of a linear string of co-moving cells, partitioned into substrings of $\geqslant 1$ adjacent cells, the left-most of which has $\langle BR \rangle = 1$, the remainder of which (if any) have $\langle BR \rangle = 0$. The mth such substring from the left represents the mth instruction of the CT-programme and contains one cell unless the mth instruction in $t \pm (n)$, in which case it contains $n \mp 1$ cells.

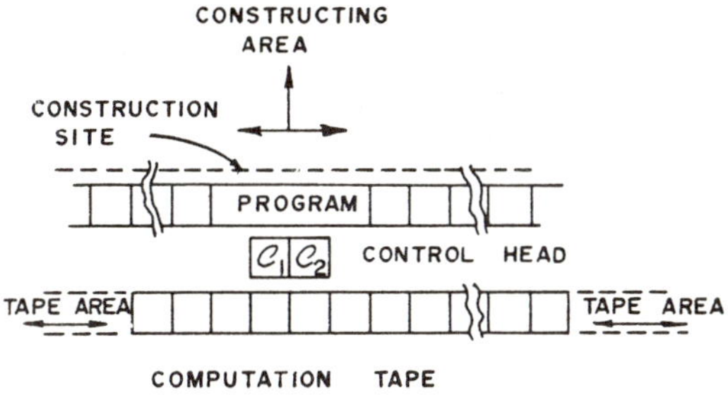

FIGURE 5. Overall plan of embedded CT-machine.

The tape head, comprising two cells, serves to read T-instructions from the programme; to execute them; to initiate construction operations above the programme cell above it; and to move the programme. The tape consists of a linear string of co-moving cells, one for each square of the tape of the simuland—with $\langle BR \rangle$ of a cell being 1 or 0 as the corresponding square of the simuland is marked or not.

We do not have a constructing arm as in the CT-machine shown in Fig. 3. Rather, we construct (or print) only in the square that is up two from C_1 and then move constructed cells about in the construction area by appropriate programming.

For the instruction code of the modules, and the detailed programming, see Arbib [2] and [2a].

5. ABSTRACT THEORY OF SELF-REPRODUCTION

This section is designed to relate our constructions to the abstract approach taken by Myhill [21].

A CT-machine may be completely specified at any stage by a quadruple:

P. its programme;
I. the instruction of the programme it is executing;
T. the state of the tape (finite support);
C. the state of the construction area (finite support).

The construction we have just outlined yields

THEOREM 1. *Any (P, I, T, C) configuration may be effectively embedded in the tessellation.*

THEOREM 2. *There is an effective procedure whereby one can find, for a given CT-automaton A, an embedded CT-automaton $c(A)$ ("constructor of A") which, when started (by telling its control head "go to 1"), will proceed to construct a copy of A in the three rows of its constructing area immediately above it, and activate that copy of A by telling its control head "go to 1."*

Now, the procedure of Arbib [2] takes perhaps three cells in the programme of $c(A)$ to code one cell of A, so it may still seem that any machine is only capable of constructing simpler machines.

We can effectively enumerate all the (P, I, T, C) configurations.

THEOREM 3. *There exists a universal constructor M_u with the property that, given the number n, coded in binary form on its tape as I_n, it will construct the nth configuration M_n in its configuration area. Symbolically $I_n : M_u \to M_n$.*

One may "prove" this by an appeal to a modified Turing hypothesis, but the reader may find some details in [2a].

Why isn't M_u self-reproducing? Because $I_u : M_u \to M_u$, but in the "second generation," $M_u \to ?$, and reproduction fails. (A cell which produces a copy of itself minus the genes is not self-reproducing.)

THEOREM. *For each recursive function h, there exists a machine M_c such that*

$$I_n : M_c \to M_{h(n)} \qquad (c \text{ depends on } h).$$

Proof. There clearly exists a programme $P(h)$ of tape instructions which will convert I_n to the tape-expression $I_{h(n)}$. The machine M_c then has for programme $P(h)$ followed by the instructions of the programme of M_u.

THEOREM (Myhill [21]). *For any computable function g, there exists a machine M_a such that*

$$M_a \to M_{g(a)}.$$

Proof. Let $M_{s(x)} = I_x : M_x$. Then s is a recursive function, and so is $g \circ s$. Thus, taking $h = g \circ s$ in the above, we have

$$I_n : M_c \to M_{h(n)} = M_{g(s(n))}.$$

Setting $n = c$, we obtain

$$M_{s(c)} = I_c : M_c \to M_{g(s(c))},$$

so that $a = s(c)$ satisfies the theorem.

For instance, g could be the function taking M into its mirror image, etc. Taking $g(x) = x$, we have the following theorem.

THEOREM (von Neumann [35]). *There exists a self-reproducing machine.*

6. SELF-REPRODUCTION AND QUANTUM MECHANICS

The scheme we have given for self-reproduction is clearly realizable with electronic circuitry which is subject to the laws of quantum mechanics. It is thus disconcerting to find Wigner [38] claiming to show that self-reproduction is virtually impossible in a quantum-mechanical system (as part of a desire to show that the explication of much of biology, as well as of consciousness, requires "biotonic" laws supplementary to those of physics; cf. Elsasser [9]).* It seems worthwhile at this point to show wherein we believe Wigner's argument fails.

6.1. Summary of Wigner's Argument

(*a*) Let there be a quantum-mechanical state ν which is the living state. (Wigner later replaces this by a whole subspace, which changes neither his conclusion nor our refutation!)

(*b*) The state at time t_0 is given by

$$\Phi = \nu \times w,$$

which is the initially non-interacting combination of the organism and environment.

(*c*) If and when self-reproduction has taken place (say at time t_1), we have two organisms present in a different environment. The state is then

$$\Psi = \nu \times \nu \times r,$$

where r describes the new environment and the relative position of the two organisms.

(*d*) *Assumption* 1. Choosing co-ordinates so that

$$\Phi_{\kappa\lambda\mu} = \nu_\kappa w_{\lambda\mu}$$

$$\Psi_{\kappa\lambda\mu} = \nu_\kappa \nu_\lambda r_\mu,$$

we assume there are only finitely many dimensions—κ and λ taking on N values, say, while μ runs through R values.

(*e*) The change of state from t_0 to t_1 is given by a collision matrix S:

$$\nu_\kappa \nu_\lambda r_\mu = \sum_{\kappa',\lambda',\mu'} S_{\kappa\lambda\mu;\kappa'\lambda'\mu'} \nu_{\kappa'} w_{\lambda'\mu'}. \qquad (*)$$

(*f*) *Assumption* 2. There are no special relationships between the components of S.

(*g*) Since (*) must be valid for any κ, λ, and μ we have N^2R complex equations.

*As scientists, I don't think we can say "biotonic" laws are unnecessary—merely that Wigner's argument does not imply their necessity.

There are several identities here, but since N^2 is enormous, this fact will be disregarded. There are N unknown ν components. r has R unknowns, and w has RN. Thus there are $N + R + NR$ unknowns. These are very much fewer than the number of equations, and so "it would be a miracle" if the equations could be satisfied.

6.2. Critique of Wigner's Argument

Wigner comments that "If one tries to confront the above with von Neumann's specific construction (of a self-reproducing automaton), one finds the confrontation impossible because the von Neumann model can assume only a discrete set of states whereas our model is continuous." This argument seems specious since (*a*) continuity was not used in Wigner's argument; and (*b*) the discreteness of the von Neumann model can be obtained to a sufficient degree of accuracy using *Q.M.* variables. The argument is not saved by the comment, "The inapplicability of his [von Neumann's] model to biological applications was realized by von Neumann," since Wigner's argument does not depend on the bioticity of ν.

We believe that our self-reproducing automaton is, in fact, a counter-example to Wigner's claim, and that the weak link (as, in fact, Wigner has conceded) lies in Assumption 2. Self-reproduction only occurs when ν is highly structured, and w is sufficiently structured to ensure rich interaction between the organism and environment. This rich interaction is represented by many relationships between the components of S which vitiate Assumption 2, and annihilate the "need for a miracle" of (*g*) above.

It is amusing to see that Wigner's argument may be applied to show that there cannot exist a pair of stable objects! Let ν describe a chair, and let $w = \nu \times r_0$ describe a second chair, and the remainder of the environment, at t_0. Let $\Psi = \nu \times \nu \times r_1$ where the two ν's describe the same two chairs, and r_1 the environment at t_1. Pursuing Wigner's argument, we would have to conclude that "it would be a miracle" if the two chairs survived from t_0 to t_1. Once again, Assumption 2 is at fault: in the reproduction example it was the development of organism-environment correlations that vitiated the choice of random S in our stability example, whereas, in Assumption 2, it is the "inter-chair" correlations that vitiate the assumption.

7. CONCLUDING COMMENTS

The remainder of this paper is given over to a consideration of questions which suggest much exciting future research. Parallel computation will be briefly discussed and then the formal model dealt with above will be related to some real biological questions.

(a) Parallel Computation in Iterative Arrays

What we have been concerned with here is iterative arrays (Hennie [10])—identical components set up in some tessellation (in our case, a square array). We have been seeing configurations that can reproduce within this. The question of reproduction of a configuration is not of interest only in terms of biology, because one can think of an embedded automaton as a programme with which we may want to carry out several computations in parallel (Holland [13], Wagner [36]). We then become interested, as computer programmers, in getting this programme to reproduce, i.e., to move up the tessellation and then to carry out its computations on appropriately modified data. Unfortunately, all the publications I have seen (up to 1966) on embedding parallel computations in arrays of such networks seem exploratory—indicating that research has not progressed too far as yet. However, the question of how programmes are copied in an array provides the link between out topic and the computer science end of systems theory, and leads us to many deep and worrying questions. Embedding computations in tessellations emphasizes the point made by Rutledge (at the Conference): in setting up parallel computers, one has to look at processes being hooked together, and to be concerned about how their interactions can be controlled.

I want to emphasize that in the construction I use here, I have slavishly followed the design of the Turing machine which looks only at one square at a time, and moves the tape back and forth. It would be much more sensible, if one were to ever build such a thing, to use the full capabilities of parallel processes.

*(b) Some Loose Analogies with Biology**

We have not yet answered the question "How can a complex multi-cellular automaton grow from a single cell, given that a finite programme can be executed within each cell?" I want to outline the method whereby our model can be modified to yield an answer, and wherein that model differs from the embryological situation.

Let us first note that the self-reproduction we have studied is a far narrower concept than the embryological one—a human zygote grows into a human, not into a replica of one of the parents. The zygote contains only an outline, a programme which, in interaction with the environment, produces an organism of a certain species (if questions of mutation and evolution are ignored). This behaviour leads to a whole series of questions—which here I can merely raise—of the "identity" of an automaton. What does it mean to say two automata belong to the same "species"? Embryological reproduction gives rise to offspring with

*The computer scientist and system theorist can probably find no better introduction to embryology than Ebert's *Interacting Systems in Development*, where he will find further references. See also Dodgson [7].

similar structure, and this implies similarity of function. What are measures of structural similarity and functional similarity, such that the first implies the second? Are there interesting classes of automata for which we may carry out decompositions into a species-dependent automaton, and an individuality-expressing automaton? Can we study reliability of reproduction for such systems, where the genetic information determines the species-dependent automaton with high probability, random influences making their appearance chiefly in modifying the individuality-expressing automaton? Incidentally, I envisage such decompositions as being at the functional, rather than at the structural, level.

We have used modules far more complex than von Neumann's 29-state elements. This seems admirable now that we are turning to biological questions (in fact, we shall shortly introduce a model with far larger modules). The living cell, with its synthetic machinery involving hundreds of metabolic pathways, can rival any operation of our module, and is under the control of DNA molecules, we believe, with far more bits of information than our cell can store. So, perhaps we lose biological significance by unduly limiting the information content of the module.

Basic to the problems of replication (perhaps a better term than self-reproduction in view of our above remarks) are the following.

(i) *Cell reproduction.* This has not been discussed. Our construction rests on the assumption that we can produce new cells at will, our only problem being to ensure that they contain the proper instructions. For a very interesting discussion (albeit with a somewhat incorrect view of algorithmic processes) of self-reproduction at the single-cell level, see Stahl [29]. Stahl does not reproduce actual cells, but string configurations—it is not clear whether the simulated cell is as autonomous in its action as a real cell.

(ii) *Organism replication.* Given cell reproduction, how do we replicate an organism? It seems sensible to subsume a lot of hard work in (i) by using complicated cells. In fact, we might hypothesize (cf. the section on the evolution of complex systems in Simon [28]*) that multicellular organisms can evolve (by whatever mechanism) only when there are complicated reproducing cells available. Contrasting our model with organism reproduction, we note the following.

(1) Our programme was embedded in a string of cells, whereas the biological programme is a string stored in each cell.
(2) We use a complete specification, whereas "Nature" uses an incomplete specification.
(3) We did not use anything like the full power of our model (i.e., the operation was sequential instead of parallel).
(4) We constructed a passive configuration—we set up all the cells with their internal programme, and only then did we activate the machine by telling the control head to execute its first instruction. Contrast the living, growing embryo with this. Our construction relied on the passivity

*This reference contains a good discussion of the importance of hierarchies in our theorizing.

of the components, and demanded that any sub-assembly would stay fixed and inactive until the whole structure was complete. The biological development depends on active interaction and induction between sub-assemblies.

Now I want merely to indicate how our model should be modified to take account of (1). The logic of taking (2)–(4) into account is harder, and so it is postponed—but note how the logic of (2)–(4) must tie up with our problems of parallel computation.

We still think of a cell of the tessellation as corresponding to a cell of the organism—with the active cells in the construction area corresponding to the embryo. But in view of (1), we are interested in a model in which each cell contains the whole programme. So, in a sense, activation of different subroutines in a cell of our modified model would correspond to differentiation of cells of embryo.

Model Mark II. Each module stores a whole programme, but with only a substring loaded into the 20 instruction registers. Biologically, we put all the available information in one cell and let it grow. In our new model, we can think of the machine as starting as a single cell with two strings (one corresponding to our Mark I programme, one to the tape). The machine secretes new cells and manipulates tape to produce a discrete aggregate of cells. Starting with one cell containing this information in our model, we could programme it to secrete extra cells for tape as well as for the growing organism. The tape cells are discarded "at birth." (We still haven't used parallel processing—for we put a large, but bounded, amount of information in the individual cell, and then used extra cells for tape; this captures nothing of the embryological organization—"A germ cell doth not a living embryo make!") A universal constructor will need tapes of arbitrary length to deal with arbitrary automata, and this could never be packed in a single biological cell. For further details, see Arbib [2a].

In our Mark I model the universal constructor is the precursor of the replicating system. In biology there are only replicating systems. In our Mark I model, given the universal constructor, we found a tape which turns it into a self-reproducing machine. But this tape is bounded, and so we could imagine a single germ cell for a self-reproducing machine in our Mark II model containing the universal constructor's programme, plus the necessary instruction tape. But we cannot imagine a biological germ cell evolving this way. We have not said anything about where cells came from—a problem which corresponds to the question of the evolution of life (cf., Pattee [22]). There are two distinct problems of evolution: (*a*) if you start from a relatively unstructured universe, how do you get cells? and (*b*) how do cells start aggregating? I think at the present moment it is relatively easy, at least qualitatively, to get the idea of cells competing for various nutrients in the environment, co-operating to form aggregates, and these aggregates then evolving in a classical domineering fashion. The question of where the cells came from is a very different and very difficult one. What I have talked about here is how to reproduce a whole multi-cellular organism, assuming you can place any component you like in the construction area. But we have not touched on the question of how to place a given component in the construction area, and the question of how the reproducing cells evolved in the first place. This question is somewhat outside the scope of the present paper, but it should be borne in mind. Codd [6] considers tessellations with even simpler components than von Neumann's. A pure automata problem is to embed our module in Codd's model, where one of our cells is simulated as an aggregate of Codd's cells with appropriate change of time scale (Holland, personal communication).

Perhaps we can approach the cellular evolution problem by imagining a sub-tessellation with components comparable to the macromolecules of biology; and we can consider reproduction of our modules as aggregates of these pseudo-macromolecules. Our constructions would then treat arrays of arrays.

Discussion

FLOOR: How big are your modules and can you choose whatever size you want?

ARBIB: You can use less instructions and more modules, or more instructions and less modules. I broke off at twenty instructions but it is quite clear that you can get down to about seven without much difficulty; and you can simplify the design by going up to forty—it's a matter of choice.

FLOOR: You contrasted two fixed-point theorems, and I would like a little more clarification, namely, on the idea of the component fixed-point theorem and the construction fixed-point theorem.

ARBIB: Given a set of components $C1$, to construct machines which build all the automata made of components from $C1$, you may need a bigger set of components $C2$. To build all machines constructed of components $C2$, you may need machines put together from a bigger set of components $C3$. The question is, is there a fixed point; can we find a set of components C such that all automata built from components of C can be constructed by automata built from the same set C?

FLOOR: The set C itself would be what you call a fixed point?

ARBIB: Yes. The set C would be the fixed point for components. Looking at this set, we can find for each automaton A, an automaton $c(A)$ which constructs A. That I can then find a machine U such that $c(U)$ is the same as U (U can "construct itself") is the construction fixed-point theorem.

FLOOR: You said that it was an easy matter to go from a component to a construction fixed point. I think what you mean is that it is easy, given the cases that have been studied.

ARBIB: Once you have got a programme for the construction of a given machine, it's easy to convert it into a tape to be fed into a universal constructor which interprets that tape and then proceeds as if it were under the guidance of the programme for the constructor of the automaton. Thus, in all such arrays as ours, the usual sort of Turing machine argument will show that you can always build the universal constructor once you have got the constructor of individual things—it is the usual process of going from a programme to an encoded description of the programme on the tape of the Turing machine.

FLOOR: You are really talking about two-dimensional models, and you mentioned the crossover problems. Can you do anything more in three dimensions than you can in two?

ARBIB: I think that three dimensions make it easier in some ways. In fact, what I have essentially done in my design is to make my cell complex enough to work in three dimensions. You can think of my module, in a sense, as a finite tier of von Neumann's components with enough depth of computation in each cell, in my single layer, to avoid the crossover problem.

FLOOR: Is there any real limitation on dimensionality?

ARBIB: I don't think you can do unbounded self-reproduction in one dimension even with complex cells. I would conjecture that you can have a limited number of self-reproductions in one dimension by having your single cells corresponding to a finite set of my cells. If a single cell in your one-dimensional string corresponded to, say, nine of my cells, you can have two self-reproductions, but after that there would be nowhere to go.

FLOOR: Can you make a Turing machine in one dimension?

ARBIB: Yes, that is trivial, because you make a single cell correspond to both a control box and a tape square.

MESAROVIČ: It seems that it is not question of existence, but a question of efficiency.

HARTMANIS: I was going to make a general comment about Turing machines. It is now known that whatever you can do with an N-dimensional tape in time t, you can do in time t^2 on a one-dimensional tape.

ARBIB: Theorem by Hartmanis and Stearns, in the *Transactions of the American Mathematical Society* (1965).

FLOOR: There is a very interesting result regarding the crossover; using and's, or's, and not's (rather than von Neumann's components), it is possible to construct a net in a planar arrangement that does the following simple thing: it sends a piece of information from any corner to its diagonally opposite corner. The number of elements is something like 18, or slightly less if you are clever.

BANERJI: If you make your cell three-dimensional, then the crossover problem could be eliminated, so it seems it is only theoretical fun rather than a real problem.

ARBIB: You get into other problems if you go into three dimensions because you have to remember what the heights of various sub-blocks are when you detour wires around them—so again it is a matter of choice.

REFERENCES

Note. This list contains a number of items related to self-reproduction, which are not cited in the text.

[1] Arbib, M. A., *Brains, Machines, and Mathematics* (New York: McGraw-Hill, 1964).

[2] ——— "Simple Self-Reproducing Universal Automata," *Information and Control*, 9 (1966), 177–89.

[2a] Arbib, M. A., "Automata Theory and Development: Part I," *J. Theoret. Biol.* 14 (1967), 131–56.

[3] Arbib, M. A., and Blum, M., "Machine Dependence of Degrees of Difficulty," *Proc. Amer. Math. Soc.*, 16 (1965), 442–7.

[4] Burks, A. W., "Historical Analysis of von Neumann's Theories of Artificial Self-Reproduction," dittoed notes, eight pages (Department of Philosophy, University of Michigan, Nov. 23, 1960).

[5] ——— "Computation, Behavior and Structure in Fixed and Growing Automata," "*Behavioral Science*, 6 (1) (Jan. 1961), 5–22. A revised version of the paper of the same title in *Self-Organizing Systems*, M. Yovits and S. Cameron, eds. (New York: Pergamon Press, 1960), 282–311.

[6] Codd, E. F., "Propagation, Computation and Construction in 2-Dimensional Cellular Spaces" (University of Michigan Technical Report, March, 1965).

[7] Dodgson, M. C. H., *The Growing Brain: An Essay in Developmental Neurology* (Bristol: John Wright and Sons, 1962).

[8] Ebert, J. D., *Interacting Systems in Development*, Modern Biology Series (New York: Holt, Rinehart and Winston, 1965).

[9] Elsasser, W. M., *The Physical Foundations of Biology* (London: Pergamon Press, 1958).

[10] Hennie, F. C., *Iterative Arrays of Logical Circuits* (Cambridge: M.I.T. Press, 1961).

[11] Holland, J. H., "Iterative Circuit Computers," *Proc. Western Joint Computer Conference* (1960), 259–65.

[12] ——— "Outline for a Logical Theory of Adaptive Systems," *Journal Assoc. Comp. Mach.*, 9 (1962), 297–314.

[13] ——— "Iterative Circuit Computers: Characterization and Resumé of Advantages and Disadvantages," in *Proc. Symp. Microelectronics and Large Systems* (Washington, D.C.: Spartan Press, 1965).

[14] Landsberg, P. T., "Does Quantum Mechanics Exclude Life?", *Nature*, 203 (Aug. 29, 1964), 928–30.

[15] Lee, C. Y., "Automata and Finite Automata," *Bell Syst. Tech. Journal*, 39 (1960), 1267–96.

[16] ——— "A Turing Machine Which Prints Its Own Code Script," in *Proc. Symp. Math. Theory of Automata*, vol. 12 of the Microwave Research Institute Symposia Series (Brooklyn: Polytechnic Press, 1963), 155–64.

[17] Moore, E. F., "Machine Models of Self-Reproduction," in "Mathematical Problems in the Biological Sciences," *Proc. Symp. Appl. Math.* (Amer. Math. Soc.), 14 (1962), 17–33.

[18] Morrison, P., "A Thermodynamic Characterization of Self-Reproduction," *Rev. Mod. Phys.* (April, 1964), 517–23.

[19] Myhill, J., "Self-Reproducing Automata," course notes, summer school course on automata theory (University of Michigan, 1963).

[20] ——— "The Converse of Moore's Garden of Eden Theorem," *Proc. Amer. Math. Soc.*, 14 (1963), 685–6.

[21] ——— "The Abstract Theory of Self-Reproduction," in *Views on General Systems Theory*, M. D. Mesarovič, ed. (New York: John Wiley & Sons, 1964), 106–18.

[22] Pattee, H. H., "Experimental Approaches to the Origin of Life Problem," in *Advances in Embryology*, 27 (New York: Interscience, 1965), 381–415.

[23] ——— "The Recognition of Hereditary Order in Primitive Chemical Systems," in *The Origins of Prebiological Systems and of their Molecular Matrices* (New York: Academic Press, 1965).

[24] Post, Emil L., "Finite Combinatory Processes—Formulation I," *Journal Symbolic Logic*, 1 (1936), 103–5.

[25] Rabin, M. O., *Degree of Difficulty of Computing a Function, and a Partial Ordering of Recursive Sets* (Jerusalem: Hebrew University, April, 1960).

[26] Ritchie, R. W., "Classes of Predictably Computable Functions," *Trans. Amer. Math. Soc.*, 106 (1963), 139–73.

[27] Shannon, C. E., "von Neumann's Contributions to Automata Theory," *Bull. Amer. Math. Soc.*, 64, 3 (1958), 123–9.

[28] Simon, H. A., "The Architecture of Complexity," *Proc. Amer. Phil. Soc.*, 106, 6 (Dec. 1962), 467–82.

[29] Stahl, W. R., "Self-Reproducing Automata," *Perspectives in Biology and Medicine*, 8, 3 (Spring, 1965), 373–93.

[30] Thatcher, J. W., "The Construction of the Self-Describing Turing Machine," in *Proc. Symp. Math. Theory of Automata*, vol. 12 of the Microwave Research Institute Symposia Series (Brooklyn: Polytechnic Press, 1963), 165–71.

[31] ——— "Universality in the von Neumann Cellular Model," in *Essays in Cellular Automata*, A. W. Burks, ed., in preparation.

[32] Turing, A. M., "On Computable Numbers," *Proc. London Math. Soc.*, 2, 42 (1936), 230–65; 43 (1937), 544–6.

[33] Ulam, S. M., *A Collection of Mathematical Problems* (New York: Interscience, 1960). See sec. II.2: "A Problem on Matrices Arising in the Theory of Automata."

[34] von Neumann, J., "The General and Logical Theory of Automata," in "Cerebral Mechanisms in Behavior," *Proc. of the Hixon Symp.*, L. A. Jeffress, ed. (New York: John Wiley & Sons, 1951), 1–31.

[35] ——— *The Theory of Self-Reproducing Automata*, A. W. Burks, ed. (Urbana, Illinois: University of Illinois Press, 1966).

[36] Wagner, E. G., "An Approach to Modular Computers: I. Spider Automata and Embedded Automata," *IBM RC 1107* (Jan. 28, 1964); "II. Graph Theory and the Interconnection of Modules," *IBM RC 1414* (June 4, 1965).

[37] Wang, H., "A Variant to Turing's Theory of Computing Machines," *Journal Symbolic Logic*, 4 (1957), 63–92.

[38] Wigner, E. P., "The Probability of the Existence of a Self-Reproducing Unit," from *The Logic of Personal Knowledge: Essays Presented to Michael Polanyi* (Glencoe, Illinois: The Free Press, 1961), 231–8; see also Landsberg, *supra* ref. [14].

Multiple Control Computer Models

C. C. Elgot, A. Robinson, and J. D. Rutledge*

1. INTRODUCTION

Unlike most of the papers presented here, this paper is not an exposition of a subject area which has reached some degree of maturity; instead, it presents the beginning of a development making an initial attempt to treat formally an area which has not previously been so studied. We would like to be able to treat, with mathematically acceptable rigour, a range of problems associated with the programming of highly parallel computers, and to make it possible to bring some of the arsenal of mathematics to bear on them. For this we need a formal model of a parallel computer (which term we must define) which is sufficiently comprehensive and sufficiently detailed to represent the principal problems of interest, and which, at the same time, is not too intricate to be manipulated. In this paper, we give a first attempt at a solution of this problem, with an indication of the motivations of the various choices which are made. We believe that the main lines of development are useful ones, and that the work is worth exposing to criticism in its present state for the purpose of this Conference.

The present development is closely related to that in [E–R], and, while this paper may be read alone, the reader who is familiar with [E–R] may well find it easier reading.

2. INFORMAL DISCUSSION

The underlying picture is that of a system consisting of a central addressable store, serving a multiplicity of processor units. The processors execute single instructions, obtaining first the instruction and then the required data from the store. The processors then carry out the indicated operation to produce a result and to determine a successor location set. They place the result in the store and designate a set, possibly empty, of successor instruction locations. These locations

*Thomas J. Watson Research Center, Yorktown Heights, New York.

are assigned to the same and/or other processors (by an assignment mechanism which we do not examine) and the cycle repeats. The processors, as such, are viewed only as the mechanisms for the execution of single instructions; nothing is retained in a processor from one instruction execution to the next, and all storage is equally available to all processors. Any number of processors can be active simultaneously, and no synchronization is assumed. Hence, we must use real time (or some approximation thereof) as the sequence background, rather than the usual sequence of discrete "distinguished instants" which usually serves in such studies.

It is assumed that the machines are deterministic in the sense that, given a total machine state, the future behaviour of the machine is determined exactly. We go even further and assume that, during the execution of an instruction, a given processor is independent of the remainder of the machine, i.e., both the resulting change to the store and successor instruction locations and the time to completion of the instruction are determined by the store state and instruction location at those times when the processor examines it (at the initiation of the instruction and at the time data is picked up, if this is different). However, execution times do depend on the values of the data, and so would be expected to change from one execution of the programme to another, for example, when the machine is being used to compute the value of a given function for a succession of argument values. The question thus arises, under what conditions does a programme (which we must define) compute a function independent of the execution times of its instructions? Put another way, how can a programmer who has incomplete information on the execution time of instructions (or does not want to bother to do detailed timing studies of his programme) be sure that his programme will do what he wants? Some preliminary results are obtained along these lines, and a direction for further investigation is indicated.

With the above general idea, the problem of the proper formalization arises. We have chosen to concentrate on a particular formal model, which we will now outline informally. In formalizations, we follow the spirit of previous work ([E–R], [AMT],* etc.). A machine at any instant is described by its "state"; a sequence of states is a computation. The specification of a machine must give its allowable states, any alphabets on which it operates, the rules which determine the state or states which may follow a given state, and possibly other information. A machine is deterministic if the successor of a given state is uniquely specified. Following [E–R], the central store is represented by a content function which maps a set of addresses or locations, A, into a set of possible contents for those addresses, B. A familiar example, corresponding to a $32K$ store of 36 bit words, would be a function k: $2^{32} \rightarrow 2^{36}$. We do not, of course, require either A or B to be finite. In the RASP model, the state of the machine was adequately described by a content function k and an instruction location $a \in A$; the simplest way of extending this to a multi-processor model would be to replace a by a set $\mathfrak{a} \subset A$,

*But note that we use the term "state" for "total system state," not internal state.

giving the set of all currently active instruction locations. However, this is inadequate, since processors take their instructions and finish them at different times, and there may even be two or more processors working on instructions (possibly different) taken from the same location. Since we want to retain our notion of "state" as a description adequate to determine all future states, we must include much more information in it. The neatest solution seems to be to include an item for each active processor, giving enough information to determine when it will next alter the main store, what alteration it will make, and what the successor instruction locations will be. There are a variety of ways of representing this information in the formal model; one has been chosen on the basis of notational simplicity.

A realistic model of the real-world situation must take into account the fact that it is impractical for each processor to make its own private copy of the content of the main store each time it starts an instruction. Instead, it must first interpret the instruction to determine what locations are relevant to that instruction, and then obtain the contents of those locations. Since the main store can be changed by other processors at any time, it is quite possible for the values in the argument locations to change between instruction access and data access. More complex instructions, of course, may require several successive accesses to the store. A model which takes the simplest case of this into account has been formulated but has not been developed, since it is much simpler to ignore this complication and to pretend that instruction and data are accessed simultaneously. For "well-behaved" programmes one would like it to be the case that the representation of an instruction in storage is not altered by an independently operating instruction while the first instruction is in process of being executed. By making this approximation, we obtain the model presented here, in which each instruction execution is accomplished with only two references to main storage, one to fetch and one to store. So far, this gives a simple, relatively realistic formal model of a machine in which a number of interacting streams of computation can proceed simultaneously, generating new streams and recombining parallel ones. However, in the re-combining of parallel streams, the model does not have all the properties one would like, since it has no explicit provision for synchronization. While it is natural in this model for an instruction to have several co-equal immediate successors, the converse situation, that it has several immediate predecessors, is not so naturally provided for. This is also the case in practical machines, where it is provided for either by programming tricks or by special instructions, e.g., the "cut" instruction of the Γ–60. In the present study, we would like to be able to deal with the problem of determinacy of computation in the absence of complete timing information; thus, we would like to have a model in which the recombination of parallel streams appears explicitly. Of a number of possibilities, we have selected the following as the neatest formalization. Each instruction, in addition to specifying its immediate successor locations, also specifies the number of immediate predecessors that it requires. It becomes active as an instruction only when its location has been specified as successor location that many times.

In a stored programme machine, of course, the instructions are subject to change, and the content of a location may be changed in the middle of this process. We have chosen a system in which, when an instruction location (say a) is designated, but the content of that location calls for multiple designations, a "waiting" state is established, including the number of additional designations required. When the required number of designations of a have been received, the instruction, the code for which is currently in a, is executed.

This yields a notion of computation. Our next idea is that of "programme," which is defined very broadly; we then must say what we mean by a programme computing a function, and more usefully, by a programme *embeddably* computing a function. This last notion is peculiar to parallel computation, since we must be able, in order to make effective use of parallelism, to carry on the computation of two functions simultaneously and without interference. It seems to be necessary, therefore, to introduce here notions of structure of a computation which also bear on the problem of computation in the absence of timing information. In order to discuss the composition of programmes in the generality permitted by the parallel structure, we require two more ideas—one corresponding to the familiar intuitive notion of re-locatability of programmes and the other, that of a "coupler" programme. The need for the latter arises from the following considerations. In a serial computation context, operations are carried out in some linear order—each operation in a computation (though *not* in a programme) has, at most, one successor and one predecessor. In a parallel context, the corresponding order is partial. If we want to compute the function $f(g_1(x_1), g_2(x_2))$, given programmes for f, g_1, g_2, we could, of course, compute first $g_1(x_1)$, then $g_2(x_2)$, and finally $f(g_1(x_1), g_2(x_2))$. This would not use the available parallelism. It would be more efficient to compute $g_1(x_1)$ and $g_2(x_2)$ simultaneously, and then compute $f(g_1(x_1), g_2(x_2))$ as soon as they are both finished. But x_1, x_2 may not become available at the same time—it may be that x_1 may be expected to be available well before x_2 is, and that g_1 takes much longer to compute than g_2, so it would be desirable to start work on g_1 immediately when x_1 is available, and start g_2 at a later time. Now the programme for $f(g_1(x_1), g_2(x_2))$ will have at least two independent entries, each with an associated subset of the total input. Obviously, the same sort of thing can happen at the exits of a programme, and in writing programmes for later combination with other not-yet-specified programmes, one will wish to make the inputs and outputs and corresponding entries and exits as independent as possible, for maximum speed. However, in combining such programmes, an exit-output set correspondence of the predecessor programme may not match the entry-input set correspondence of the successor. To correct this mismatch we introduce the notion of "coupler" programme, which computes the identity function on storage, and activates each of its exits as soon as a corresponding subset of its entries has been activated. There are alternative formulations which accomplish this purpose, one being to require of the "relocation" notion that, in addition to providing a programme able to compute a function at given locations using a portion of the store which does not interfere with a given reserved area, it must also provide one with specified entry-input (or exit-output) correspondence.

One additional comment on the "relocation" notion. In specifying the form in which storage is reserved, we must allow for the fact that a programme, while its "code" requires only a finite amount of storage, as in [E–R], may well require an unbounded amount of "working storage" which must be protected from interference by concurrent programmes. It seems desirable, since the possibility of an effective re-location function should not be ruled out, that this reserved area be simple in some appropriate sense. We have, more or less arbitrarily, adopted the criterion "primitive recursive" as the formalization of simplicity. The reader will observe that this could be replaced by any other reasonable formalization of "simple" without altering the argument.

The final result of the present preliminary discussion is a "plausibility" theorem—we can indeed compose programmes in a reasonable way, so our elaborate apparatus of definitions at least begins to work. This leaves the way open for further development toward a treatment of programming languages for this sort of machine.

3. FORMAL DEVELOPMENT

A multiple control RASP (MCR) is specified by a quintuple,

$$M = \langle A, B, K_0, h, \tau \rangle,$$

where A is a set, finite or infinite, of locations;
B is a set, finite or infinite, of words;
$K_0 \subseteq B^A$ is the set of permissible *content functions*

(a frequently interesting example is the set of all assignments of members of B to members of A such that all but a finite number of elements of A are assigned a special member of B denoting "empty" or "cleared"; these are called the *finitely supported* content functions);

h: $K_0 \times A \times B \rightarrow \mathfrak{P}(K_0) \times P'(A) \times N$, where $\mathfrak{P}(K_0)$ is the set of partial functions having completions in K_0, $P'(A)$ is the set of finite (including empty) subsets of A, and N is the natural numbers (the components of h are denoted h^1, h^2, and h^3);

τ: $K_0 \times A \times B \rightarrow R$, the real numbers greater than $\epsilon > 0$ (this determines the time of execution of instructions).

A *State* of an MCR is a triple $\langle k, \alpha, \beta \rangle$, $k \in K_0$, α a (finite) set of ordered triples of form $\langle k, a, t \rangle$, and β a (finite) set of ordered pairs of form $\langle a, n \rangle$ where

$k \in K_0$,
$a \in A$,
$t \in$ Real numbers > 0,
n an integer.

Each of the members of α may be thought of as representing an active processor and will be referred to as a processor state. For formal convenience we represent the processor as having a copy of the entire content of main store at the time of its reference to same, although its behaviour will normally depend on only a portion of this content. The second element a is the location of the instruction being executed by this processor, and the real number t is the time to go until the completion of this execution. Various alternatives were considered for this time component, such as the actual time of completion, time of initiation, etc., all of which would provide adequate information; however, only the one chosen here allows for the repetition of the same state at different times in a computation, a condition which we would like to make formally simple and easily recognizable.

The elements of β similarly represent instructions which have been designated as successors, but which require further such designation before they can become active. They are thus *quasi-active*, but have no definite termination time. The number of designations required by an instruction (location) is determined by the memory content at the time of first designation; the operation actually performed depends on the content at the time of the final designation.

The machine exists in continuous time, but since the state only changes at discrete instants, we can represent a computation as usual by a discrete sequence of states, each indexed with the time at which it comes into existence. Let C be a computation. Then

$$\langle 0, \langle k_0, \alpha_0, \beta_0\rangle\rangle \in C$$

where k_0 may be any member of K_0, α_0 is a finite set of triples of form

$$\langle k_0, a, \tau(k_0, a, k_0(a))\rangle,$$

and β_0 is a finite set of pairs of form $\langle a, n\rangle$. Let $\langle t_i, s_i\rangle \in C$, and $s_i = \langle k_i, \alpha_i, \beta_i\rangle$, and let

(1) $$\Delta t = \min\{t|(\exists a, k)\ \langle k, a, t\rangle \in \alpha_i\}.$$

Then $t_{i+1} = t_i + \Delta t$. Let $\alpha_i = \gamma \cup \delta$, where

$$\langle k, a, t\rangle \in \gamma \Leftrightarrow t = \Delta t \wedge \langle k, a, t\rangle \in \alpha_i, \qquad \delta = \alpha_i - \gamma.$$

We require that

(2) $$\langle t_i, s_i\rangle \in C \wedge \langle k_1, a_1, t_1\rangle, \langle k_2, a_2, t_2\rangle \in \gamma \Rightarrow$$
$$\mathfrak{D}[h^1(k_1, a_1, k_1(a_1))] \cap \mathfrak{D}[h^1(k_2, a_2, k_2(a_2))] = \emptyset.$$

With this requirement, the following is well defined, for $x \in A$:

(3) $$k_{i+1}(x) = \begin{cases} k_i(x) & \text{for } x \notin \bigcup_{\langle k,a,\Delta t\rangle \in \gamma} \mathfrak{D}[h^1(k, a, k(a))], \\ [h^1(k, a, k(a))](x) & \text{for } x \in \mathfrak{D}[h^1(k, a, k(a))] \text{ and } \langle k, a, \Delta t\rangle \in \gamma. \end{cases}$$

To simplify notation slightly, we define the function $\#_\gamma(a)$, or simply $\#a$, to denote the number of elements of γ which designate a as successor location, or more formally

$$\#a = \text{card}\{\langle k, a', \Delta t\rangle | \langle k, a', \Delta t\rangle \in \gamma \wedge a \in h^2(k, a', \Delta t)\};$$

also, we let

$$\beta(a) = \begin{cases} n & \text{if } \langle a, n\rangle \in \beta \\ 0 & \text{otherwise.} \end{cases}$$

Now we can define α_{i+1} and β_{i+1} by

$$(4) \qquad \alpha_{i+1} = \{\langle k, a, t - \Delta t\rangle | \langle k, a, t\rangle \in \delta\} \cup \{\langle k_{i+1}, a, \tau(k_{i+1}, a, k_{i+1}(a))\rangle | \#a \geqslant \beta(a) > 0 . \vee . \#a > \beta(a) = 0 \wedge h^3(k_{i+1}, a, k_{i+1}(a)) = 1\},$$

$$(5) \qquad \beta_{i+1} = \{\langle a, n\rangle | n \text{ is least positive } n' \text{ such that } n' \equiv (\beta(a) - \#a) \bmod[h^3(k_{i+1}, a, k_{i+1}(a))] \wedge n \neq 0\}.$$

(Note that we are ignoring the possibility of private storage in the processors.)

It is possible that $h^2(k, a, b)$ may be empty—hence a computation may be finite or infinite.

A few explanatory comments are in order. t_{i+1} is determined as the next instant after t_i, at which a processor completes its job and refers to main store again. Note that while the range of τ is bounded away from zero, we can only say $\Delta t > 0$, so that it does not immediately follow that $t_i \to \infty$ as $i \to \infty$. A proof of this will be given after we have developed a required bit of structure. All the processors which finish at exactly the same time are considered as acting on the main store simultaneously; this set is just that represented by γ, with δ representing those not yet completed. Since in a practical machine some more or less complicated priority device would prevent two processors from simultaneously storing results in the same location, we prefer to leave this implicit in the present model, merely requiring by (2) that no such conflict occur in a proper computation. The range of h^1 is specified as consisting of a partial content function, with the intention that the domain of $h^1(k, a, b)$ should be just those locations in which this operation makes changes. Thus (3) properly specifies the new content function.

The set α_i is composed of states representing processors which are still running, with their respective times to completion updated, together with the processor states newly activated as successors to those which have just finished. These all operate on the main store state as produced in step $i + 1$, and so incorporate k_{i+1}.

β_{i+1} contains all the quasi-active instruction locations, together with a record of the number of remaining designations required to activate each; those in β_i not designated as successors by processor states of γ; those in β_i which are named, but which have not had their required number of designations; and those newly designated locations, the instruction code in which requires more designations than are received at instant t_{i+1}. Note that while the same instruction location may figure in any number of elements of α, it can appear in only one member of β; only activations directed to a given address while that address is quasi-active are counted as designations of that particular execution.

A location may receive multiple activations in a single computation step. These are all counted, but only one active processor state (member of α) may be initiated. However, for instructions requiring multiple designation, all designations in a given step are counted, and any surplus over that required for initiation is saved toward a subsequent initiation.

This completes our definition of the notion of MCR and its computation. We must now see how to apply it, that is we must state what it means for such a machine to compute a function or relation, how and when it is possible, given that a machine can compute two functions, for the machine to compute their composition, and see what is required for an MCR to be able to compute all recursive functions. We could follow the programme of [E–R] almost exactly—in fact it is easy to show that the set of RASP's may be naturally embedded in the MCR's, with the addition of the special passive instruction code $b_0 \in B$, preserving computations. However, we want a development which takes account of the parallelism available in the MCR. Our first objectives are notions of programme and programme schema which will be analogous to the notions of instruction and instruction schema, and which could be used for definitional extension.

A computation is said to *terminate* if it is finite, and if its final quasi-active list, β, is empty. This will occur if and only if for some i, $\alpha_i = \emptyset = \beta_i$. A notion of more interest is that of a computation terminating on a specified set (of locations). A computation is said to *reach* a set $e = \{e_1, \ldots, e_n\} \subset A$ when each of the members of that set appears as instruction location in some processor state of the computation, i.e., $e \in \Pi_2{}^3 C$. We will say that a computation C *terminates on set* $e = \{e_1, \ldots, e_n\}$ if and only if C reaches e, C terminates, and

$$\langle t_i, s_i\rangle \in C \wedge \langle k_i, e_j, t\rangle \in \alpha_i \Rightarrow h^2(k_i, e_j, k_i(e_j)) = \emptyset,$$

that is, the instruction activated whenever a member of e appears as an instruction location has no successors.

We are now ready for a (preliminary) definition of the notion of programme. A *programme* $\mathfrak{P}$ is an $(m + 2)$-tuple $\langle k, a, e_1, e_2, \ldots, e_m\rangle$ where k is a (finite) partial content function, a and the e_i are finite ordered subsets of A, and $e_i \cap \mathfrak{D}k = \emptyset$, $i = 1, \ldots m$. A *synchronous-start* (*s-s*) *computation of a programme* is just a computation with initial state $s_0 = \langle k_0, \alpha_0, \emptyset\rangle$ such that $k \subseteq k_0$ and $\Pi_2(\alpha_0) = a$ (unordered). The computation is successful if, and only if, it terminates on one of the sets e_i.

This notion of programme is an extremely broad one, allowing the term to be applied to any partial content function together with entry and exit sets. On the other hand, this is just the case in programming practice at the machine language level. To get a more manageable notion, we will introduce various restrictions and auxiliary conditions.

A programme $\mathfrak{p} = \langle k, a, e \rangle$ *synchronous-start computes a function* $\varphi : B^{m_1} \to B^{m_2}$ at datum locations $d = \langle d_1, \ldots, d_{m_1} \rangle$ and value locations $v = \langle v_1, \ldots, v_{m_2} \rangle$, $d_i, v_i \in A$, if and only if for every computation of $\mathfrak{p}$ with initial state $s_0 = \langle k_0, \alpha_0, \emptyset \rangle$ such that $k \subseteq k_0$ and $\Pi_2(\alpha_0) = a$, if $\varphi(k_0(d))$ is defined, then the computation is successful, and terminates (on e) with state $s_t = \langle k_t, \emptyset, \emptyset \rangle$ such that $k_t(v) = \varphi(k_0(d))$; if $\varphi(k_0(d))$ is undefined, the computation is unsuccessful.

We will have an entirely analogous definition for the notion of a programme computing a relation; a programme will often, of course, compute both a relation and a function. It is clear that not all programmes compute functions—in fact, a programme which computes a function is, so far as the terminal content of v is concerned, insensitive to how k is completed. Thus $\mathfrak{D}(k) \cup d$ is, in a sense, the domain of dependence of $\mathfrak{p}$, or at least contains it. The converse notion we will now define, just as in [E–R].

The (*terminal*) *range of influence* of a programme $\mathfrak{p}$, $tRI(\mathfrak{p})$, is defined by $d \in tRI(\mathfrak{p})$ if and only if for some s_0, s_0 is initial state of successful *s-s* computation of $\mathfrak{p}$, s_t is final state, and $k_0(d) \neq k_t(d)$.

We also require a stronger notion.

The (*dynamic*) *range of influence* of a programme $\mathfrak{p}$, $dRI(\mathfrak{p})$ is that set of locations whose contents may be changed during any *s-s* computation of $\mathfrak{p}$, even though they may be later restored, i.e., $d \in dRI(\mathfrak{p})$ if and only if for some s_0, s_0 is the initial state of a s-s computation C of $\mathfrak{p}$, $s_t \in \Pi_2 C$, and $k_0(d) \neq k_t(d)$.

We are working toward a notion of composition of programmes, which should result in the composition of the functions and relations which they compute, if any. Two major difficulties remain. First, we have defined programme as including a particular partial content function, and thereby being fixed in location in memory. This will mean that two programmes may very well require the same part of memory, and thus be unable to coexist. The second one, of more interest since it is peculiar to multiple control machines, arises from the fact that a programme terminates on a set of locations and may reach members of that set at different times; a subsequent programme starts on a set of locations; the coupling of programmes, which constitutes composition, may therefore result in a composite computation in which all the starting locations of the second programme appear as instructions locations, but at different times. While we could require that a programme have a unique exit for each value of the relation which it computes, this seems to us an unnatural and unnecessary restriction to put on the general idea of programme. The alternative seems to be to define an especially well-behaved class of programmes which can tolerate this rather cavalier treatment and will compute their functions and/or relations independent of the relative timing of their various starts. This kind of independence of timing will be of considerable interest in any case, and nothing is lost by bringing it in at this point. Even more is required, however, since we may have parts of two or more programmes running simultaneously, so the requirements on programmes to be so used are even more stringent.

For this purpose we need to look at the structure of a computation in a slightly different way. Informally, a computation consists of a number of executions of instructions, some pairs going on concurrently; others constrained by the logic of successor-predecessor specifications to occur in a fixed sequence. This is what we wish to reflect now in the formalism.

An *instruction execution* in a computation C consists of two components, a "waiting" phase and an "active" phase.

An *active instruction execution* is a maximal sequence of processor states $E_c = \langle k, a, t_0\rangle, \ldots, \langle k, a, t_m\rangle$ satisfying:

(1) if $\langle k, a, t_0\rangle \in \Pi_2^2(t_j, s_j)$, $\langle t_j, s_j\rangle \in C$, then

$\langle k, a, t_i\rangle \in \Pi_2^2(t_{j+i}, s_{j+i})$ for all $i \leqslant m$,

(2) $\langle k, a, t_i\rangle \in \Pi_2^2(t_k, s_k)$ and $\langle k, a, t_{i+1}\rangle \in \Pi_2^2(t_{k+1}, s_{k+1})$,

then $t_{i+1} - t_i = t_{k+1} - t_k$.

The *instruction location* of E_c is a.

A *waiting instruction execution*, similarly, is a maximal sequence of quasi-active states $E_w = \langle a, n_0\rangle, \ldots, \langle a, n_l\rangle$ satisfying:

(1) if $\langle a, n_0\rangle \in \Pi_3\,\Pi_2(t_j, s_j)$, $(t_j, s_j) \in C$, then

$\langle a, n_i\rangle \in \Pi_2\,\Pi_3(t_{j+i}, s_{j+i})$ for $i \leqslant l$.

(2) No active execution with instruction location a begins in any of the elements of C, $\langle t_j, s_j\rangle, \ldots, \langle t_{j+l}, s_{j+l}\rangle$ of condition 1.

The *instruction location* of E_w is a.

An *instruction execution* is an ordered pair $\langle E, t\rangle$, such that E is a maximal sequence obtained by following a waiting instruction execution E_w (which may be of length 0) by an active instruction execution E_c, with the same instruction location, and such that the last element of E_w occurs in the member of C immediately preceding that containing the first element of E_c, and t is the first component of the member of C which contains the first component of E_c.

We define a relation $\mathfrak{S}_I$ (immediate successor) on instruction executions by $\mathfrak{S}_I(\langle E_1, t_1\rangle, \langle E_2, t_2\rangle)$ if and only if

(1) The last element of E_1 occurs in s_k of C and either some element of $E_{w,2}$ or the first element of $E_{c,2}$ occurs in s_{k+1} of C.

(2) $\langle k_1, a_1, t_1\rangle \in E_1 \Rightarrow$ instruction location of E_2 is in $h^2(k_1, a_1, k_1(a_1))$.

Let $\mathfrak{S}$ be the transitive closure of $\mathfrak{S}_I$. It is immediate that $\mathfrak{S}$ partially orders the set of instruction executions in C. We can index the instruction executions by the set of natural numbers, in some way consistent with the ordering $\mathfrak{S}$, i.e., $\mathfrak{S}(\langle E, t\rangle_n, \langle E', t'\rangle_m) \Rightarrow n < m$. Now replace each (indexed) instruction execution $\langle E, t\rangle_n$ by the (indexed) ordered pair containing the first two components of the last component of E. These are all distinct by virtue of the indexing, and we will use "$\mathfrak{S}$" also for the partial order relation induced by this mapping. The partially ordered set thus obtained will be called the *instruction execution form* of C, denoted $I(C)$. Actually, it is only a skeleton of C, since all the timing information has been dropped, and the relative timing of non-$\mathfrak{S}$-related executions is not explicitly specified (although much of it could be inferred from the content functions). Where no confusion will result, we will omit the index.

We will use I as a symbol for this mapping and for the obvious induced mappings, so that for $(t, s) \in C$,

$$s = \langle k, \langle k_1, a_1, t_1\rangle, \ldots, \langle k_m, a_m, t_m\rangle\rangle, \beta\rangle\rangle,$$

$I(\langle k_1, a_1, t_1\rangle)$ is the element of $I(C)$ representing the instruction execution to which $\langle k_1, a_1, t_1\rangle$ belongs. Note that $(I(\langle k_i, a_i, t_i\rangle), I(\langle k_j, a_j, t_j\rangle)) \notin \mathfrak{S}$ for $i \neq j$, where the two processor states belong to the same machine state.

The instruction execution form of a computation is the structure needed for the proof of the proposition stated earlier, namely:

PROPOSITION. *For a computation* $C = (t_0, s_0), \quad (t_1, s_1), \ldots, (t_i, s_i), \ldots$ $\lim_{i\to\infty} t_i = \infty$.

Proof. Consider the directed graph with nodes the instruction executions of C and edges just the elements of $\mathfrak{S}_I$. This graph has a finite root, the minimal elements of $I(C)$ under $\mathfrak{S}$, and a finite number of outgoing branches at each node. Since the number of nodes becomes infinite with i, a well-known argument (Königs Lemma) tells us that the graph contains an infinite path $\langle E_0, t_0\rangle$, $\langle E_1, t_1\rangle$, $\langle E_2, t_2\rangle, \ldots$. In this path, $t_{i+1} - t_i$ is just the time required for the operation E_i, which is greater than ϵ; hence $t_k > k\epsilon$ for all k, and the proposition is proved.

It is a consequence of this proposition that an active instruction execution has a finite number of elements; that is, during the finite time between the initiation and the termination of an active instruction execution, only a finite number of computation steps can occur. This number may be arbitrarily large for a given time interval. The same does not, of course, hold for waiting instruction executions which have no time limit.

Now we can proceed to define the notion of well-behaved programme referred to earlier.

A programme $\mathfrak{p}(K, a, e)$ *embeddably computes* (*E-computes*) *a function* φ: $B^{m1} \to B^{m2}$ at datum locations $d = \langle d_1, \ldots, d_{m1}\rangle$ and value locations $v = \langle v_1, \ldots, v_{m2}\rangle$ if and only if there exist mappings

$$\begin{aligned}
&\rho_1: a \to P(d),\\
&\rho_2: e \to P(v),\\
&\rho_3: a \to P(\mathfrak{D}(K)),\\
&\rho_4: a \to \mathfrak{P}(D), \text{ where } D = dRI(p) \cup d \cup v \cup \mathfrak{D}(K),
\end{aligned}$$

such that all computations C satisfy:

(1) if C contains a set S of instruction executions $\langle k_i, a_i\rangle$, $a_i \in a$, with locations exactly the members of a, which are not ordered by $\mathfrak{S}$, and if

$$\langle k_i, a_i\rangle \in S \Rightarrow k_i(\rho_3(a_i)) = K(\rho_3(a_i)),$$

and if every instruction execution in C whose h^1-domain intersects $\rho_4(a_i)$ for $a_i \in a$ is $\mathfrak{S}$-related to the execution of a_i is S, then $d^* = \bigcup\{k_i\lceil\rho_1(a_i)\,|\,\langle k_i, a_i\rangle \in S\}$ is a function and if $\varphi(d^*)$ is defined, then there is a set T of instruction executions at locations exactly the members of e such that no execution at a member of e is $\mathfrak{S}$-between members of S and of T, T is a lower $\mathfrak{S}$-bound for S, and S an upper $\mathfrak{S}$-bound for T, $v^* = \bigcup\{k_j\lceil\rho_3(e_j)\,|\,\langle k_j, e_j\rangle \in T\}$ is a function and $v^* = \varphi(d^*)$; also, for any instruction execution $\langle k', a'\rangle$ such that $\mathfrak{S}(S, \langle k', a'\rangle)$ and $\mathfrak{S}(\langle k', a'\rangle, T)$, $\mathfrak{D}(h^1(k', a', k'(a'))) \cap \rho_2(e_j) \neq 0 \Rightarrow \langle k', a'\rangle$ $\mathfrak{S}$-precedes the execution of e_j in T. If $\varphi(d^*)$ is not defined, then no such set T exists.

The notion of a programme E-computing a relation is again similar, and the extensions to this case and that of a programme computing both a function and a relation simultaneously are left to the reader.

To go with this restricted notion of E-computing, we need a corresponding notion of range of influence. Note that, while the easiest way to design a programme to E-compute a function would be to seal it off entirely from any concurrent computation, this is not required by the definition. In fact, any amount of interaction between a programme which E-computes a function and other concurrent activity can take place. Our previous definitions of range of influence took account only of the special computations in which there was no concurrent activity. To define the area of storage in which an "embeddable" programme may make changes, we need the following.

The *embedded range of influence* $R(eRI(\mathfrak{p}))$ of a programme $\mathfrak{p}$ which embeddably computes a function φ and/or a relation is defined by: $d \in eRI(\mathfrak{p})$ if and only if for some computation C satisfying (1) and some instruction execution $\langle k, a\rangle$ in $I(C)$ such that $\mathfrak{S}(S, \langle k, a\rangle)$ and $\mathfrak{S}(\langle k, a\rangle, T)$, $d \in \mathfrak{D}[h^1(k, a, k(a))]$. If $\varphi(d^*)$, $(R(d^*))$, is undefined, read "$(\exists e_i)e_i \in e \wedge \mathfrak{S}(\langle k, a\rangle, (k, e))$" in place of "$\mathfrak{S}(\langle k, a\rangle, T)$."

Another notion which is needed before we are ready to discuss composition is that of relocation. Two programmes which are to be composed must be capable of coexisting in storage without interference, and the inputs and outputs must be suitably related, none of which can be expected to happen for two arbitrary programmes. We can adopt the definition in [E–R] almost verbatim, as follows.

An MCR P *computes a function* f: $B^n \rightarrow B^m$ if and only if for some p, q, for every set $d_1, \ldots, d_n, v_1, \ldots, v_m, a_1, \ldots, a_p, e_1 \ldots, e_q$ of distinct members of A, and every pair of co-infinite, primitive recursive subsets A_0, $A_1 \subset A$, such that $a \cap A_0 = \emptyset$ and $A_1 \subseteq A_0$, there exists a programme $\mathfrak{p} = \langle K, a, e\rangle$ satisfying:

(1) $\mathfrak{p}$ E-computes f at datum locations d and value locations v.
(2) $\rho_4(a) \cap A_0 = \emptyset$.
(3) $\mathfrak{D}(K) \cap (A_0 \cup \{d_1, \ldots, d_n\} \cup \{v_1, \ldots v_m\}) = \emptyset$.
(4) If $A_1 \cap v = \emptyset$, then $dRI(\mathfrak{p}) \cap A_1 = \emptyset$.
(5) $(\exists A_2)A_0 \cup eRI(\mathfrak{p}) \cup \rho_4(a) \subset A_2$ and A_2 is co-infinite, primitive recursive.

An alternative course would be to introduce the notion of a programme $\mathfrak{p}$ embeddably and relocatably computing a function, requiring that a relocation function be given which will generate the programmes required in the above definition as a function of $\mathfrak{p}$, d, v, A_0, A_1. The method chosen here is possibly more general, easier to use, and clearly includes the alternate.

We must now consider the meaning of composition in the present context. We can easily parallel the development in [E–R] to show that if an MCR computes functions f and g, then it computes $f \cdot g$ (at least in the case where the appropriate entry and exit sets are singletons), but it seems that we should ask for more—that, for example, $f(g_1, g_2)$ (as defined below) can be computed, with a partial order $\begin{matrix} g_1 \searrow \\ g_2 \nearrow \end{matrix} f$, or perhaps, that $f(g_1(h), g_2(h))$ can be similarly computed with only a single computation of h. More complex situations also require analysis. Even this case, however, is more complex than it appears to be on the surface, since the simple arrow from g_1 to f will correspond in the programmed form to a sequencing from the members of the terminal set of the programme for g_1 to the members of the initial set of the programme for f. If $\mathfrak{p}_i = \langle K_i, a_i, e_i \rangle$ is a programme for g_i, $i = 1, 2$ and f, $i = 0$, we must arrange that all the members of a_0 are activated, but that each is activated only when all the data which it requires is present, i.e., identifying $v_1 \cdot v_2$ and d_0, we require that $a_{0,i} \in a_0$ be executed in a computation as successor to some minimal $\alpha \subseteq e_1 \cup e_2$ such that $(\rho_{1,2} \cup \rho_{2,2})(\alpha) \supseteq \rho_{0,1}(a_{0,i})$.

Except in special cases, then, we cannot simply use the procedure of identifying the initial instruction locations of a successor programme with the terminal locations of its predecessors. For many purposes we can restrict ourselves to such special cases, and this would be one way to proceed. It appears that any alternate requires either greatly strengthening the requirements for an MCR to be said to compute a function, or else to make some assumptions about the availability of a class of "coupler" programmes which can be used to implement the required relation between terminal and initial location sets. While we could distinguish subclasses of more or less complex and powerful couplers, for the present we will simply define one class which seems to be adequate for all purposes.

A *coupler programme* $C(n, m, \psi)$ where n, m are integers, and $\psi : P(n) \rightarrow P(m)$ is a programme satisfying $C(n, m, \psi) = \langle K, a, e \rangle$ E-computes the identity function at any set of datum and value locations; $|a| = n$, $|e| = m$; in any computation of $C(n, m, \psi)$, if $\psi(\{i_1, \ldots, i_l\}) = \{j_1, \ldots, j_k\}$ then the executions of $e_{j_1}, \ldots, e_{j_k}$ are $\mathfrak{S}$-successors to the executions of $a_{i_1}, \ldots, a_{i_l}$.

An MCR which can compute any C, we will call an MCR_C. Now at last we are ready to say something about composition.

PROPOSITION. *If an MCR_C P computes functions g_1, g_2, and f, then it also computes $\varphi = f(g_1, g_2)$, where*

$$f\colon B^{n_0} \rightarrow B^{m_0}, \quad g_1\colon B^{n_1} \rightarrow B^{m_1}, \quad g_2\colon B^{n_2} \rightarrow B^{m_2}, \qquad n_0 = m_1 + m_2.$$

Proof. A proper proof of this proposition involves a rather lengthy and detailed construction given in the appendix. At least the outlines of this should be clear to the reader. Briefly, the hypotheses allows us to choose programmes $\mathfrak{p}_1$, $\mathfrak{p}_2$, C, $\mathfrak{p}_0$ successively for g_1, g_2, $C(q_1 + q_2, p_0, \psi)$, and f such that each E-computes its function, making use only of storage disjoint from A_0 and from that required by the previously chosen programmes. Outputs of $\mathfrak{p}_1$, $\mathfrak{p}_2$ are assigned to a set of distinct locations, exits of $\mathfrak{p}_1$, $\mathfrak{p}_2$ are identified with entries of C and exits of C with entries of $\mathfrak{p}_0$, and ψ is defined from the input-entry and output-exit relations of $\mathfrak{p}_1$, $\mathfrak{p}_2$, $\mathfrak{p}_0$ by: $a_{0,i} \in \psi(e')$, $e' \subseteq e_1 \cup e_2$, if and only if $e' = (\rho_{1,2} \cup \rho_{2,2})^{-1}(\rho_{0,1}(a_i))$ —i.e., if and only if e' is the set of all exits of $\mathfrak{p}_1$ and $\mathfrak{p}_2$ ρ_2-associated with outputs of $\mathfrak{p}_1$, $\mathfrak{p}_2$ which as inputs of $\mathfrak{p}_0$ are ρ_1-associated with a_i. The total programme $\mathfrak{p} = \langle K_0 \cup K_1 \cup K_2 \cup K_C, a, e\rangle$ then E-computes $f(g_1, g_2)$ at d, v, for the given a, e, d, v, and satisfies (4) of the definition with A_2 equal to the set A_2 for the last chosen programme, say $\mathfrak{p}_0$, since this set includes A_0, and all the reserved areas of programme $\mathfrak{p}_1$, $\mathfrak{p}_2$, C as well as that for $\mathfrak{p}_0$.

This brings us to the end of this phase of the development. It seems clear that we can exhibit an MCR which can compute any recursive function, for example, and that real use can be made of the parallelism of the model. We have defined E-computing in such a way that the programmer need have little knowledge of the timing functions provided in the basic model—in fact, it is clear how one could write successful programmes without such knowledge.

APPENDIX

PROOF OF PROPOSITION, p. 72

Given: $a = a_1, \ldots, a_p,$
$e = e_1, \ldots, e_q,$
$d = d_1, \ldots, d_n,$
$v = v_1, \ldots, v_m,$
A_0, A_1 co-infinite, primitive recursive sets $a \cap A_0 = \emptyset$, $A_1 \subseteq A_0$.

1. For $p_1, q_1, \boldsymbol{a}_1 = \{a_1, \ldots, a_{p1}\}, \boldsymbol{e}_1 = \{e_1, \ldots, e_{q1}\}$
$\boldsymbol{d}_1 = \{d_1, \ldots, d_{n1}\}, \boldsymbol{v}_1 = \{w_1, \ldots, w_{m1}\}$
$A_{1,0} = A_0 \cup \boldsymbol{a}_2 \cup \boldsymbol{e}_2 \cup \boldsymbol{d}_2 \cup v \cup \boldsymbol{w}_2 \cup e$
$A_{1,1} = A_{1,0}$

$\exists \mathfrak{p}_1$: (1) $\mathfrak{p}_1$ E-computes g, at $\boldsymbol{d}_1$, $\boldsymbol{v}_1$,
(2) $\rho_{1,4}(\boldsymbol{a}_1) \cap A_{1,0} = \emptyset$,
(3) $\mathfrak{D}(K_1) \cap (A_{1,0} \cup \boldsymbol{d}_1 \cup \boldsymbol{v}_1) = \emptyset$,
(4) $A_{1,1} \cap dRI(\mathfrak{p}_1) = \emptyset$,
(5) $(\exists A_{1,2})$, $A_{1,0} \cup eRI(\mathfrak{p}_1) \cup \rho_{1,4}(a_1) \subseteq A_{1,2}$ and $A_{1,2}$ is co-infinite, primitive recursive.

2. For $p_2, q_2, \boldsymbol{a}_2 = \{a_{p_1+1}, \ldots, a_p\}, \boldsymbol{e}_2 = \{e_{q_1+1}, \ldots, e_q\}$

$$\boldsymbol{d}_2 = \{d_{n_1+1}, \ldots, d_n\}, \boldsymbol{v}_2 = \{w_{m_1+1}, \ldots, w_m\}$$

$$A_{2,0} = (A_{1,2} - (\boldsymbol{a}_2 \cup \boldsymbol{e}_2 \cup \boldsymbol{v}_2))$$

$$A_{2,1} = A_{2,0} - \boldsymbol{v}_2$$

$\exists \mathfrak{p}_2$: (1) $\mathfrak{p}_2$ E-computes g_2 at $\boldsymbol{d}_2, \boldsymbol{v}_2$,
(2) $\rho_{2,4}(\boldsymbol{a}_2) \cap A_{2,0} = \emptyset$,
(3) $\mathfrak{D}(K_2) \cap (A_{2,0} \cup \boldsymbol{d}_2 \cup \boldsymbol{v}_2) = \emptyset$,
(4) $A_{2,1} \cap dRI(\mathfrak{p}_2) = \emptyset$,
(5) $(\exists A_{2,2})$, $A_{2,0} \cup eRI(\mathfrak{p}_2) \cup \rho_{2,4}(\boldsymbol{a}_2) \subseteq A_{2,2}$ and $A_{2,2}$ is co-infinite, primitive recursive.

3. For $q_1 + q_2, p_0, \boldsymbol{a}_c = \boldsymbol{e}_1 \cup \boldsymbol{e}_2, \boldsymbol{e}_c = a_0$

$$\boldsymbol{d}_c = w, \boldsymbol{v}_c = w$$

$$A_{c,0} = A_{2,2}, A_{c,1} = A_{2,2} - w$$

$\exists C(a, e, \psi)$: (1) C E-computes identity at w_1 and for $e' \subseteq \boldsymbol{e}_1 \cup \boldsymbol{e}_2$, $a_{0,i} \in \psi(e'), \Leftrightarrow e' = (\rho_{1,2} \cup \rho_{2,2})^{-1}(\rho_{0,1}(a_i)$,
(2) $\rho_{c,4}(\boldsymbol{a}_c) \cap A_{c,0} = \emptyset$,
(3) $\mathfrak{D}(K_c) \cap (A_{c,0} \cup w) = \emptyset$,
(4) $dRI(C) \cap A_{c,1} = \emptyset$,
(5) $(\exists A_{c,2})$, $A_{c,0} \cup eRI(C) \cup \rho_4(\boldsymbol{a}_c) \subseteq A_{c,2} \wedge A_{c,2}$ is co-infinite, primitive recursive.

4. For $p_0, q_0, \boldsymbol{a}_0 = \boldsymbol{a}_0, \boldsymbol{e}_0 = \boldsymbol{e}$

$$\boldsymbol{d}_0 = w, \boldsymbol{v}_0 = v$$

$$A_{0,0} = A_{c,2}; A_{0,1} = A_{c,1} - v$$

$\exists \mathfrak{p}_0$: (1) $\mathfrak{p}_0$ E-computes f at w and v,
(2) $\rho_4(\boldsymbol{a}_0) \cap A_{0,0} = \emptyset$,
(3) $\mathfrak{D}(K_0) \cap (A_0 \cup w \cup v) = \emptyset$,
(4) $dRI(\mathfrak{p}_0) \cap A_{0,1} = \emptyset$,
(5) $(\exists A_{0,2})$, $A_{0,0} \cup eRI(p_0) \cup \rho_4(a_0) \subseteq A_{0,2} \wedge A_{0,2}$ is co-infinite, primitive recursive.

5. Now, let $p = p_1 + p_2, q = q_0$

$$\mathfrak{p} = \langle K_1 \cup K_2 \cup K_3 \cup K_c, a, e\rangle$$

(1) $\mathfrak{p}$ E-computes $f(g_1, g_2)$ at d and v.

$$\rho_1 = \rho_{1,1} \cup \rho_{2,1}$$

$$\rho_2 = \rho_{0,2}$$

$$\rho_3(a_i) = \begin{cases} \rho_{1,3}(a_i) \cup K_c \cup K_0 & \text{for } 1 \leqslant i \leqslant p_1 \\ \rho_{2,3}(a_{i-p}) \cup K_c \cup K_0 & \text{for } p_1 < i \leqslant p_1 + p_2 \end{cases}$$

$$\rho_4(a_i) = \begin{cases} \rho_{1,4}(a_i) \cup \rho_{c,4}(\boldsymbol{a}_c) \cup \rho_{0,4}(\boldsymbol{a}_0), & 1 \leqslant i \leqslant p_1, \\ \rho_{2,4}(a_i) \cup \rho_{c,4}(\boldsymbol{a}_c) \cup \rho_{0,4}(\boldsymbol{a}_0), & p_1 < i \leqslant p_1 + p_2 \end{cases}$$

(ρ_3 and ρ_4 are not the most economical but will serve).

Let a computation Γ contain a set of unordered instruction executions $S = \{\langle k_i, a_i\rangle | a_i \in a\}$ at locations a_i, such that $k_i(\rho_3(a_i)) = K(\rho_3(a_i))$, and such that every instruction execution of Γ whose h^1-domain intersects $\rho_4(a_i)$ is $\mathfrak{S}$-related to $\langle k_i, a_i\rangle$. Then:

(*a*) this condition holds for $\mathfrak{p}_1$ and $\mathfrak{p}_2$ as they appear embedded in $\mathfrak{p}$, since the choices of $A_{1,0}$, $A_{2,0}$, $A_{c,0}$, $A_{0,0}$ protect these programmes from other components of $\mathfrak{p}$. Hence Γ contains unordered sets of executions T_1, T_2 which are lower bounds for the two components of S respectively, and such that

$$\Big(\bigcup_{e_j \in e_1} (k_j \lceil \rho_{1,3}(e_j))\Big)(v_1) = g_1\Big[\Big(\bigcup_{a_i \in a_1} (k_i \lceil \rho_{1,1}(a_i))\Big)(d)\Big].$$

and similarly for g_2.

(*b*) Since $\boldsymbol{a}_c = \boldsymbol{e}_1 \cup \boldsymbol{e}_2$, and the set of executions $T_1 \cup T_2$ at $\boldsymbol{e}_1 \cup \boldsymbol{e}_2$ is unordered by the above, Γ contains an unordered set of executions at a_c; by choice of $A_{1,1}$, $A_{2,1}$ the ρ_3- and ρ_4-conditions are satisfied, so Γ contains a set T_c of executions at $\boldsymbol{e}_c = \boldsymbol{a}_0$, at which appropriate members of w contain the outputs from $\mathfrak{p}_1$ and $\mathfrak{p}_2$, and by the definition of ψ an execution at a member of $\boldsymbol{a}_0$ occurs exactly as successor to the set of members of $T_1 \cup T_2$ which, by $\rho_{1,2}$, and $\rho_{2,2}$, supply the inputs which it, by $\rho_{0,1}$, requires.

(*c*) As in 1, Γ contains an unordered set (T_c) of executions at a_0, and the other conditions are satisfied to give the existence of an unordered set of executions T, lower bound to T_c and hence to S, such that

(1) $\Big(\bigcup_{e_j \in e_0} (k_j \lceil \rho_{0,2}(e_j))\Big)(v) = f\Big[\Big(\bigcup_{a_i \in a_0} (k_i \lceil \rho_{0,1}(a_i))\Big)(d_0)\Big]$

$\qquad = f\Big[g_1\Big[\Big(\bigcup_{a_i \in a_1} (k_i \lceil \rho_{1,1}(a_i))\Big)(d_1)\Big], g_2\Big(\Big[\bigcup_{a_i \in a_2} (k_i \lceil \rho_{2,1}(a_i))\Big)(d_2)\Big]$

(2) $\rho_4(a) \cap A_0 = \emptyset$ by 1.2, 2.2, 3.2, 4.2,

(3) $\mathfrak{D}(K) \cap (A_0 \cup d \cup v) = \emptyset$ by 1.3, 2.3, 3.3, 4.3,

(4) if $A_1 \cap v = \emptyset$, then $eRI(\mathfrak{p}) \cap A_1 = \emptyset$, by 1.4, . . . , 4.4 and choice of $A_{1,1}, \ldots, A_{0,1}$,

(5) take A_2 as $A_{0,2}$.

NOTATION

We collect here definitions of a few notational conventions which might be unfamiliar to the reader.

$\mathfrak{P}$ power set operator. For A a set, $\mathfrak{P}(A)$ is the set of all subsets of A.

P finite subset operator. For A a set, $P(A)$ is the set of all *finite* subsets of A.

$\mathfrak{D}$ domain operator. For f a function or relation, $\mathfrak{D}(f) = \{x | \langle x, y\rangle \in f\}$.

card cardinality, the number of elements in a set.

Π_i^j iterated projection. For C a set of ordered n-tuples, $n \geqslant i$, $\Pi_i(C)$ is the set of ith components of elements C. $\Pi_i^j = \underbrace{\Pi_i \cdot \Pi_i \cdot \ldots \cdot \Pi_i}_{j \text{ times}}$.

$k|a$ domain restriction. For k a relation and a a set,

$$k \upharpoonright a = \{x, y | \langle x, y \rangle \in k \wedge x \in a\}.$$

Thus, $\mathfrak{D}(k \upharpoonright a) = \mathfrak{D}(k) \cap a$.

REFERENCES

[AMT] Turing, A. M., "On Computable Numbers, with an Application to the Entscheidungsproblem," *Proc. London Math. Soc.*, Ser. 2, 42 (1936–37), 230–65.

[E-R] Elgot, Calvin C., and Robinson, Abraham, "Random-Access Stored-Program Machines, an Approach to Programming Languages," *Journal Assoc. Comp. Mach.*, 11, (1964), 365–99.

Explicit Definitions and Linguistic Dominoes

Saul Gorn*

There are many categories of mechnical languages in which all expressions can be "parsed" by exhibiting hierarchical orderings among their sub-expressions. When we speak of a category of languages here, we mean it in the algebraic sense in which each language is an object of the category; for any two languages of a category, $\mathfrak{L}_1$ and $\mathfrak{L}_2$, there is a unique one-to-one "translator" T_{12} which transforms any expression of $\mathfrak{L}_1$ into its corresponding expression in $\mathfrak{L}_2$; furthermore, these translators have the following compositional relations:

1. For any two languages of the category, $\mathfrak{L}_1$ and $\mathfrak{L}_2$, T_{12} and T_{21} are inverse transformations one of the other, so that $T_{12}\,T_{21}$ will translate any expression of $\mathfrak{L}_1$ back into itself, as will $T_{21}\,T_{12}$ for the language $\mathfrak{L}_2$.
2. For any three languages of the category, $\mathfrak{L}_1$, $\mathfrak{L}_2$, and $\mathfrak{L}_3$, $T_{13} = T_{12}\,T_{23}$.

(In other words, the translators form a grouppoid, in the sense of Brandt.)

Those categories of languages in which simple parsing may be done include the "constituent analysis" or "phrase structure models" of portions of natural languages, the notations in mathematics involving such "control characters" as parentheses, commas, etc., the two-dimensional languages in which expressions are presented in "labelled trees," functor notations generally, and the "combinator" notations of the Polish prefix type, and, programming languages in which each elementary expression contains one operator and its requisite operands.

Now any system of representation languages for deductive systems whose axioms are categorical will form such a category of languages. For example, the Peano axiom system for the natural numbers gives rise to a large category of "natural number representation languages." Each language of a category has its own peculiar syntactic properties, but those properties and theorems definable or expressible in terms of the primitive concepts of the axioms alone will be uniquely translated among the languages of the category. Thus, if we have chosen a language of the category whose individual syntactic properties make it easy to prove

*The Moore School of Electrical Engineering. The research behind this paper was made possible by the joint support of the Army Research Office (DA-31-124-ARO(D)-98), the National Science Foundation (NSF-GP-1476), and the Public Health Service Research Grant (1 RO1GM-13, 494-01) from the National Institute of General Medical Science.

such a theorem, we have none the less proved the theorem for all languages of the category. I have elsewhere [4] called this the "principle of syntactic invariance."

There is a strong suspicion that large and important categories of languages exist for the handling of ramified structures, each such category containing a language of the "prefix type" [5].

What we have said so far applies also to systems of "language functions," each of which produces a language when applied to any one of a domain of "alphabets." We will therefore mainly restrict our attention to the one-dimensional language function, called the "complete prefix language function," $\mathfrak{P}$, which we have defined as follows.

Let $\mathfrak{A}$ be any "simply-stratified" alphabet (often infinite), by which we mean that for each character $c \in \mathfrak{A}$, there is a uniquely determined natural number, nc, called the stratification of c, and presented in some natural number representation language. Thus $\mathfrak{A}$ is partitioned into a possibly infinite set of subalphabets $\mathfrak{A}^{(n)}$, $n = 0, 1, 2, \ldots$. A string of characters of $\mathfrak{A}$ is a word of $\mathfrak{P}\mathfrak{A}$ if it is constructed recursively as follows. If $nc = n$, and $\alpha_1, \ldots, \alpha_n \in \mathfrak{P}\mathfrak{A}$, then $c\alpha_1 \ldots \alpha_n \in \mathfrak{P}\mathfrak{A}$. (If $n = 0$, this is interpreted to mean that the string $c \in \mathfrak{P}\mathfrak{A}$.)

In this paper we shall discuss the effect produced in "complete prefix languages" undergoing extension at the introduction of new characters into their alphabets by a process of "explicit definition."

The effect on the formal side is a rather simple algebraic one. The words in the extended language fall into classes of words "equivalent by definition," and each equivalence class possesses a natural partial ordering whose structure is simply related to a fairly obvious analysis of the definition.

The effect on the non-formal side is even more interesting. A number of processors of several varieties become available to all prefix languages (and hence, by translation, to all languages of their category)—processors which recognize patterns in which the defined symbols may be introduced, processors which eliminate such symbols until a "normal form" is generated, and some processors which recognize when words in the extended languages are equivalent.

Before introducing the concepts and their processors, I would like to make a few remarks about the connections between this subject and problems in other parts of information science, and in logic and linguistics.

(*a*) Mechanical languages may "grow" for a number of reasons and by a number of methods. One class of methods of growth is that in which a family of languages is involved, all of which involve the "same processors," the variations being only in the "alphabets" to which they are applied; the method of growth is by extension of the alphabet without changing the bulk of the processors. However, when a language in such a family grows due to an extension of its alphabet, some processors will be needed which are concerned precisely with the relationship between the original language and its extension. When we use such "growing languages," we will want to "control the growth" or "control the effects of the growth." Processors of the type discussed in this paper will be obvious candidates for "controllers" of growth (see Gorn [6]).

(*b*) Many problems in artificial intelligence may be interpreted in the following way. A language is specified for which a recognizer is not known (or for which none can exist because the language is undecidable); a "heuristic" processor is to be constructed by which we can "often" determine whether a word belongs to this language by attempting to solve the problem of setting up a derivation for it. If the processor is halted before this happens, the recognition is left undecided.

The investigations of Newell, Shaw, and Simon in regard to the logic theorist [7] and the general problem solver [8] can be formulated in this manner. For example, theorems in logic belong to a language generated from its "axioms" by appropriate "transformation rules" such as rules for substitution, *modus ponens*, and others (see [9]). The heuristic programme recognizes patterns of applicability of the transformation rules in order to set up "subgoals"; often such pattern searches can be made efficient by complete algorithms rather than heuristic ones, thereby making the main heuristics more efficient. This is the approach of the master's theses of Christine Boyer [1], and Gerhard Chroust [2]; the latter generalizes the problem from theorem proving to derivation seeking, and both eliminate definitions down to a "normal form" in order to simplify the pattern search.

(*c*) The first four chapters of Curry and Feys [3] are concerned with the extension of formal systems by definitions; a main result, which they needed for the problems of λ-conversion, is the Church-Rosser theorem which states that if two expressions are equivalent by definition, then there is an expression to which they both can be reduced. The corresponding result in this paper is much easier to obtain because there is more structural information about the languages involved, and the problem is restricted to "explicit definitions." However, the question is transformed from one about formal systems to one about languages, and the derivation rules of the formal systems are changed into the "transformation rules" of the language, in the sense of Carnap.

(*d*) The restriction to prefix languages in this paper has the effect that every word involved corresponds to a "parsing tree" related to the "phrase marker" used by natural linguists when they discuss phrase-structure models, as we have already remarked. The classification of explicit definitions which will naturally arise would apply to a classification into types of the "linguistic transformations" in the sense of Harris and Chomsky.

(*e*) Suppose we have two programming languages, such as, for example, the assembly languages of two different binary general purpose machines of the von Neumann type. We can consider that each instruction of either language is defined in terms of a common "microlanguage" of bit-manipulations. Some of these definitions are recursive, but many can be made explicit. Among the difficulties of machine-to-machine translation is the fact that many instructions, like "*add*," do not really have the same definition in the different machines.

The same problem arises if a community of users, beginning with a common language, have the freedom of introducing macrodefinitions independently of one another. Subcommunities with common problems will develop different sublanguages by different systems of macrodefinitions, and if these different subcommunities do not remain in constant communication, their distinct "dialects" will, in effect, grow into distinct languages, and the translations, even with only explicitly defined macro-instructions, will become difficult because of what, in this paper, is called the domino effect among definitions.

RAMIFICATION CONCEPTS

In the categories of languages which have parsing structures, or which represent such structures, we would like to define syntactic relations which are applicable to all the languages of their categories. If, for example, such a language is one-dimensional, the relation in which a certain character is the fifth in a word, reading from left to right, is in that language a syntactic relation between the character and the word which we would *not* expect to be an important one in the category; but the relation in which a character in a word is "the root" of that word's parsing structure *is* a syntactic relation which we would expect to be maintained in all corresponding words and characters in the languages throughout the category.

Thus, a category of languages for algebraic manipulation might have the following corresponding expressions in six of its languages.

1.

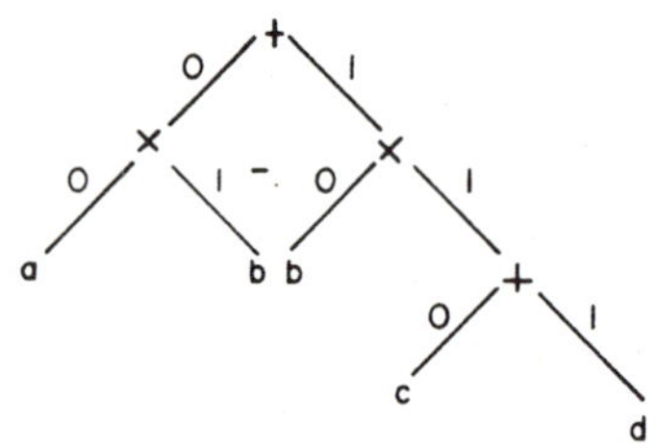

2. $ab + b(c + d)$
3. $+ * * a\ b\ b + c\ d$
4. $(d + cb) + (ba)$
5. $+(*(a, b), *(b, + (c, d)))$
6. $+ * a\ b * b + c\ d$

In all of these, the second occurrence of "multiplication," where the scan is from left to right in all but 4, has a *depth* of 1, a *height* of 2, and a *scope* which involves b, c, and d; furthermore, it is the second element at depth 1 beyond the root. Similarly, the second occurrence of b is at depth 2, at height 0, has no scope beyond itself, and is the first element at depth 2 in the scope of the second element at depth 1 in the scope of the root. Furthermore, in all of them, the first character at depth 2 in the scope of the second character at depth 1 in the scope of the root is the same as the second character at depth 2 "under" the first character at depth 1 "under" the root: more briefly, the characters at addresses 1.0 and 0.1 are both *b*.

All are representations of the same "labelled tree," or simply "tree," in a number of languages of a category, and the concepts of root, node, end-point, depth, height, subtree (or scope), address of a node, or a scope, etc., in the whole category determine the unique translations from language to language in the category, though there may be many translators of different types of efficiency for the same resulting translations.

Similarly, there may be many processors for each language of a category, and they may be "recognizers," "scope analyzers," "depth analyzers," "deconcatenators," "replacers," "substitutors," "address generators," "form recognizers," etc. We are interested in those which can be transformed from language to language in the category.

It is therefore useful to have in each category a language built out of these concepts alone; we will choose one in which the phrase structuring of each word is expressed as a mapping from a universal tree-addressing language to an alphabet. In such a language the corresponding expression to the six we had above might be:

7. $*, +; 0, \times; 00, a; 01, b; 1, \times; 10, b; 11, +; 110, c; 111, d.$

And, as before, there might be a variety of internal orderings for different languages of this type in the category. However, the main idea, mathematically, is that we will define a tree, not as a partial ordering, but as a subset of certain partially ordered set; and a "labelled tree" is not defined like a "coloured graph" but like a mapping. The information science component of the definition is the following: a labelled tree is a ramified system of storage areas, endowed with an appropriate "addressing system," and such that each storage element corresponds to a node of the tree and contains the label in storage.

The main utility of such a choice of representation is that, as we modify words (for example, by introducing or eliminating defined equivalents), we are able to keep track of the identity or the loss of identity of its components throughout a series of modifications.

Definition. "A universal tree-addressing language."

Let $\mathfrak{N}$ be any natural number representation language (we shall use, for example, the usual Arabic notation), and let $\{*,.\}$ be two characters not in the alphabet of the language $\mathfrak{N}$. We shall effectively use $\mathfrak{N} \cup \{*,.\}$ as our alphabet, and the set of all finite sequences of elements of $\mathfrak{N}$ (variously designated $\sum\mathfrak{N}$, or $\mathfrak{N}^*$, or $\mathfrak{N}^\omega$), with certain conventions as our addressing language: $*$ will represent the "null-chain" in $\mathfrak{N}^\omega$, and "." will represent the chaining operation. Thus, the character "$*$" will have a descriptive interpretation (namely, the identity element in a partially-ordered semi-group we are about to define), and a prescriptive (or command, or programming) interpretation (namely, "perform a no-op" in the chaining operation we will define). Similarly, the character "." will have an object interpretation (representing itself, and used as a separator of characters of $\mathfrak{N}$), a descriptive interpretation (representing the semi-group's binary functor), a prescriptive interpretation (representing a chaining or concatenation operation), and a control interpretation (representing the allocation of some storage in a ramified storage system, and the extension of an "address selector" in order to be able to retrieve its contents). When the language $\mathfrak{N}$ is uniquely deconcatenable, the object "." will be dispensed with, so that we would write, for example, "021013" instead of "0.2.1.0.1.3."

With this multiplicity of interpretations we shall define a specific partial ordering for these addresses, and we shall call this ordered system of addresses "A."

1. $*$ is called the root address. If $a \in A$, then $*.a = a.* = a$. The "depth" of $*$ is zero.

2. If $a \in A$ and $i \in \mathfrak{N}$, then $a.i$ is "an immediate successor of a" with relative address i, and a is "the immediate predecessor of $a.i$," and the "depth of $a.i$" is one more than the depth of a (hence, if $a \in A$, the depth of a is the number of characters of $\mathfrak{N}$ contained in a), and we begin the definition of the partial ordering by writing $a < a.i$ and $a.i > a$. Thus, $* < i$ for every $i \in \mathfrak{N}$, we do not use the ordering of $\mathfrak{N}$ to define that in A.

3. Thus, for every $a \in A$, either $a = *$ or $a = i_1.i_2.\ldots.i_n$ is of depth n for $n > 0$ and determines a unique chain of immediate predecessors

$$a > i_1.i_2.\ldots.i_{n-1} > \ \ldots > i_1.i_2 > i_1 > *.$$

We make the relation ">" the minimal transitive extension of the definition in 2. Thus, if $a_1, a_2 \in A$, then $a_2 < a_1$ if and only if there is an $a_3 \in A$ such that $a_3 \neq *$ and $a_1 = a_2.a_3$; in this case, the depth of a_1 is the sum of the depths of a_2 and a_3, a_2 is "a predecessor" of a_1, $a_1 > a_2$, a_1 is "a successor" of a_2, and the "relative address of a_1 with respect to a_2 is a_3." In fact, we can introduce the usual meanings for "$\geqslant$" and "$\leqslant$," so that the following are equivalent: $a_1 \geqslant a_2$, $a_2 \leqslant a_1$, there exists an $a_3 \in A$ such that $a_1 = a_2.a_3$; a_1 has relative address a_3 with respect to a_2 even if $a_3 = *$ and $a_1 = a_2$.

Notice that if $a_1 = a_2.a_3$, $a_2 \leqslant a_1$, but there need be no such relation between a_3 and a_1; all "heads" of a_1 are predecessors, and all "tails" of a_1 are relative addresses of a_1 with respect to the corresponding heads. This addressing system will be used on all labelled trees; in effect, we are mapping all finite trees root to root and as far left as possible into the infinite tree of figure 1.

This addressing system will be recognized as a restricted form of the Dewey decimal system used by librarians.

The following notation will be useful. Let A_1 and A_2 be two sets of addresses, $A_1 \subseteq A$ and $A_2 \subseteq A$. Then by $A_1 \cdot A_2$ we will mean the set of all addresses whose relative address with respect to some address of A_1 is some address of A_2, i.e., all

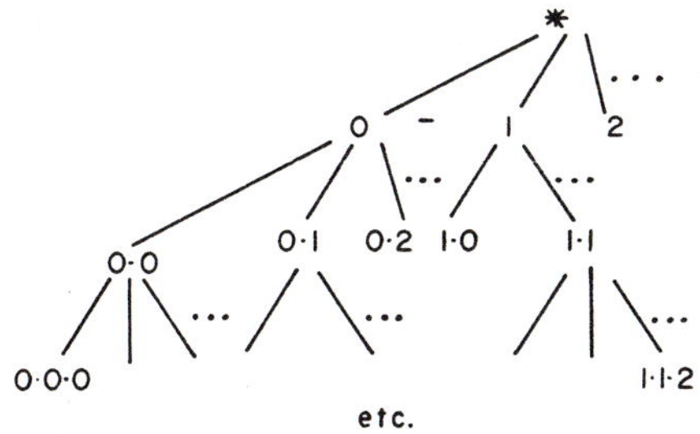

FIGURE 1. The universal finite tree and its addresses.

addresses of the form $a_1 \cdot a_2$ where $a_1 \in A_1$ and $a_2 \in A_2$. If $A_1 = \{a_1\}$, we simply write $a_1 \cdot A_2$. It is easy to verify that

$$(A_1 \cup A_2) \cdot A_3 = A_1 \cdot A_3 \cup A_2 \cdot A_3,$$

and that

$$(A_1 \cap A_2) \cdot A_3 = A_1 \cdot A_3 \cap A_2 \cdot A_3.$$

For the last identity, it is convenient to accept the convention that, if A_1 or A_2 is null, so is $A_1 \cdot A_2$; in fact,

$$A_1 \cdot A_2 = \Lambda \Leftrightarrow \text{either } A_1 = \Lambda \text{ or } A_2 = \Lambda.$$

The most important subsets of A are those giving all end-points of particular trees, or subsets of all end-points of particular trees, or sets of all addresses of particular trees.

A set of addresses $A_1 \subset A$ will be called an "independent set" if no address in A_1 is predecessor of another. Any independent set forms a uniquely deconcatenable code because no element is head of another. For example,

$$\{0 \cdot 1, 2 \cdot 0 \cdot 1 \cdot 1, 2 \cdot 0 \cdot 2\}$$

is an independent set. Clearly the set of all maximal elements of any set is independent.

An independent set of addresses is called a "complete end-point set" if it fulfils the following condition. If $a \cdot i$ is an element or a predecessor of an element of A_1, where $i \in \mathfrak{N}$, and if $j < i$ in $\mathfrak{N}$, then either $a \cdot j$ or some successor is also an element of A_1. Any independent set generates a unique minimal complete end-point set. For example, the minimal complete end-point set generated by $\{0 \cdot 1, 2 \cdot 0 \cdot 1 \cdot 1, 2 \cdot 0 \cdot 2\}$ is obviously the following:

$$\{0 \cdot 0, 0 \cdot 1, 1, 2 \cdot 0 \cdot 0, 2 \cdot 0 \cdot 1 \cdot 0, 2 \cdot 0 \cdot 1 \cdot 1, 2 \cdot 0 \cdot 2\}.$$

A finite set of addresses, $D \subset A$, is called a tree domain if it fulfils the following two conditions:

(*a*) If $a_1 \in D$ and $a_2 < a_1$, then $a_2 \in D$.
(*b*) If $a_1 \cdot i \in D$ and $j < i$ in $\mathfrak{N}$, then $a_1 \cdot j \in D$.

Clearly, the maximal elements of a tree domain form a complete end-point set; conversely, a complete end-point set and all its predecessors form a tree domain determining it.

By a scope-set we mean any address set of the form $a \cdot D$ where D is a tree domain. If $a \in D$, where D is a tree domain, then clearly a and all its successors in D form a scope-set; we will call it the scope of a in D, and designate it by SaD; if $a \notin D$, we will take $SaD = \Lambda$. $SaD = \{a\}$ if and only if a is maximal in D; we shall call it an end-point of D. If $a = a_1 \cdot a_2$ is an element of a tree domain D, and if $Sa_1D = a_1 \cdot D_1$, and if $Sa_2D_1 = a_2 \cdot D_2$, then $SaD = a \cdot D_2 = a_1 \cdot Sa_2D_1$. In other words, if $Sa_1D = a_1 \cdot D_1$, then $Sa_1 \cdot a_2D = a_1 \cdot Sa_2D_1$.

If D is a tree domain and $a \in D$, we shall define the height of a in D (written: $h(a; D)$) recursively as follows.

(*a*) If a is an end-point of D, its height in D is zero.

(*b*) If $a \cdot 0, a \cdot 1, \ldots, a \cdot n$ are all the immediate successors of a in D, then

$$h(a; D) = 1 + \max_{0 \leqslant i \leqslant n} h(a \cdot i; D).$$

More generally, if D is a tree domain and $A_1 \subseteq D$, the relative height of $a \in A$, is defined recursively as follows:

(*a*) if no $x \in A_1$ is a successor of a, then $ht(a; A_1) = 0$.

(*b*) otherwise $ht(a; A_1) = 1 + \max_{\left\{\substack{a < x \\ a \in A_1}\right\}} ht(x, A_1)$.

By the depth of a tree domain D we mean the maximum of the depths (i.e., $\mathfrak{N}$-lengths) of its addresses. We write $d(a)$ and $d(D)$ for the depths of a and D respectively, so that $d(D) = \max_{a \in D} d(a)$. By the depth of the scope $SaD = a \cdot D_1$ we mean $d(D_1)$. Thus $h(a; D) = d(SaD)$, as follows immediately by recursion from the definition of h.

By the breadth of a tree domain D we mean the number of its end-points; we designate it by $b(D)$. If $SaD = a \cdot D_1$, then we define $b(SaD)$ to be $b(D_1)$. Thus $b(\{*\}) = 1$. If $D = \{*, 0, 1, \ldots, n-1\}$, then $b(D) = n$. If $n \in \mathfrak{N}$, where n is the maximum in $\mathfrak{N}$ of the elements of depth 1 in D, then

$$b(D) = \sum_{i=0}^{n-1} b(SiD).$$

More generally, let D_0 be a tree domain, let S_0 be the complete end-point set of D_0, which therefore contains $n = b(D_0)$ elements, and let $D_1, \ldots, D_n$ be tree domains, and let $a_0, \ldots, a_{n-1}$ be the end-points of D_0 in left-to-right order, then by the "sum tree domain" we mean

$$D_0 \cup \bigcup_{i=0}^{n-1} a_i \cdot D_{i+1},$$

which we designate by "$+D_0D_1 \ldots D_n$." Then

$$b(+D_0D_1 \ldots D_n) = \sum_{i=1}^{n} b(D_i).$$

It is clear that iterated use of "$+$" allows us to represent any tree domain as a sum of tree domains of depth one, except for $\{*\}$.

Finally, let us define "the ramification of an address a in a tree domain D" to be the number of immediate successors of a in D. If D has depth one, then the ramification of $*$ in D is the breadth of D. The ramification of any end-point in D is zero. If $a \notin D$, then the ramification of a in D is undefined. If $SaD = a \cdot D_1$, then the ramification of a in D is $r(a; D) = r(*, D_1)$. We extend this definition to tree domains and scopes by defining $r(D) = b(D)$ and $r(SaD) = b(D_1)$.

Now let D_1 and D_2 be two tree domains; then a mapping, f, from D_2 "into" D_1, is called monotone if

(*a*) $f(*) = *$;
(*b*) if a_2 is an end-point of D_2, then $f(a_2)$ is an end-point of D_1;
(*c*) if $a_{21} \leqslant a_{22}$ in D_2, then $f(a_{21}) \leqslant f(a_{22})$ in D_1;
(*d*) if $a_{11} \leqslant f(a_{22})$ in D_1, then there is an $a_{21} \leqslant a_{22}$ in D_2 such that $a_{11} = f(a_{21})$.

The address map is said to be "onto" D_1 if every $a_1 \in D_1$ has a pre-image in D_2, $f^{-1}(a_1) \neq \Lambda$. Thus if a mapping f: $D_2 \rightarrow D_1$ is monotone, then every complete chain in D_2 maps into a complete chain (no gaps) in D_1. If we let $a_1 \cap a_2$ mean the "greatest common head of a_1 and a_2," so that it designates the maximum predecessor of both, then condition c is equivalent to c^1: if $a_1, a_2 \in D_2$, then $f(a_1 \cap a_2) \leqslant f(a_1) \cap f(a_2)$.

If we take the usual interpretations of $f(A_1)$, $f^{-1}(A_2)$, and $f^{-1}(a_1)$, then f is a monotone address map of D_2 *onto* D_1 only if

$$a_2 \in f^{-1}(a_1) \Rightarrow Sa_2D_2 \subseteq f^{-1}(Sa_1D_1);$$

in other words, f is a monotone address map of D_2 onto D_1 only if each restriction of f to a scope-set is a map into the corresponding scope-set of D_1, and every scope-set of D_1 has pre-image in D_2.

If f_2: $D_2 \rightarrow D_1$ and f_1: $D_1 \rightarrow D_0$ are two monotone address maps, then their composite is also a monotone address map which we will designate by $f_1 f_2$: $D_2 \rightarrow D_0$; $f_2(D_2) \subseteq D_1$ and $f_1(D_1) \subseteq D_0$, whence $f_1 f_2(D_2) \subseteq D_0$. The existence of a monotone address map is an important relation between tree domains; because of condition d it is clear that there can be no such map if the image domain is deeper than the pre-image, i.e., if $f(D_2) \subseteq D_1$, then

$$d(f(D_2)) \leqslant d(D_2).$$

Another important relation between tree domains is the following. We say that D_1 is an initial pattern of D_2, and write $D_1 \leqslant_i D_2$ if $D_1 \subseteq D_2$ and every non-ending node of D_1 has the same ramification in D_1 as in D_2; in other words, $D_1 \leqslant_i D_2$ if and only if $a \in D_1 \Rightarrow a \in D_2$ and either $r(a; D_1) = r(a, D_2)$ or $r(a; D_1) = 0$. Suppose $D_0 \leqslant_i D$ and $b(D_0) = n$; then there are n tree domains, $D_1, \ldots, D_n$, such that $D = +D_0D_1 \ldots D_n$. The converse is also obvious: $D_0 \leqslant {}_iD_2$.

We say that D_1 "dominates" D_2 if there is a scope of D_1 which is non-trivial (i.e., not just an end-point) and which is an initial pattern of D_2; in other words, there is an address a, not an end-point of D_1, such that $SaD_1 = a \cdot D_0$, and $D_2 = +D_0 D_2' \ldots D_2^n$. (Note that the condition of non-triviality is important,

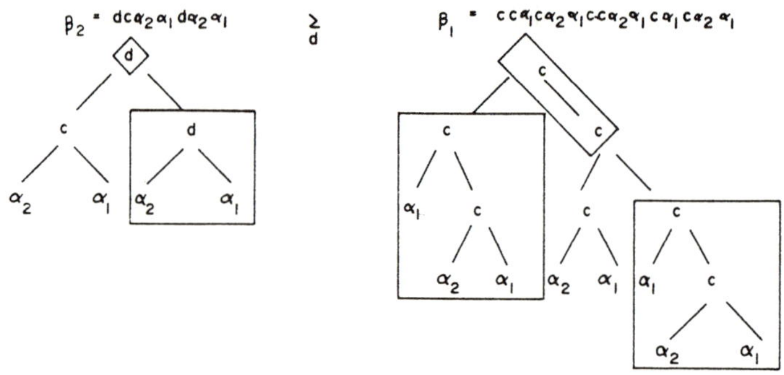

β_2	
$D\beta_2$	$Ca\beta_2$
*	d
0	c
1	d
00	α_2
01	α_1
10	α_2
11	α_1

β_1			
$D\beta_1$	$Ca\beta_1$	$f(D)$	$Cf(a)\beta_2$
*	c	*	d
0	c	1	d
1	c	*	d
00	α_1	11	α_1
01	c	1	d
10	c	0	c
11	c	1	d
010	α_2	10	α_2
011	α_1	11	α_1
100	α_2	00	α_2
101	α_1	01	α_1
110	α_1	11	α_1
111	c	1	d
1110	α_2	10	α_2
1111	α_1	11	α_1

FIGURE 2. Example of a *d*-map and its monotone address map.

because otherwise every D_1 would dominate every D_2 because every $D = +*D$ and $D = +D* \ldots *$, and we could use $D_0 = \{*\}$.) In such a case, we call D_0 the "intermediary of the $D_1 D_2$-domination": $D_1 = +D_1'* \cdots *D_0* \cdots *$, $D_2 = +D_0 D_2' \cdots D_2^{(n)}$. The tree domain

$$+D_1'* \cdots *D_2* \cdots * = +D_1'* \cdots *+D_0 D_2' \cdots D_2^{(n)}* \cdots *$$

is then called a "$D_1 D_2$-domino" (with intermediary D_0). It is, of course, quite possible for a tree domain to dominate itself, and for a D_1D_1-domino to exist; the intermediary may be as small as $D_0 = \{*, 0\}$, or more generally of depth one and arbitrary breadth, or of depth and breadth as large as we please, and we can still find a suitable D_1, say $D_1 = +D_0 D_0 \cdots D_0$, to yield a $D_1 D_1$-domino with intermediary D_0.

The intermediary, D_0, in a $D_1 D_2$-domination will also be an "internal pattern" of the domino, in accordance with the following definition. D_0 is an "internal pattern" of D if, roughly speaking, it is the initial pattern of some scope of D, thereby making D a domino with D_0 as an intermediary. More precisely, D_0 is an internal pattern of D if and only if there is an $a \in D$ such that, if $SaD = a \cdot D_2$, then $D_0 \leqslant_i D_2$. Thus, let $D = +D_1'* \cdots *+D_0 D_2' \cdots D_2^{(n)}* \cdots *$, where

$D_2 = +D_0 D_2' \cdots D_2^{(n)}$; if we take, for example, $D_1 = +D_1' * \cdots * D_0 * \cdots *$, then D is a $D_1 D_2$-domino with intermediary D_0. (See Fig. 3.)

The addressing system we have discussed generalizes, for partially ordered sets, the property of well-ordering among fully ordered sets; in fact, the relation $\leqslant_i$ among all finite tree domains has this well-ordering property; furthermore, every tree domain has a finite number of predecessors in the sense of $\leqslant_i$.

Another well-ordering for finite chains of integers, one which is a full-ordering, is of the lexicographic ordering type; to distinguish it from the addressing system, we will present its elements in another notation. Instead of writing $n_0 \cdot n_1 \cdot \cdots \cdot n_m$, we will write $(n_0, n_1, \ldots, n_m)$. The particular ordering of this type which we will have occasion to use is the following.

$$(n_0, n_1, \ldots, n_m) \leqslant_l (n_0', n_1', \ldots, n_{m'}'),$$

if and only if

(*a*) either the "degree" of $(n_0, n_1, \ldots, n_m)$ is less than the degree of $(n_0', n_1', \ldots, n_{m'}')$ where degree $(n_0, n_1, \ldots, n_m) = \max_{n_i \neq 0} i$ (if $n_i = 0$ for all i, we call the degree -1),

(*b*) or degree $(n_0, n_1, \ldots, n_m) =$ degree $(n_0', n_1', \ldots, n_{m'}') = d$ and $n_p \leqslant n_p'$ where $p = \max_{n_i \neq n_i'} i$ (if $n_i = n_i'$ for all i, p does not exist, but then we say $(n_0, \ldots, n_m) = (n_0', \ldots, n_{m'}')$; thus

$$(n_0, \ldots, n_m) = (n_0, \ldots, n_m, 0, \ldots, 0),$$

like the coefficients of polynomials).

In this full-ordering, every element of degree greater than zero has infinitely many predecessors. For example $(n_0) \leqslant_l (n_0', n_1')$ if $n_1' > 0$, no matter how big n_0 may be. Nevertheless, even though most elements have infinitely many predecessors, any strictly descending chain from a given element must be finite.

As examples of the use of this ordering for tree domains, let D be any tree domain, and let $A_1 \subseteq D$.

1. Let n_i be the number of nodes of A_1 of ramification i in D; $(n_0, n_1, \ldots)$ is then the "ramification distribution of A_1 in D."

2. Let n_i be the number of nodes of A_1, of depth i in D; $(n_0, n_1, \ldots)$ is then the "depth distribution of A_1 in D."

3. Let n_i be the number of nodes of A_1 of height i in D; $(n_0, n_1, \ldots)$ is then the "height distribution of A_1 in D."

4. Let n_i be the number of nodes of A_1 of relative height i in A_1; $(n_0, n_1, \ldots)$ is then the "absolute height distribution of A_1 in A_1."

Suppose, for example, that we have a stratified alphabet for use in algebra and analysis, $\mathfrak{A}$, in which

$$\mathfrak{A}^{(0)} = \{x, y, z, \ldots, a, b, c, \ldots\} \cup \mathfrak{N},$$
$$\mathfrak{A}^{(1)} = \{C, S, E, \ldots\},$$
$$\mathfrak{A}^{(2)} = \{+, -, \times, \div, \ldots\}, \text{ etc.},$$

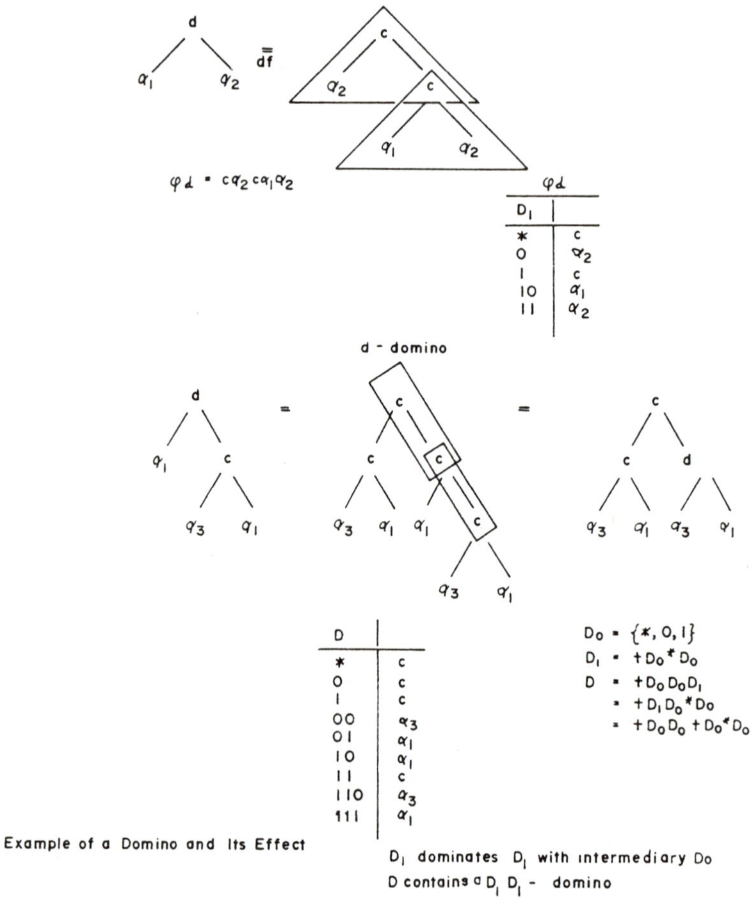

FIGURE 3. Example of a domino and its effect.

and we extend $\mathfrak{A}$ by definining d_x by recursion as follows:

$$(1) \qquad d_x\,\alpha, \quad \text{where } \alpha \in \mathfrak{A}^{(0)} = \begin{cases} 1 & \text{if } \alpha = x, \\ 0 & \text{if } \alpha \neq x. \end{cases}$$

$$(2) \qquad \begin{aligned} d_x\,C\alpha &= -0 \times S\alpha\, d_x\,\alpha, \\ d_x\,S\alpha &= \times\; C\alpha\, d_x\,\alpha, \\ d_x\,E\alpha &= \times\; E\alpha\, d_x\,\alpha, \quad \text{etc.,} \end{aligned}$$

for α any word of the extended $\mathfrak{P}\mathfrak{A}$.

$$(3) \qquad \begin{aligned} d_x + \alpha\beta &= +d_x\,\alpha d_x\,\beta, \\ d_x \times \alpha\beta &= +\times d_x\,\alpha\beta \times \alpha d_x\,\beta, \quad \text{etc.,} \end{aligned}$$

for α and β any words of the extended $\mathfrak{P}\mathfrak{A}$.

For any word γ of the extended $\mathfrak{PA}$ we let A_1 be the set of nodes of the tree domain of γ at which d_x occurs; furthermore, we notice that each rule in the recursive definition of d_x has the property that the height distribution of the occurrences of d_x on the right is less than that on the left, even though the depth, breadth, and number of A_1-occurrences increases. It therefore follows immediately from the well-ordering of height distributions that any sequence of relevant applications of the rules for the recursive definition of d_x must end in a finite number of steps with a "total elimination" of d_x, no matter what the initial expression of the extended $\mathfrak{PA}$ may have been.

Although we will be considering "explicit definitions" rather than recursive ones, our application of the well-ordering of these distributions can still be somewhat more complicated.

LABELLED TREES, FORM LANGUAGES, AND EXPLICIT DEFINITIONS

We shall now define labelled trees. When no confusion is possible, we shall call them simply trees, though, to every labelled tree there will be a corresponding one which we shall call "the tree of the labelled tree" which will serve as a generalized "syntactic type" of that labelled tree.

Let $\mathfrak{A}$ be any simply stratified alphabet; by a "labelled tree over $\mathfrak{A}$" we shall mean a mapping, t, from a tree domain D into $\mathfrak{A}$, such that if $a \in D$ and $r(a; D) = r$, then $t(a) \in \mathfrak{A}^{(r)}$. In other words, the character of $\mathfrak{A}$ "at a node" must have a stratification equal to the ramification of that node. If $\mathfrak{A}$ is a natural number representation language $\mathfrak{N}$, where the stratification is the identity function, then the labelled tree is simply a tree. If $s(c)$ is the stratification of $\mathfrak{A}$, i.e., a mapping of $\mathfrak{A}$ into $\mathfrak{N}$, then $s(t(a))$ is "the tree" of the labelled tree $t(a)$.

As we have already noted, there is a whole category of languages for representing labelled trees over $\mathfrak{A}$, and another for representing trees. In this paper we shall use three in each category:

(1) the two-dimensional type;

(2) the complete prefix type, $\mathfrak{PA}$ and $\mathfrak{PN}$;

(3) the mathematical mapping type just defined.

For each language of such a category we can now refer to addresses, roots, nodes, end-points, scopes, address sets, address domains, depths, heights, depth distributions, height distributions, stratifications, ramifications, initial and internal patterns, domination relations, dominoes, etc. We shall broaden each of these categories by first extending their basic stratified alphabets to include variables, and will therefore extend all these concepts to "forms" of expressions rather than to expressions alone.

When the alphabet of a particular language of a category employs "control characters" such as parentheses, commas, connecting lines, etc., these are not considered to be in the basic alphabet, have no stratification number, and serve mainly as elements in the particular processors designed to perform the appropriate symbol manipulations belonging to the category. Apart, then, from the role of control characters, which may differ from language to language in a category, it is clear what is meant by "the character at address a of labelled tree β" (written $Ca\beta$), "the tree of β" ($\tau\beta$), "the scope at address a of β" ($Sa\beta$), "the domain of β" ($D\beta$); we also note that "an occurrence of the character $c \in \mathfrak{A}$ in β" is an address a such that $c = Ca\beta$; in other words, $a \in C^{-1}c\beta$ is "an occurrence of c in β."

Another way to express these concepts, a formal way, is as follows. If $\mathfrak{N}$ is a number representation language, $\mathfrak{A}$ a "simply-stratified alphabet" of stratification s (so that $\mathfrak{A} \xrightarrow{s} \mathfrak{N}$, and $\mathfrak{A}^{(r)} = s^{-1}r$), and if $\mathfrak{D} \subset 2^A$ is the set of tree domains in a representation language employing the address language A, then the category of "labelled tree languages over $\mathfrak{A}$" includes a "mapping representation language," $T\mathfrak{A}$; a labelled tree $\beta \in T\mathfrak{A}$ is therefore a mapping $D \xrightarrow{\beta} \mathfrak{A}$ such that $a \in D \Rightarrow s \circ \beta(a) = r(a; D)$. The domain function, which we have also designated by D, is therefore a mapping $T\mathfrak{A} \xrightarrow{D} \mathfrak{D}$, and "$D\beta = D$" represents a useful and very usual ambiguity in the use of the character "D." Because of the category, we will not hesitate to identify β in $T\mathfrak{A}$ with the corresponding words in $\mathfrak{P}\mathfrak{A}$, $\mathfrak{P}_d\mathfrak{A}$, $\mathfrak{T}\mathfrak{A}$ (the last two language functions are illustrated on p. 80 by items 1 and 3), etc., throughout the category, and this identification carries over to the domain function D, independent also of the representation $\mathfrak{D}$ in its category. Then the "character function C" may be identified as that mapping of $A \times T\mathfrak{A}$ onto $\mathfrak{A}$ such that $Ca\beta$ is identical to $\beta(a)$, and again we can consider it independently of the A or T in their categories. Similar remarks apply to τ ("the tree of"), and S (the scope at $\cdots$ of $\cdots$). We shall define the "scope at address a of tree β" as follows. If $D \xrightarrow{\beta} \mathfrak{A}$, and $a \in D$, then β also maps $SaD = a \cdot D_1$ into $\mathfrak{A}$; $Sa\beta$ will mean the map $Ca\beta$, or $\beta(a)$, restricted to SaD, which we will not distinguish from the tree whose domain is D_1: $D_1 \xrightarrow{Sa\beta} \mathfrak{A}$, the restriction of β to SaD. Thus it now makes sense to consider, for each $a \in A$, the mappings Ca and Sa of $T\mathfrak{A}$ onto $\mathfrak{A}$; if $a \notin D\beta$, we can take $Ca\beta$ to be "the null-character of $\mathfrak{A}$" or "the null-word of the language of trees over $\mathfrak{A}$ which we have chosen from the category," and we can take $Sa\beta$ to be "the null-tree" with null address domain (not to be confused with trees of depth zero, whose address domain is $\{*\}$). Thus, if $a = a_1 \cdot a_2$ in A, then $Ca = Ca_2\, Sa_1$; the meaning is perfectly clear if $a_1 = *$ or $a_2 = *$. Similarly, $Sa_2\, Sa_1 = Sa_1 \cdot a_2$.

If β_1 is a tree over $\mathfrak{A}$, and if $D\beta_1 = D_1$, and if $a \in D_1$ is an address in the domain of β_1, then by the tree β_2 "obtained from β_1 by replacing the scope at a by β_3" we mean one whose domain is $D_2 = (D_1 - SaD_1) \cup a \cdot D_3$ and defined by

$$\beta_2\, a_1 = \begin{cases} \beta_1\, a_1 & \text{if } a_1 \in D_2 - a \cdot D_3, \\ \beta_3\, a_3 & \text{if } a_1 = a \cdot a_3 \in a \cdot D_3 \end{cases}$$

(in $\mathfrak{P}\mathfrak{A}$, replacement is achieved by the replacement of one connected string by another), and we write "$\beta_2 = R\beta_3\, a\beta_1$" as well as "$\beta_2 = \beta_1(\beta_3 \rightarrow Sa)$."

If a_1 and a_2 are independent addresses, then $R\beta_2\, a_2\, R\beta_1\, a_1\, \beta_0 = R\beta_1\, a_1\, R\beta_2\, a_2\, \beta_0$; in other words, replacements at independent addresses commute, and may in fact be achieved by simultaneous action; a particularly important example is that in which a_1 and a_2 are end-points of $D\beta_0$. This generalizes, for $n > 2$, to the concept of n simultaneous replacements, from which the concept of "free substitution" is easily defined. If $\beta_2 = \beta_1(\beta_3 \rightarrow Sa)$, then $C^{-1}c\beta_2 = (C^{-1}c\beta_1 - C^{-1}cSa\beta_1) \cup a \cdot C^{-1}c\beta_3$; this describes the modification of occurrences by replacements.

Now let $s(c) = 0$, whence $C^{-1}c\beta_1$ must be included in the end-point set of $D\beta_1$, i.e., "c must be end-points of β_1." Then we write "$\beta_2 = \beta_1(\beta_3 \rightarrow c)$" for the result of all simultaneous replacements at each address of $C^{-1}c\beta_1$, and say that "β_2 is obtained from β_1 by free substitution of β_3 for c"; in this case, if $d \in \mathfrak{A}$, $C^{-1}d\beta_2 = (C^{-1}d\beta_1 - C^{-1}c\beta_1) \cup C^{-1}c\beta_1 \cdot C^{-1}d\beta_3$. This definition of free substitution as a simultaneous replacement at end-points may obviously be generalized for many characters of stratification zero to "simultaneous substitution." Simultaneous replacements and substitutions are possible in the categories of tree languages because "scopes do not overlap."

Now, it is easy to see how the concept of addition of tree domains may be extended to addition of trees where $s(c) = 0$ only for $c = 0$ in $\mathfrak{N}$: in $\mathfrak{P}\mathfrak{N}$ we may write $+\tau_0\, \tau_1 \cdots \tau_n$, where τ_0 has n end-points and the sum refers to simultaneous replacement of the i'th zero from left-to-right in τ_0 by τ_i.

By the depth (height, ramification) of a character we will mean the depth (height, ramification) of the occurrence, i.e., the address in the address domain of that tree. It is also clear what we mean by the depth distribution, height distribution, etc., of characters in a tree.

Now let us consider categories of "form languages for trees."

If $\mathfrak{A}_c$ is a simply-stratified alphabet (the constants), and if $\mathfrak{A}_v = \{\alpha_1, \alpha_2, \ldots\}$ is an alphabet (which we will call variables) such that $\mathfrak{A}_c \cap \mathfrak{A}_v = \Lambda$, then extend the stratification mapping from $\mathfrak{A}_c$ to $\mathfrak{A}_c \cup \mathfrak{A}_v$ by taking $s(\alpha_i) = 0$ for each i; the "descriptive semantics" of $\mathfrak{A}_v$ consist in the fact that each α_i is a variable whose domain is "words" of $\mathfrak{P}\mathfrak{A}'$ or any other language of its category for "suitable extensions $\mathfrak{A}'$ of $\mathfrak{A}$"; the "command semantics" of $\mathfrak{A}_v$ consist in the fact that each α_i refers to the contents of some storage area, while the "pragmatics" of $\mathfrak{A}_v$ have each object "α_i" being the name of some storage area adaptable to contain any word of $\sum\mathfrak{A}$. When we want to emphasize our choice of the command interpretation of α_i, we shall write "(α_i)," and read "the content of α_i." Thus "$\alpha_i \rightarrow \alpha_j$" is a word in a programming language which means "place the contents of α_i into α_j"; but "$\alpha_i \rightarrow (\alpha_j)$" means "place the contents of α_i into the address which is the contents of α_j" (this is called "indirect addressing").

This extension of $\mathfrak{A}_c$ to $\mathfrak{A}$ has therefore also extended "the trees over $\mathfrak{A}_c$" to "the trees over $\mathfrak{A}$," with representations in all languages of the category, such as $\mathfrak{P}\mathfrak{A}$, $\mathfrak{P}_d\mathfrak{A}$, $\mathfrak{T}\mathfrak{A}$, $T\mathfrak{A}$, etc. We shall call these trees over the extended alphabet $\mathfrak{A}$ "tree forms," or simply "forms" over $\mathfrak{A}_c$, designate their class by $\mathfrak{F}\mathfrak{A}_c$, and most often represent an element of $\mathfrak{F}\mathfrak{A}_c$ by a correspondent in $\mathfrak{P}\mathfrak{A}$, $\mathfrak{I}\mathfrak{A}$, or $T\mathfrak{A}$. By the rank of a form we mean the number of distinct α_i in it; a form has rank zero if and only if it is a tree over $\mathfrak{A}_c$, i.e., a tree without variables.

It is not unusual for two distinct forms to have the same (descriptive) meaning. For example, α_1 and α_2 both mean (descriptively) any word because their domains of values (or contents, in the command interpretation) are identical, even though they are distinct both as objects and as names (addresses) of storage areas. There are, consequently, semantic equivalence classes for which we might like to choose standard representatives; we will call them "normal forms." For example, we could define a form as normal if its prefix representation has the following property: if α_j occurs in the form, and $j > 1$, then α_{j-1} also occurs, and the first occurrence of α_{j-1} is left of the first occurrence of α_j. For example, "α_1" and "$c\alpha_1\, c\alpha_2\, \alpha_1$" are normal forms from the classes of "α_3" and "$c\alpha_3\, c\alpha_1\, \alpha_3$."

By a simple form we mean a normal form of depth zero or one; thus, the elements of $\mathfrak{A}_c^{(0)}$ and the depth one trees over $\mathfrak{A}_c$ are simple forms of rank 0, α_1 is the only "non-null" simple form of depth zero (its rank is one), and any simple form of depth one and rank $n > 0$ has the prefix representation $c\alpha_1 \ldots \alpha_n$ where $s(c) = n$. This means that there is an extra condition which we have imposed on a normal form of depth one before we will call it simple, namely, that end-points be variable (i.e., none will be constant; for example $\times 0 \alpha_1$, $(0 \times \alpha_1)$ or

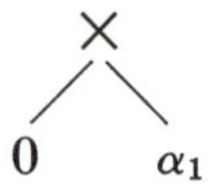

are *not* representations of simple forms).

Explicit definitions have traditionally been presented in the form $\alpha_1 =_{df} \alpha_2$ where (the contents of) α_1 is a form called the "definiendum" and (the contents of) α_2 is a form called the "definiens." Examples would be $\sim \alpha_1 =_{df} 1\alpha_1\alpha_1$, $0 =_{df} (\alpha_1 - \alpha_1)$, $\text{axiom}_1 =_{df} \supset \alpha_2 \supset \alpha_1\, \alpha_2$, $\text{axiom}_1\alpha_1\, \alpha_2 =_{df} \alpha_2 \supset (\alpha_1 \supset \alpha_2)$, $\text{theorem}_1 =_{df} \alpha_1 \supset \alpha_1$,

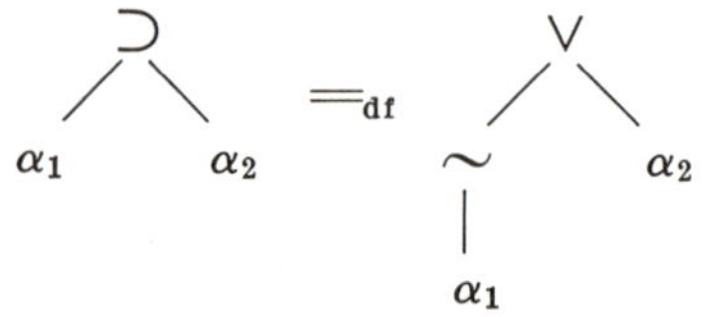

$$\begin{array}{c}\supset \\ \alpha_1 \quad \alpha_2\end{array} =_{df} \begin{array}{c}\vee \\ \sim \quad \alpha_2 \\ | \\ \alpha_1\end{array}, \quad \alpha_1 + 1 =_{df} S(\alpha_1),$$

$\alpha_1 + \alpha_2 + \alpha_3 =_{df} ((\alpha_1 + \alpha_2) + \alpha_3)$, $\langle \alpha_1, \alpha_2 \rangle =_{df} \alpha_1(\alpha_2)$, $a\alpha_3\, \alpha_1\, \alpha_2\, \alpha_3 =_{df} \alpha_1(\alpha_2)$, $a\alpha_1\, \alpha_2 =_{df} F\alpha_1\, c_1\, \alpha_2\, c_2$, etc.

We shall restrict our definition of "explicit definition" in such a way that if an alphabet is "extended" by an explicit definition, the prefix language in the category, which was "complete" before the extension in that $s(c) = n$ permits c to be concatenated with *any* n good words, remains complete after the extension by definition. This requires that the definiendum be a simple form.

Definition. By an explicit definition of "*d*" we mean an expression of the form $\alpha_1 =_{df} \alpha_2$, where

(1) (the contents of) α_1 is a simple form whose root is d;

(2) (the contents of) α_2 is a form such that every variable in α_2 occurs in α_1 (thus the rank of $\alpha_1 \geqslant$ that of α_2).

Explicit definitions will be used to establish an equivalence relation in the extended language and thereby permit us to process "simpler expressions" such as ones with reduced depth, or reduced breadth; their function is therefore to permit "transformations" in the language.

We shall therefore assume available to us a number of processors in a command syntax (i.e., a programming meta-language). Among such might be recognizers of when a form is normal, or simple, or generators which translate any form into its unique equivalent normal form (an efficient one for the prefix representation works in a single left-to-right scan). Less obvious processors, which we can assume to be available for each representation of the category of languages, are "form recognizers" or "tree-pattern recognizers," which do the following.

Given a normal form contained in the storage α_1 and any form in the storage α_2, the form recognizer will tell whether α_2 is or is not "of form α_1," and, if so, what scopes of α_2 replace the variables of α_1 (we shall define these terms shortly). In general, it will recognize when a form α_2 has α_1 as its form (Figure 4). The component in Figure 4 called the "scope analyzer" is specified further in Gorn [6].

Definition. The form β_1 is said to be "of the form β_0," and β_0 is called "an initial pattern of β_1," and, as with tree domains, we write $\beta_0 \leqslant_i \beta_1$ (but we also write $\beta_1 \in \beta_0$), if there is a complete independent set of scopes $\sigma_1, \ldots, \sigma_n$ of β_1, where n is the number of end-points of β_0, such that if $\alpha_i = \alpha_j$ in β_0, then $\sigma_i = \sigma_j$; and such that $\beta_1 = \beta_0(\sigma_1 \rightarrow \alpha_1, \ldots, \sigma_n \rightarrow \alpha_n)$.

The notation is consistent in that $\beta_0 \leqslant_i \beta_1 \Rightarrow D\beta_0 \leqslant_i D\beta_1$. In fact, $\beta_1 \leqslant_i \beta_2$ if and only if $D\beta_1 \leqslant_i D\beta_2$ and $Ca\beta_1 = Ca\beta_2$ for each $a \in D\beta_1$ which is not a variable, while $Ca\beta_1 = Ca'\beta_1 \Rightarrow Sa\beta_2 = Sa'\beta_2$ where a and a' are end points of β_1.

We accept it as obvious that all substitution processors with the same component substitutions for the individual α_i in β_0 will produce the "same" result throughout the category of $\mathfrak{F}\mathfrak{A}$ and, indeed, within any language or form language of the category, independent of the sequencing order or even simultaneity of the individual α_i substitutions or replacements. This is a pragmatic assumption about the behaviour of the storage processors, namely, that the separate storage cells are independent of one another with respect to the process of "binding," in spite of any dependence they may have with respect to the process of selecting among them.

β_0 is called an internal pattern of β_1 it it is the initial pattern of a scope of β_1, i.e., if there is a scope of β_1 of the form β_0; an address of β_0 as an internal pattern of β_1 is called "an occurrence of β_0 in β_1," and we extend the semantics of our notation to write $a \in C^{-1}\beta_0\,\beta_1$ (note that if $s(c) = n$ and $\beta_0 = c\alpha_1 \ldots \alpha_n$, then

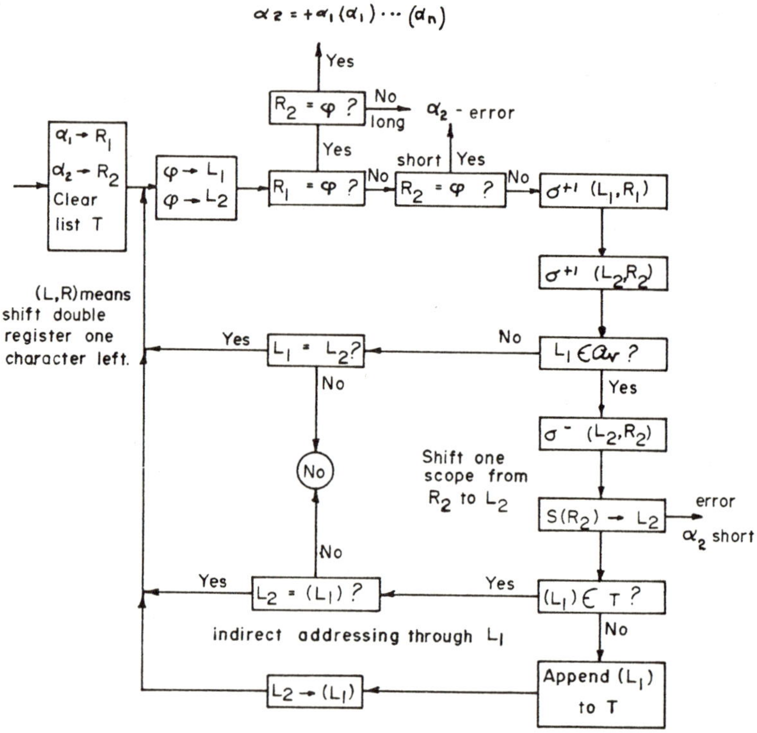

FIGURE 4. Form recognizer—prefix language; left-to-right scan.

$C^{-1}\beta_0\,\beta_1 = C^{-1}c\beta_1$). The form recognizer of Figure 4 is evidence that is is an "effective" definition.

We can now also extend the definition of domination and dominoes from tree domains to trees. The form β_1 is said to dominate the form β_2 with intermediary β_0 if $d(\beta_0) > 0$ and β_0 is a scope of β_1, say at address a, and if, finally, $\beta_0 \leqslant_i \beta_2$. In this case, the replacement (not substitution) of β_0 at address a in β_1 by β_2 is called a $\beta_1\,\beta_2$-domino: $\beta_1(\beta_2 \longrightarrow Sa)$, which we shall also write as $\beta_1(\beta_2 \overset{a}{\longrightarrow} \beta_0)$. It is distinguished from the substitution symbol by the presence of "a." It is obvious that these relations entail the corresponding ones for the tree domains of β_0, β_1, and β_2, and that there is a one-to-one correspondence between dominoes with intermediary β_0 and trees with internal pattern β_0.

We can now specify the various kinds of replacements for which explicit definitions were intended. Let d be defined, with $s(d) = n$, by an explicit definition of the form: $d\alpha_1 \ldots \alpha_n =_{\text{df}} \phi_d$, where $\phi_d \in \mathfrak{F}\mathfrak{A}_0$, d is not a character of $\mathfrak{A}_0$, and every variable end-point of ϕ_d occurs among $\alpha_1, \ldots \alpha_n$ (the rank of $\phi_d \leqslant n$). In the remainder of this paper we shall be studying the structure within $\mathfrak{F}\mathfrak{A}_1$, where $\mathfrak{A}_1 = \mathfrak{A}_0 \cup \{d\}$, generated by this explicit definition.

If $\beta_2 \in \mathfrak{F}\mathfrak{A}_1$, there may be a number of occurrences of d and ϕ_d in β_2. The definition is intended to allow us to replace any occurrence of one type by an occurrence of the other, and such replacements are called "d-introductions" and "d-eliminations."

Definition. If $\beta_2 \in \mathfrak{F}\mathfrak{A}_1$ and $a \in C^{-1}d\beta_2$, and if $\sigma_i = Sa \cdot i\beta_2$ and we call $\psi_1 = \phi_d(\sigma_1 \to \alpha_1, \ldots, \sigma_n \to \alpha_n)$ (so that $\phi_d \leqslant_i \psi_1$), then the tree $\beta_1 = \beta_2(\psi_1 \to Sa)$ is said to be obtained from β_2 by "elimination of d at a," and we write: $\beta_1 = E_d\, a\beta_2$, and also

$$\beta_2 \xrightarrow[a]{E_d} \beta_1.$$

In anticipation of the relation of d-descendence which we will define in $\mathfrak{F}\mathfrak{A}_1$, we shall also say that β_1 is "an immediate d-descendent of β_2" and write $\beta_2 \geqslant_d \beta_1$. (This only begins the definition of "$\geqslant_d$"; note that the definition may be what we will call "expansive," so that not every α_i need occur in ϕ_d, so that some of the replacement commands may be vacuous.)

Conversely, if $\beta_1 \in \mathfrak{F}\mathfrak{A}_1$ and $a \in C^{-1}\phi_d\, \beta_1$, so that $\phi_d \leqslant_i Sa\beta_1$, then there is a complete independent set of scopes $\sigma_1, \ldots, \sigma_n$, occurring with the repetitions corresponding to the α_i in ϕ_d, at addresses in β_1 whose relative addresses with respect to a equals the addresses of the end-points α_i in ϕ_d; let

$$\beta_0 = +d\alpha_1 \ldots \alpha_n\, \sigma_1 \ldots \sigma_n$$

and $\beta_2 = \beta_1(\beta_0 \to Sa)$. Then we shall say that β_2 is obtained from β_1 by "introduction of d at a," and we write $\beta_2 = I_d\, a\beta_1$, and

$$\beta_1 \xrightarrow[a]{I_d} \beta_2;$$

β_2 is "an immediate d-predecessor of β_1," $\beta_1 \leqslant_d \beta_2$.

Obviously, if $a \in C^{-1}d\beta_2$, then $a \in C^{-1}\phi_d\, E_d\, a\beta_2$, and $\beta_2 = I_d\, aE_d\, a\beta_2$; conversely, if $a \in C^{-1}\phi_d\, \beta_1$, then $a \in C^{-1}dI_d\, a\beta_1$, and $\beta_1 = E_d\, aI_d\, a\beta_1$. Not only do we have a category of languages, but each language is now a category of expressions with respect to the operations of d-introduction and d-elimination for appropriate occurrences.

$$\beta_1 \xrightarrow[a]{I_d} \beta_2 \xrightarrow[a]{E_d} \beta_1;\ \beta_2 \xrightarrow[a]{E_d} \beta_1 \xrightarrow[a]{I_d} \beta_2.$$

We can now define the partial ordering and the equivalence relation generated by an explicit definition in any language of the category of $\mathfrak{F}\mathfrak{A}_1$ or $T\mathfrak{A}_1$ exactly as it is done for recursive definitions and the extension of formal systems (Curry and Feys [3]).

$\beta_2 \geqslant_d \beta_1$ if either $\beta_2 = \beta_1$ or if there is a chain of d-eliminations

$$\beta_2 = \beta_{01} \xrightarrow[a_1]{E_d} \beta_{02} \xrightarrow[a_2]{E_d} \ldots \beta_{0n-1} \xrightarrow[a_{n-1}]{E_d} \beta_{0n} = \beta_1.$$

We say, in this case that β_2 is a d-predecessor of β_1 and β_1 is a d-descendent of β_2. We also say that β_1 is obtained from β_2 by d-reduction.

β_2 is said to be d-equivalent to β_1, and we write $\beta_2 \equiv_d \beta_1$, if the chain called a "derivation chain" contains any combination of d-introductions and d-eliminations. It is easy to see that $\geqslant_d$ is indeed a partial ordering, and $\equiv_d$ an equivalence relation. We shall designate the set of all d-descendents of β_0 (the principal dual-ideal, or filter, generated by β_0, as it is called in algebra) by $L(\beta_0)$, and the d-equivalence class of β_0 by $E(\beta_0)$. The remainder of this paper will be concerned with the structures of the various $L(\beta_0)$ and $E(\beta_0)$.

THE NORMAL d-DESCENDENT

If $\beta_2 \geqslant_d \beta_1$, the evidence would be the presentation of a particular d-reduction chain; there might be many such chains for the same pair β_2 and β_1. We would like to be able to identify the occurrences in β_2 "responsible for" each occurrence in β_1, and we would like to know whether such "maintenance of identity" through a particular "reduction history" is changed by an alternative reduction history. This we do by extending the concept of monotone address map from tree domains to trees.

Let f be a monotone address map, $D_2 \xrightarrow{f} D_1$. If $a_1 \in D_1$, then $f^{-1}(a_1)$ possesses a finite number of minimal elements which constitute an independent address subset: $Ma_1 f =_{\text{df}} \{a_2 : f(a_2) = a_1,\ a_2' < a_2 \Rightarrow f(a_2') < a_1\}$, $Ma_1 f \subset D_2$, $Ma_1 f$ independent. From this definition it is obvious that

$$a_2 \in Ma_1 f \Rightarrow f(Sa_2 D_2) \subseteq Sa_1 D_1 \quad \text{and} \quad f^{-1}(Sa_1 D_1) = \bigcup_{a_2 \in M} Sa_2 D_2,$$

where a_2, $a_2' \in Ma_1 f$ and $a_2 \neq a_2' \Rightarrow Sa_2 D_2 \cap Sa_2' D_2 = \Lambda$, because scopes do not overlap. We therefore call the scope sets $Sa_2 D_2$ for $a_2 \in Ma_1 f$ the "component pre-image scopes of D_2 determined by f from a_1."

By a "component of D_2 determined by f from a_1" we mean an address set $Sa_2 D_2 \cap f^{-1}(a_1)$ where $a_2 \in Ma_1 f$; in other words, it is that part of the pre-image of a_1 contained in a single component pre-image scope. A primitive example of this concept extended to tree maps is provided by "the generating d-map" of a definition:

Definition. If $s(d) = n$, and d is defined by explicit definition by the form ϕ_d, so that $d\alpha_1 \cdots \alpha_n =_{\text{df}} \phi_d$, then let $A_{di} = C^{-1}\alpha_i \phi_d$ (the occurrences, possibly null, of α_i in ϕ_d), and let

$$A_{d0} = D\phi_d - \bigcup_{i=l}^{n} A_{di}$$

(the "interior" of ϕ_d). Then by the "generating d-map g_d" we mean that determined by the address map

$$g_d(a) = \begin{cases} i-1 & \text{if } a \in C^{-1}\alpha_i \phi_d, \text{ i.e., if } a \in A_{di}, \\ * & \text{otherwise, i.e., if } a \in A_{d0}. \end{cases}$$

Because α_i occurs in ϕ_d only if $1 \leqslant i \leqslant n$, it is obvious that g_d is a monotone address map of $D\phi_d$ into $Dd\alpha_1 \cdots \alpha_n$; it will be a map "onto $d\alpha_1 \cdots \alpha_n$" if and only if the rank of ϕ_d is n, so that α_i occurs in ϕ_d if and only if $1 \leqslant i \leqslant n$. Let the number of elements of A_{di} be m_i, and call it the multiplicity of α_i in ϕ_d.

In this example the component pre-image of $*$ is A_{do}, and the component pre-image of α_i is the A_{di}, either null (if α_i is not an end-point of ϕ_d, the map is not "onto") or depth 0 trees $\{\alpha_i\}$ of characters at relative addresses in A_{di}.

Suppose we let $\beta_{20} = d\alpha_1 \cdots \alpha_n$, and $\beta_{10} = \phi_d$, and

$$D\beta_{20} = \{*, 0, \ldots, n-1\} = D_{20},$$

$D\beta_{10} = \{*, \ldots, a_{11}, \ldots, a_{1m_1}, \ldots, a_{n1}, \ldots, a_{nm_n}\} = D_{10}$, where $Ca_{ij}\beta_{10} = \alpha_i$, and either a_{ij} is null (if α_i does not occur in ϕ_d), or is a component pre-image of α_i. Then the following diagram schematizes both the symbol manipulation of transformation from β_{10} to β_{20}, and the generating d-map.

$$\begin{array}{ccc} D_{10} & \xrightarrow{\beta_{10}} & \mathfrak{A} \\ g_d \downarrow & \downarrow & (Ca\beta_{10} \to g_d(a)) \\ D_{20} & \xrightarrow[\beta_{20}]{} & \mathfrak{A} \end{array}$$

We can therefore consider the mapping β_{10} a kind of composite of β_{20} and g_d and write

$$\beta_{10} \overset{E}{\sim} \beta_{20} \circ g_d.$$

A more detailed description would be $(Ca\beta_{10} \to g_d(a)) \circ \beta_{10} = \beta_{20} \circ g_d$.

Consider now any $\beta_2 \in \mathfrak{F}\mathfrak{A}_1$, and suppose d occurs in β_2 at a: $a \in C^{-1}d\beta_2$, and let $\beta_1 = E_d a \beta_2$, so that

$$\beta_2 \xrightarrow[a]{E_d} \beta_1;$$

then g_d generates a natural monotone address map from $D\beta_1 = D_1$ to $D\beta_2 = D_2$ at $a \in C^{-1}\phi_d\beta_1$ as follows.

$$g_d(a; a_1) = \begin{cases} a_1 & \text{if } a \nleqslant a_1, \\ a & \text{if } a_1 = a \cdot a_1' \text{ and } a_1' < a_{ij} \\ a \cdot i - 1 \cdot a_1'' & \text{if } a_1 = a \cdot a_1' \text{ and } a_1' = a_{ij} \cdot a_1''. \end{cases}$$

If we remember that

$$\beta_1 = R\phi_d(\cdots, Sa \cdot i - 1\beta_2 \to a_{ij}, \cdots) a\beta_2,$$

we see that

(1) β_2 and a uniquely determine both $g_d(a; a_1)$ and the immediate d-descendent β_1;

(2) β_1 and $a \in C^{-1}\,\phi_d\,\beta_1$ uniquely determine both $g_d(a; a_1)$ and the immediate d-predecessor β_2;

(3) if $\beta_1 = E_d a \beta_2$, or $\beta_2 = I_d a \beta_1$, then $g_d(a; a_1)$ is uniquely determined.

$g_d(a; a_1)$ is the simplest example, including $g_d(a_1) = g_d(*; a_1)$ for $\beta_{20} = d\alpha_1 \cdots \alpha_n$, and $\beta_{10} = \phi_d$, of the type of monotone address map yielding what we will call a "d-map for trees." As before, we write

$$\beta_1 \overset{E}{\sim} \beta_2 \circ g_d(a; a_1).$$

Because the compositions of monotone address maps are also monotone address maps, every d-elimination which is a derivation of $\beta_2 \geqslant_d \beta_1$ yields a monotone address map $\beta_1 \overset{E}{\sim} \beta_2 \circ f$ which is a d-map according to the following definition.

Definition. A monotone address map $D_1 \xrightarrow{f} D_2$, where $D_1 = D\beta_1$ and $D_2 = D\beta_2$ is called a d-map of β_1 into β_2 if and only if for every $a_2 \in D_2$ and for every non-null (in case the mapping is not "onto") component of β_1 determined by f at a_2, say with minimal element $a_1 \in D_1$, the following conditions hold:

(1) either $Ca_2\beta_2 = Ca_1\beta_1$ (i.e., $\beta_2(a_2) = \beta_1(a_1)$) and $f^{-1}(a_2) \cap Sa_1\beta_1 = \{a_1\}$, and, if a_2 is not an end-point of D_2, $f(a_1 \cdot i) = a_2 \cdot i$ for $0 \leqslant i \leqslant r(a_2; D_2)$;

(2) or $\beta_2(a_2) = d \neq \beta_1(a_1)$ and the component $f^{-1}(a_2) \cap Sa_1\beta_1 = \{a_1 \cdot Ad_0\}$, and if $a \in A_{d0}$, then $\beta_1(a_1 \cdot a) = \phi_d(a)$, while, if $a \in A_{di}$, then $f(a_1 \cdot a) = a_2 \cdot i - 1$. (In other words, f maps characters of β_1 into β_2 either identically, possibly even when d occurs, or "the interior of" ϕ_d into d, and in any event the permutation of immediate descendents is maintained or dictated by ϕ_d.) In the case of a d-map "onto," $a_2 \in Ma_1 f \Rightarrow f(Sa_2D_2) = Sa_1D_1$, as against the weaker condition for the more general monotone address map. In this case, each scope is mapped onto a scope. We have, then, $\beta_1 \overset{E}{\sim} \beta_2 \circ f$ from the diagram

$$\begin{array}{ccc} D_1 & \xrightarrow{\quad f \quad} & D_2 \\ \beta_1 \Big\downarrow & \longrightarrow & \Big\downarrow \beta_2 \\ \mathfrak{A} & & \mathfrak{A} \end{array}$$

It also follows, as indicated in the diagram

$$\begin{array}{ccccc} D_1 & \xrightarrow{\quad f_1 \quad} & D_2 & \xrightarrow{\quad f_2 \quad} & D_3 \\ \beta_1 \Big\downarrow & \longrightarrow & \beta_2 \Big\downarrow & \longrightarrow & \beta_3 \Big\downarrow \\ \mathfrak{A} & & \mathfrak{A} & & \mathfrak{A} \end{array}$$

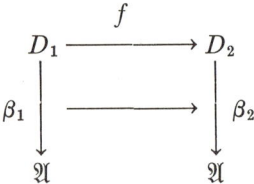

that if f_1 is a d-map from β_1 into β_2 and if f_2 is a d-map from β_2 into β_3, then the composite $f_2 \circ f_1$ is also a d-map; $\beta_1 \overset{E}{\sim} \beta_3 \circ (f_2 \circ f_1)$.

We shall now show that the compositions of maps generated by g_d, already defined, effectively exhaust all d-maps, and that any two forms such that $\beta_2 \geqslant_d \beta_1$ determine a unique d-map of β_1 into β_2 independent of the particular chain of eliminations from β_2 to β_1. Furthermore, given any $A_2 \subseteq C^{-1}d\beta_2$, a unique d-descendent of β_2, β_1 is determined by "complete elimination at A_2." (However, not every d-descendent need be such a complete eliminant. One of our tasks will be to find conditions that make all elements of $L(\beta_2)$ complete eliminants.)

LEMMA 1. *If there is a d-map of* β_1 *into* β_2, *then it is unique.*

Proof. The method is essentially one of recursion in depth; we prove the lemma by induction on the depth of β_2. Let f and f' be two d-maps of β_1 into β_2. If $d(\beta_2) = 0$ and $\beta_2 \neq d$, then $\beta_1 = \beta_2$ and both f and f' are the identity mapping. If $d(\beta_2) = 0$ and $\beta_2 = d$, then ϕ_d must be a null-form (no variable), and either $\beta_1 = d$ or $\beta_1 = \phi_d$; in the first case f and f' are again the identity, and in the second case they are both g_d. This is sufficient to begin the induction when either $\beta_2(*) \neq d$, or $\beta_2 = d$ and ϕ_d is a null-form. If $\beta_2(*) = d$ and ϕ_d are not null-forms, but $d(\beta_2) = 1$, then either $\beta_1(*) = d$ and both f and f' are the identity or $\beta_1 = E_d{*}\beta_2$ and both f and f' are g_d.

Now suppose the lemma to be true for all trees of depth less than $d(\beta_2)$. If $\beta_2(*) = \beta_1(*)$, then, because $f(*) = f'(*)$, we need only prove that $f = f'$ when restricted to each $Si\beta_1$ to map into $Si\beta_2$ for each i between 1 and $r(*, D_1) = r(*, D_2)$; this is true by the induction hypothesis. If, however, $\beta_2(*) \neq \beta_1(*)$, then $\beta_2(*) = d$ and β_1 begins with the nodes of ϕ_d which are either not end-points or not variable. Because all these initial nodes of β_1 must map into $*$ under both f and f', we need only prove that $f = f'$ when restricted to each $Sa_{ij}\beta_1$ to map into $Si\beta_2$ for each i and j; again this is true by the induction hypothesis because $Si\beta_2$ have depths less than $d(\beta_2)$. Thus the lemma is proved.

Definition. If f is the d-map of β_1 into β_2, and if a_2 is a d-occurrence in β_2, $Ca_2\beta_2 = d$, then by the d-multiplicity of f at a_2 we mean the number of components in $f^{-1}(a_2)$ which are not identical with d, i.e., which are the "interior" of ϕ_d. For example, the d-multiplicity of g_d at $*$ is one, as is the d-multiplicity of $g(a; a_1)$ at a for $\beta_1 = E_d\, a\beta_2$. We designate it by $m(f, a_2)$.

By the d-multiplicity of f, $m(f)$ we mean the sum of the multiplicities at all occurrences of d in β_2:

$$m(f) = \sum_{a_2 \in C^{-1}d\beta_2} m(f, a_2).$$

More generally, if $A_2 \subseteq C^{-1}d\beta_2$, then the d-multiplicity of f on A_2 is the sum

$$m(f, A_2) = \sum_{a_2 \in A_2} m(f, a_2).$$

Thus $m(f, \{a_2\}) = m(f, a_2)$, and $m(f) = m(f, C^{-1}d\beta_2)$.

For example, suppose $A_2 = \{a_2', a_2''\} \subseteq C^{-1}d\beta_2$; if a_2' and a_2'' are independent, then $E_d\, a_2'\, E_d\, a_2''\beta_2 = E_d\, a_2''E_d\, a_2'\beta_2 = \beta_1$, say, and either reduction yields the same d-map of multiplicity 2 on A_2. However, if a_2' and a_2'' are not independent, say $a_2' < a_2''$, or, more specifically, $a_2'' = a_2'\cdot i\cdot a_2''$, then

$$\beta_1 = E_d\, a_2'E_d\, a_2''\beta_2 = E_d a_2'\cdot a_{(i+1)1}\cdot a_2'''\cdot\cdots\cdot E_d\, a_2'\cdot a_{(i+1)}m_i\cdot a_2'''E_d\, a_2'\beta_2,$$

and either reduction yields the same d-map of β_1 into β_2; but this time, its multiplicity on A_2 is $m_i + 1$.

The two reduction equivalences just stated will be called "the fundamental commutation relations" of d-eliminations (or d-introductions, when rewritten) or of g_d maps.

LEMMA 2. *If there is a d-map, f, of* β_1 *into* β_2, *then* $\beta_2 \geqslant_d \beta_1$.

Proof. The proof is effectively by recursion in height, using induction on $m(f)$. If $m(f) = 0$, then $\beta_2 = \beta_1$, and the lemma is trivial. Suppose the lemma is true for all d-maps of multiplicity less than $m(f)$. Let a_2 be a d-occurrence of minimum height in β_2 among all those of multiplicity >0, so that no descendent of a_2 in β_2 is an occurrence of an "active d" in f. Now let $a_1 \in f^{-1}(a_2)$ such that $\phi_d \leqslant_i Sa_1\,\beta_1$; a_1 must exist because the multiplicity at $a_2 \geqslant 1$, and any non-identical component of $f^{-1}(Sa_2\,\beta_2)$ has the pattern ϕ_d. Now let $\beta_1' = I_d\, a_1\, \beta_1$ and define the d-map f_1 of β_1' into β_2 as follows.

$$f_1(a) = \begin{cases} a_2\cdot a_1' & \text{if } a = a_1\cdot a_1' \\ f(a) & \text{otherwise.} \end{cases}$$

In other words, f_1 is the identity map on $Sa_1\beta_1'$ and is the same as f otherwise; in fact, $\beta_1 \overset{E}{\sim} \beta_1' \circ g_d(a_1;)$, $\beta_1' \overset{E}{\sim} \beta_2 \circ f_1$, whence $\beta_2 \circ f \overset{E}{\sim} \beta_1 \overset{E}{\sim} \beta_2 \circ f_1 \circ g_d(a_1;)$, so that we have factored f into the composition of f_1 with $m(f_1) = m(f) - 1$, and $g_d(a_1;)$ with $m(g_d) = 1$. By the induction hypothesis we now have $\beta_2 \geqslant_d \beta_1' \geqslant_d \beta_1$, and the lemma is proved.

THEOREM 1. *The following conditions are equivalent*:

(1) $\beta_1 \in L(\beta_2)$.

(2) $\beta_2 \geqslant_d \beta_1$.

(3) *There is a d-map of* β_1 *into* β_2.

(4) *There is a unique d-map of* β_1 *into* β_2.

COROLLARY. *A necessary and sufficient condition that* $\beta_2 \geqslant_d \beta_1$ *is that either* $\beta_2(*) = \beta_1(*)$ *and* $Si\beta_2 \geqslant_d Si\beta_1$, *or* $\beta_2(*) = d$, *the "interior of* ϕ_d *heads* β_1," *and* $Si\beta_2 \geqslant_d Sa_{ij}\,\beta_1$ *for every i and j.*

Note. We could have used this as a recursive definition (in depth) of $\beta_2 \geqslant_d \beta_1$; it can be used to design a recognizer for $\geqslant_d$. In a programming language like LISP, it could be done in essentially one recursive conditional expression. The processor of Figure 5, designed by Martin Roberts employs two pairs of shifting double registers (or tapes).

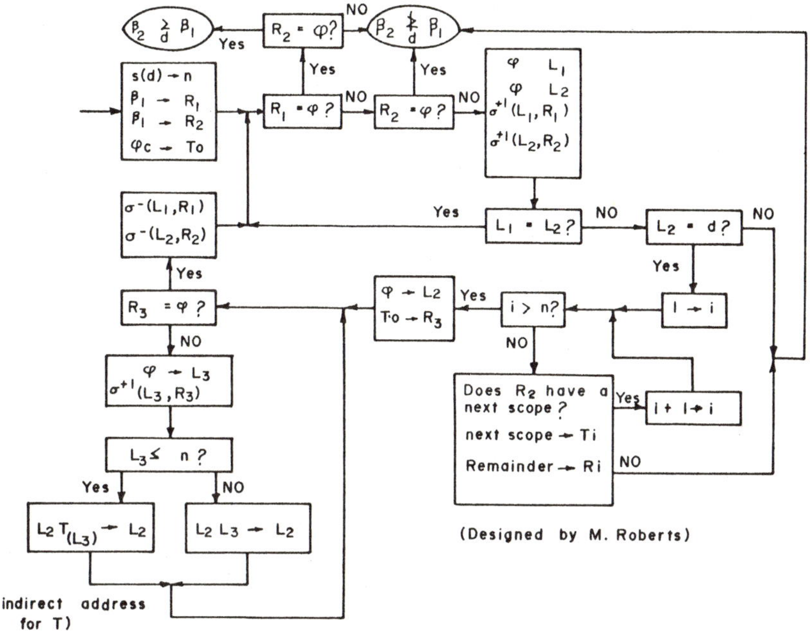

FIGURE 5. Recognizer for the relation $\beta_2 \geqslant_d \beta_1$.

Let us now consider the reductions of a $\beta_2 \in \mathfrak{F}\mathfrak{A}_1$ from a pre-assigned $A_2 \subseteq C^{-1}d\beta_2$.

All reductions of β_2 restricted to the occurrences of d from A_2 may be prescribed by the following selection game: choose any $a \in A_2$; if $\beta_2' = E_d\, a\beta_2$, so that $\beta_2 \geqslant_d \beta_2'$, then $\beta_2' \circ g_d(a;) \overset{E}{\sim} \beta_2$ exhibits the d-map of β_2' into β_2; let the minimal elements of all the components of $g_d^{-1}(a; A_2)$, excluding a itself, constitute the address set A_2'. Now iterate this procedure by choosing any $a' \in A_2'$, taking $\beta_2'' = E_d\, a'\beta_2' = E_d\, a' E_d\, a\beta_2$, etc., to yield an address set A_2'', and so on.

Before we show that any such selection sequence is finite (after all, A_2' may have many more elements than A_2), let us first remark that it is independent of the tree β_2 with which we began, provided only that $A_2 \subseteq C^{-1}d\beta_2$; only A_2 and the explicit definition (or g_d) are needed to permit the selection of a from A_2, the determination of A_2' to permit the selection of a', etc. The reason is that $E_d\, a$ is a functor applicable to any tree β_2 such that $Ca\beta_2 = d$, and the address computation required to yield β_2' is always effectively the same because of g_d. The same then applies to any elimination reduction chain, it yields a functor applicable to any tree with a proper distribution of d's among its addresses. We should therefore be able to exhibit a pure address calculus for elimination reductions when we are interested only in occurrences of certain d's without concerning ourselves about any other occurrences. This we shall now do. The situation for d-introductions is not so simple due to the possible $\phi_d\, \phi_d$-dominoes.

Because we are only concerned in the remainder of this section with eliminations and a single explicit definition, we can introduce a simplified notation. Instead of the functor $E_d\, a_1\, E_d\, a_2 \cdots E_d\, a_n$, assuming that the address set $\{a_1, \cdots, a_n\}$ is such that it is meaningful (i.e., has a non-null domain of trees in $\mathfrak{F}\mathfrak{A}_1$ to which it is applicable), we shall represent the functor by $\langle a_1, \cdots, a_n\rangle$. Not every sequence of addresses determines a functor, nor is such a representation unique, as we shall soon see. However, a simple set of rules will allow us to compute the resulting address set when $\langle a_1, \cdots, a_n\rangle$ is applied to a given address set $\{b_1, \cdots, b_n\}$. When we want to ignore the particular representation of such a functor whose domain is a set of address sets, we shall designate it by F with a subscript, just as we used f for d-maps of trees. Just as all d-maps f are generated as composites of g_d and $g_d(a;\,)$, so are all the address reduction functors F generated by $F_d = \langle * \rangle$ and $F_d(a) = \langle a \rangle$, which we will now define, if we use the composition

$$F_2 \circ F_1(A_1) = F_2(F_1\, A_1).$$

Definitions. If we remind ourselves of the pattern of repetitions of the variable end-points in an explicit definition, each address i in $\beta_{10} = d\alpha_1 \cdots \alpha_n$ is the occurrence of α_{i+1}, which is the g_d map of $\{a_{i1}, \cdots, a_{im_i}\} = A_{di}$ among the end-points of ϕ_d; we recall that m_i is the multiplicity of α_i; when g_d is not a map "onto" β_{10}, there will be at least one i less than $s(d)$ for which $m_i = 0$ and the pre-image address set $g_d^{-1}(\alpha_i) = \Lambda$, the null address set. We therefore write:

Rule 1. $\langle * \rangle \{*\} = \Lambda$ to indicate that $\langle * \rangle$ has eliminated an occurrence of d at $*$.

Rule 2. For each $i < s(d)$ and for every address a,

$$\langle * \rangle \{*, i \cdot a\} = \{a_{i1} \cdot a, \cdots, a_{im_i} \cdot a\},$$

interpreting this to mean the address set Λ when $m_i = 0$.

In general, an elimination at address a is applicable if and only if there is an occurrence of d at a. We shall therefore restrict $\langle a \rangle$ to be applicable only to those address sets A_1 for which $a \in A_1$, as we have already done in Rules 1 and 2. The definition of $g_d(a;\,)$ therefore requires the following computation rules:

Rule 3. If $\{*\} \subseteq A_1$, then $\langle a \rangle \{a \cdot A_1\} = \{a \cdot \langle * \rangle A_1\}$.

Rule 4. If $a \nless b$, then $\langle a \rangle \{a, b\} = \{b\}$.

Let us now define $a_1 \geqslant_d a_2$ among addresses to mean that there is an a_{ij} such that $a_1 \geqslant a_2 \cdot a_{ij}$.

Let us also define, recursively, $\langle a_1, \cdots, a_k, b\rangle$, when it is meaningful, to mean $\langle a_1, \cdots, a_k \rangle \circ \langle b \rangle$. Any address set to which it is applied must therefore include b. By iterating this condition, it is then fairly obvious that $\langle a_1, a_2 \rangle$ is meaningful whenever $a_1 \geqslant a_2$ implies that $a_1 \geqslant_d a_2$, so that $a_1 \geqslant a_2 \cdot a_{ij}$ will yield

$$\begin{aligned} &\langle a_1, a_2 \rangle \{a_2, a_2 \cdot i\} \\ &= \langle a_1 \rangle \langle a_2 \rangle a_2 \cdot \{*, i\} = \langle a_1 \rangle a_2 \cdot \langle * \rangle \{*, i\}, \\ &= \langle a_1 \rangle a_2 \cdot \{a_{i1}, \cdots, a_{im_i}\} = \langle a_1 \rangle \{\cdots, a_1, \cdots\}, \end{aligned}$$

which is meaningful.

It follows that the commutation rules for E_d may now be represented as follows:

Rule 5. If a and b are independent, then $\langle a, b\rangle = \langle b, a\rangle$.

Rule 6. $\langle a, a\cdot i\cdot a'\rangle = \langle a\cdot a_{i1}\cdot a', \cdots, a\cdot a_{imi}\cdot a', a\rangle$, if $m_i \neq 0$; otherwise we can only say $\langle a, a\cdot i\cdot a'\rangle \subseteq \langle a\rangle$.

It also follows that, if F_1 has the representation $\langle a_1, \cdots, a_k\rangle$ and F_2 has the representation $\langle b_1, \cdots, b_l\rangle$, then $F_1 \circ F_2$ will have the representation

$$\langle a_1, \cdots, a_k, b_1, \cdots, b_l\rangle$$

if and only if, for every $1 \leqslant i \leqslant k$ and $1 \leqslant j \leqslant l_l$, $a_i > b_j$ implies $a_t \geqslant_d b_j$.

The final rule for reduction computations on address sets is also fairly obvious.

Rule 7. If F is applicable to A_1 and to A_2, then it is applicable to

$$A_1 \cup A_2 \text{ and } F(A_1 \cup A_2) = FA_1 \cup FA_2.$$

Rules 2, 4, and 7 may be stretched a bit in application; when F is applicable to $A_1 \cup A_2$, then during the application to $A_1 \cup A_2$, we will apply it to A_2, even if not strictly applicable, by writing Rule 2 as

$$\langle *\rangle\{i\cdot a\} = \langle *\rangle\{i\}\cdot a = \{a_{i1}, \cdots, a_{im_i}\}\cdot a, \text{ and Rule 4 as } a \not< b \Rightarrow \langle a\rangle\{b\} = \{b\}.$$

An example of the calculus in application to the functor $F = \langle 111, 0, *\rangle$ and the address set $A_1 = \{*, 1, 11\}$ for the definition (in prefix) $d\alpha_1\alpha_2 = c\alpha_2 c\alpha_1\alpha_2$ is the following. First $\langle *\rangle\{*, 0\} = \{10\}$ and $\langle *\rangle\{*, 1\} = \{0, 11\}$, whence

$$\begin{aligned}
FA_1 &= \langle 111, 0, *\rangle\{*, 1, 11\}\\
&= \langle 111, 0\rangle\{\langle *\rangle\{*\} \cup \langle *\rangle\{1\} \cup \langle *\rangle\{1\cdot 1\}\}\\
&= \langle 111, 0\rangle\{\Lambda \cup \{0, 11\} \cup \{0, 11\}\cdot\{1\}\}\\
&= \langle 111, 0\rangle\{0, 11, 01, 111\}\\
&= \langle 111\rangle\{\langle 0\rangle\{0\} \cup \langle 0\rangle\{0\cdot 1\} \cup \langle 0\rangle\{11, 111\}\}\\
&= \langle 111\rangle\{\Lambda \cup 0\cdot\langle *\rangle\{1\} \cup \{11, 111\}\}\\
&= \langle 111\rangle\{0.\{0, 11\} \cup \{11, 111\}\}\\
&= \langle 111\rangle\{00, 011, 11, 111\}\\
&= \langle 111\rangle\{111\} \cup \langle 111\rangle\{00, 011, 11\}\\
&= \{00, 011, 11\}.
\end{aligned}$$

This is an example in which F does not provide a complete reduction of A_1. F is said to provide a complete reduction of A_1 if $FA_1 = \Lambda$. We also note in this example that $\langle 111, 0, *\rangle A_1 = \langle 111, 0\rangle A_2 = \langle 111\rangle A_3 = A_4$,

where
$$\begin{aligned}
A_1 &= \{*, 1, 11\},\\
A_2 &= \{0, 11, 01, 111\},\\
A_3 &= \{00, 011, 11, 111\},
\end{aligned}$$
and
$$A_4 = \{00, 011, 11\}.$$

If we examine the height distributions of the generated tree domains in this example, we would not find a clear relationship; but if we look at the "absolute" height distributions of A_i with respect to A_i, we find (1, 1, 1) for A_1, (2, 2) for A_2, (3, 1) for A_4, and (3) for A_3. In other words, the absolute height distributions decrease. This behaviour is typical, as is shown in the next few lemmas.

LEMMA 3. *If $FA_1 = A_2$, then the absolute height distribution of A_2 is less than that of A_1.*

Proof. The proof follows by recursion merely by checking that this height-decreasing property is maintained by Rules 1 to 7. Rule 1 yields a reduction from (1) to (0); Rule 2 shows a reduction from (1, 1) to (m_i); Rule 3 refers us back to Rules 1, 2, and 7; Rule 4 shows a reduction from (2) to (1) if $b \nless a$, and from (1, 1) to (1) if $b < a$. The commutation Rules 5 and 6, and Rule 7 are irrelevant as far as this property is concerned because they place no direct restriction on the resulting address set, and the lemma is proved.

We note that an indirect effect of Rule 7 is that if $A_1 \subseteq A_2$ and F is applicable to A_1, then it is applicable to A_2. Note also that if $A_1 \subseteq A_2$, then it is obvious that the absolute height distribution of A_1 is less than that of A_2.

Definition. A representative of F, $\langle a_1, \cdots, a_n \rangle$ is said to be in right normal form if the addresses $a_1, \cdots, a_n$ are in reverse prefix order when read off in their minimal generated tree domain. See also p. 107 for "redundant definition."

LEMMA 4. *Every reduction functor for a non-redundant definition has a unique right normal representation.*

Proof. The existence is shown by iterated application of the commutation Rules 5 and 6. The unicity is shown as follows. If $\langle a_1, \cdots, a_k \rangle$ and $\langle b_1, \cdots, b_l \rangle$ are both in right normal form and $a_k \neq b_l$, say a_k precedes b_l in prefix ordering, then $\langle b_1, \cdots, b_l \rangle$ is applicable to a "minimal address" set constructed by including b_l, an address of least depth which, after application of $\langle b_l \rangle$ produces b_{l-1}, etc.; then, clearly $\langle a_1, \cdots, a_k \rangle$ is not applicable to this address set. If $a_k = b_l$, apply this construction to the first $a_{k-i} \neq b_{l-i}$.

LEMMA 5. *For a non-redundant definition, every address set A_1 possesses a unique complete reduction functor having a right normal form.*

Sketch of Proof. Arrange the addresses of A_1 in prefix order:

$$A_1 = \{a_1, a_1', \cdots, a_1^{(k)}\}.$$

Then $\langle a_1 \rangle$ is applicable to A_1 and $\langle a_1 \rangle A_1 = A_2$ has a lower height distribution than A_1 by Lemma 3; iterate this process with $A_2 = \{a_2, \cdots\}$ to obtain $\langle a_2 \rangle A_2 = A_3$ so that $\langle a_2, a_1 \rangle A_1 = A_3$, etc. This process must end in a finite number of steps because the height distributions are well-ordered. Any other selection method would have to end for the same reason, but must yield the same functor because of the commutation rules.

LEMMA 6. *If $F_1 A_1 = A_2$, and the definition is non-redundant, there is a unique extension of F_1 which completely reduces A_1 and has a right normal form.*

For if F_2 is the unique reducer of A_2 (Lemma 5), then $F_2 \circ F_1$ is the extension sought, and is unique (again by Lemma 5, for A_1).

Example. We saw that

$$\langle 111, 0, *\rangle A_1 = \{00, 011, 11\},$$

where $A_1 = \{*, 1, 11\}$. It therefore follows, since $\langle 00, 011, 11\rangle\{00, 011, 11\} = \Lambda$ (because $A_4 = \{00, 011, 11\}$ is an independent set), that

$$\langle 00, 011, 11, 111, 0, *\rangle A_1 = \Lambda.$$

The right normal form of this functor is derived as follows:

$$\begin{aligned}\langle 00, 011, 11, 111, 0, *\rangle &= \langle 011, 11, 111, 00, 0, *\rangle \\ &= \langle 011, 1111, 110, 11, 00, 0, *\rangle \\ &= \langle 1111, 011, 110, 11, 00, 0, *\rangle \\ &= \langle 1111, 110, 11, 011, 00, 0, *\rangle.\end{aligned}$$

THEOREM 2. *If $\beta_2 \geqslant_d \beta_1$ and if A_2 is the set of d-occurrences in β_2 whose multiplicity in the d-mapping from β_1 into β_2 is greater than zero, then there is a unique tree β_0 obtained from β_2 by complete reduction on A_2 and $\beta_1 \geqslant_d \beta_0$.*

For let f_1, the d-map from β_1 into β_2, be factored into a composition of $g_d(a;\ )$ as in the proof of Lemma 2: $f_1 = g_d(a_k;\ ) \circ \cdots \circ g_d(a_1;\ )$; then $\langle a_k, \cdots, a_1\rangle$ is applicable to A_2, and has a unique extension $\langle a_l, \cdots, a_k, \cdots, a_1\rangle$ which completely reduces A_2. Then $f_0 = g_d(a_l;\ ) \circ \cdots \circ g_d(a_1;\ )$ is the d-map of a d-elimination chain from $\beta_0 = E_d\, a_l \cdots E_d\, a_k \cdots E_d\, a_1\, \beta_2$ into β_2, and $\beta_1 = E_d\, a_k \cdots E_d\, a_1\, \beta_2$. Hence, $\beta_2 \geqslant_d \beta_1 \geqslant_d \beta_0$, and the theorem is proved.

THEOREM 3 (*the restricted Church–Rosser theorem*). *If $\beta_3 \geqslant_d \beta_1$ and $\beta_3 \geqslant_d \beta_2$, and the definition is non-redundant, then there is a β_0 such that $\beta_1 \geqslant_d \beta_0$ and $\beta_2 \geqslant_d \beta_0$.*

Proof. Let the "active address sets" of β_3 (i.e., the d-occurrences of multiplicity > 0) with respect to β_1 and β_2 be A_1 and A_2 respectively. Let $A_0 = A_1 \cup A_2$, let F_0 be the complete reduction of A_0, and let β_0 be the reduction from β_3 generated by F_0. Because $A_i \subseteq A_0$, F_0 is an extension of each F_i completely reducing A_i; whence, if β_{10} and β_{20} are the complete eliminants guaranteed by theorem 2 for A_1 and A_2, clearly $\beta_{i0} \geqslant_d \beta_0$. But, by Theorem 2, $\beta_i \geqslant_d \beta_{i0}$, and the theorem is proved.

Conjecture. *For every $\beta \in \mathfrak{F}\mathfrak{A}_1$, $L(\beta)$ is a lattice.*

COROLLARY 1 (*the normal form theorem*). If $\beta_1 \in \mathfrak{F}\mathfrak{A}_1$, *and the definition is non-redundant, then there is one and only one $\beta_0 \in \mathfrak{F}\mathfrak{A}_0$ such that $\beta_1 \geqslant_d \beta_0$.*

Proof. Let $A_1 = C^{-1}d\beta_1$, and let F_1 be the complete eliminant of A_1, $F_1 A_1 = \Lambda$, and let f_1 be the generated d-map of F_1, and $\beta_1 \circ f_1 = \beta_0$. Then $\beta_1 \geqslant_d \beta_0$ and $C^{-1}d\beta_0 = \Lambda$, i.e., $\beta_0 \in \mathfrak{F}\mathfrak{A}_0$. Suppose β_{01} and β_{02} were two such; $\beta_1 \geqslant_d \beta_{01}$, β_{02} and $\beta_{0i} \in \mathfrak{F}\mathfrak{A}_0$. By Theorem 3, there is a β_0 such that $\beta_{0i} \geqslant_d \beta_0$; but $\beta_{0i} \in \mathfrak{F}\mathfrak{A}_0$ implies that there are no occurrences of d, i.e., $C^{-1}d\beta_{0i} = \Lambda$, whence, $A_i = \Lambda$ and the reductions from β_{0i} to β_0 are the identity, i.e., $\beta_{01} = \beta_0 = \beta_{02}$. Thus the theorem is proved.

Note: it is conjectured that these theorems are true without the restriction to non-redundant definitions. Certainly the processors work without such a restriction.

THE STRUCTURE OF d-EQUIVALENCE CLASSES

THEOREM 4 (*the restricted Church–Rosser theorem—usual form*). *If* $\beta_1, \beta_2 \in \mathfrak{F}\mathfrak{A}_1$ *and* $\beta_1 \equiv_d \beta_2$, *then there is a* $\beta_0 \in \mathfrak{F}\mathfrak{A}_1$ *such that* $\beta_1 \geqslant_d \beta_0$ *and* $\beta_2 \geqslant_d \beta_0$.

Proof. There exists a chain of d-introductions and d-eliminations leading from β_1 to β_2. If we consolidate the d-eliminations and the d-introductions, we have a chain $\beta', \beta', \cdots, \beta^{(n)}$, which we will call an equivalence chain, beginning with β_1 and ending with β_2, such that each successive triple has $\beta^{(i-1)} \geqslant_d \beta^{(i)} \leqslant_d \beta^{(i+1)}$ or $\beta^{(i-1)} \leqslant_d \beta^{(i)} \geqslant_d \beta^{(i+1)}$. In the first case, we might call $\beta^{(i)}$ an (internal) trough, and in the second an (internal) crest.

If the number of internal crests is zero, then either $\beta_1 = \beta' \geqslant_d \beta'' = \beta_2$ or $\beta_1 = \beta' \leqslant_d \beta'' = \beta_2$ or $\beta_1 = \beta' \geqslant_d \beta'' \leqslant_d \beta''' = \beta_2$, in which cases $\beta_0 = \beta_2, \beta_1$ or β'' respectively.

The theorem says, in effect, that if there is an equivalence chain at all, then we can find one with zero crests. We prove this by "infinite descent," i.e., by showing that if there is an equivalence chain with k (internal) crests, where $k > 0$, then there is one with $k - 1$. In effect, we do this by using Theorem 3 to transform a crest into a trough; only the question of how close the crest is to an end of the equivalence chain obliges us to consider several cases. Let

$$\beta^{(i-1)} \leqslant_d \beta^{(i)} \geqslant_d \beta^{(i+1)}, i > 1$$

be a crest in a given equivalence chain between β_1 and β_2. By Theorem 3, there is a $\beta_0{}^{(i)}$ such that $\beta^{(i-1)} \geqslant_d \beta_0{}^{(i)} \leqslant_d \beta^{(i+1)}$.

Case 1. If $\beta_1 = \beta^{(i-1)}$ and $\beta^{(i+1)} = \beta_2$, i.e., if $k = 1$, then we have reduced k to zero.

Case 2. If $\beta_1 = \beta^{(i-1)}$ and $\beta^{(i+1)} \neq \beta_2$, then $\beta_1 \leqslant_d \beta'' \geqslant_d \beta''' \leqslant_d \beta^{iv} \cdots$ has become $\beta_1 \geqslant_d \beta_0'' \leqslant_d \beta''' \leqslant_d \beta^{iv} \cdots$ or $\beta_1 \geqslant_d \beta_0'' \leqslant_d \beta^{iv} \cdots$ and we have reduced the number of crests.

Case 3. Similarly, if $\cdots \beta^{(i-2)} \geqslant_d \beta^{(i-1)} \leqslant_d \beta^{(i)} \geqslant_d \beta^{(i+1)} = \beta_2$, we get a reduction to $\cdots \beta^{(i-2)} \geqslant_d \beta_0{}^{(i)} \leqslant_d \beta_2$.

Case 4. Otherwise, in general, $\cdots \beta^{(i-2)} \geqslant_d \beta^{(i-1)} \leqslant_d \beta^{(i)} \geqslant_d \beta^{(i+1)} \leqslant_d \beta^{(i+2)} \cdots$ reduces to $\cdots \beta^{(i-2)} \geqslant_d \beta_0{}^{(i)} \leqslant_d \beta^{(i+2)} \cdots$. The theorem is proved.

COROLLARY (*the normal form*). *If* $\beta_1 \in \mathfrak{F}\mathfrak{A}_1$, *then there is one* (*and only one*) β_0 *such that* $E(\beta_1) \cap \langle \mathfrak{F}\mathfrak{A}_0 \rangle = \{\beta_0\}$.

By the corollary to Theorem 3, we know that $L(\beta_1) \cup \langle \mathfrak{F}\mathfrak{A}_0 \rangle = \text{some } \{\beta_0\}$. Let $\beta_2 \in E(\beta_1) - L(\beta_1)$, so that $\beta_1 \equiv_d \beta_2$. By Theorem 4, there is a $\beta_0' \leqslant_d \beta_1, \beta_2$, whence $\beta_0' \in L(\beta_1)$, so that $\beta_0 \leqslant_d \beta_0' \leqslant_d \beta_2$. If $\beta_0' \in \mathfrak{F}\mathfrak{A}_0$, the argument of that same corollary yields $\beta_0' = \beta_0$.

Definitions. If $\beta_1 \in \mathfrak{F}\mathfrak{A}_1$, and if β_0 is the d-normal form of β_1, so that $\beta_0 \in \mathfrak{F}\mathfrak{A}_0$, then by "the d-pattern of β_1" we will mean the address set of occurrences of the ϕ_d pattern in β_0: $D_d\,\beta_1 = C^{-1}\phi_d\,\beta_0$.

Note that $E(\beta_1) = \{\beta_1\}$ if and only if $D_d\,\beta_1 = \Lambda$.

The structure of $E(\beta_1) = E(\beta_0)$ as a partially ordered set is completely determined by $D_d\,\beta_1$ and g_d; in other words, the structure $E(\beta_1)$ is determined by the properties of the definition and the occurrences, both apparent and hidden, of either d or its points of application in β_1.

We now define those properties of the definition which are relevant to the structures of the d-equivalence classes in $\mathfrak{F}\mathfrak{A}_1$. The structural properties of the d-equivalence classes so determined are combinations of the following.

(*a*) The partially ordered set $E(\beta_1)$ may be finite or infinite.

(b) The partially ordered set $E(\beta_1)$ may be a lattice, or not a lattice. Because of the normal form theorem, which states that all these equivalence classes possess a minimal element, in the finite cases this structural distinction reduces to the existence or non-existence of a maximum.

(c) The partially ordered set $E(\beta_1)$ may be modular or non-modular; a discrete partially ordered set, as all our examples are because of the quantized nature of the alphabets, is *modular* if, whenever $\beta_2 \geqslant_d \beta_1$, any two complete chains between β_2 and β_1 have the same length.

(d) When $E(\beta_1)$ is a modular lattice, it may be Boolean or non-Boolean. This means that $E(\beta_1)$ may be isomorphic with the set of all subsets of a set of n elements, where the set of subsets is ordered by the relation of set-inclusion. For example, in most cases, if $D_d\,\beta_1$ is an independent set of n addresses, $E(\beta_1)$ is completely isomorphic with the set of all subsets of $D_d\,\beta_1$ because each element is the complete eliminant from the maximal element for some $A_1 \subseteq D_d\,\beta_1$.

The relevant properties of the explicit definition of d are the following.

(a) An explicit definition will be called "redundant" if the definiendum either has greater depth than the definiens or if there is a variable in the definiendum which does not occur in the definiens. We shall call the first type of redundant definition a "depth expander"; because the definiendum must be a simple form (hence of depth one), the definiens of a depth expander must consist of a simple character, a constant, or a variable.

We shall call the second type of redundant definition a "scope expander." Examples of "pure" depth expanders are: $d\alpha_1 =_{\mathrm{df}} \alpha_1$; examples of "pure" scope expanders are: $d\alpha_1\,\alpha_2 =_{\mathrm{df}} c\alpha_1$, or $d\alpha_1 =_{\mathrm{df}} cc_1\,c_2\,c_3$; examples of "pure" redundant definitions which are mixed depth and scope expanders are: $d\alpha_1\,\alpha_2 =_{\mathrm{df}} \alpha_2$, or $d\alpha_1\,\alpha_2 =_{\mathrm{df}} c_1$.

Obviously, a definition is a scope expander if and only if some variable has multiplicity zero.

(b) An explicit definition will be called a "depth contraction" if the definiens has a depth greater than one; it will be called "self-dominating" if ϕ_d dominates itself. Clearly, every self-dominating definition must be a depth contraction.

(c) An explicit definition will be called a "breadth contraction" if at least one address i of the definiendum has multiplicity greater than one.

Figures 6, 7, 8, and 9 illustrate these concepts and their effects on the structures of equivalence classes.

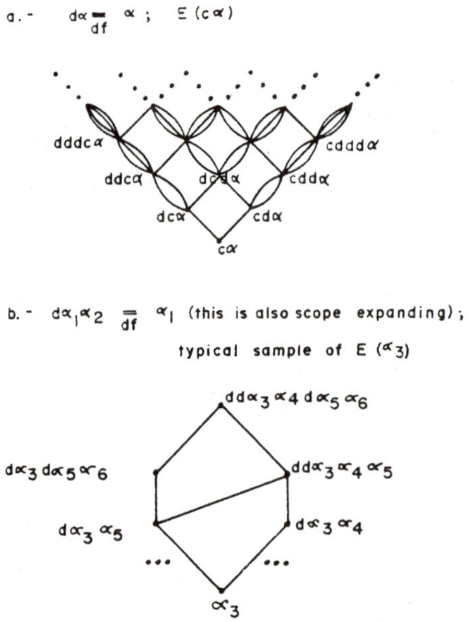

FIGURE 6. Depth expanding redundant definitions and modular, non-Boolean equivalence classes.

LEMMA 7. *A necessary and sufficient condition that* $\beta_1 \in \mathfrak{F}\mathfrak{A}_1$ *have infinitely many immediate d-predecessors is that* $D_d\,\beta_1 \neq \Lambda$ *and that the definition be a scope expander.*

For if β_1 has a d-predecessor, there is a ϕ_d occurrence in β_1, and $D_d\,\beta_1 \neq \Lambda$. But $C^{-1}\phi_d\,\beta_1$ must be finite, whence at least one (and hence all) ϕ_d-occurrence provides infinitely many d-introductions, which can happen if and only if the definiendum possesses an arbitrary variable in addition to any that require fixed substitutions; but this is precisely the definition of a scope expander.

THEOREM 5. *A necessary and sufficient condition that* $E(\beta_1)$ *be infinite is that* $D_d\,\beta_1 \neq \Lambda$ *and the definition be redundant.*

In view of Lemma 7, we need only show that if $D_d\,\beta_1 \neq \Lambda$ and the definition is not scope expanding, then $E(\beta_1)$ is infinite if and only if the definition is depth expanding. Consider the normal form β_0. Since every element of $E(\beta_1)$ has only a finite number of immediate d-predecessors, $E(\beta_1)$ can be infinite if and only if

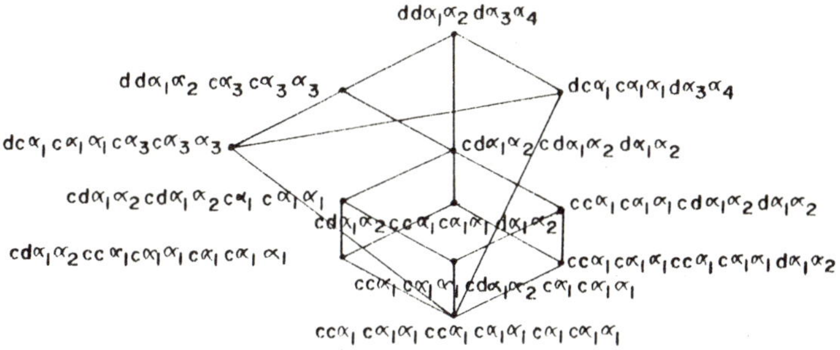

$d\alpha_1\alpha_2 \ \overline{\overline{df}}\ c\alpha_1 c\alpha_1\alpha_1$

Depth Reducer & (possible) breadth reducer with (possible) domino

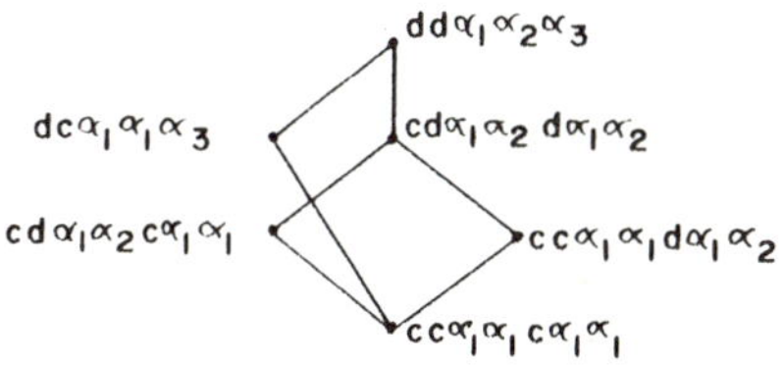

$d\alpha_1\alpha_2 \ \overline{\overline{df}}\ c\alpha_1\alpha_1$

Depth Maintainer and (possible) breadth reducer

FIGURE 7

$d\alpha_1\alpha_2 \ \overset{=}{df}\ c\alpha_2\, c\alpha_1\alpha_2$

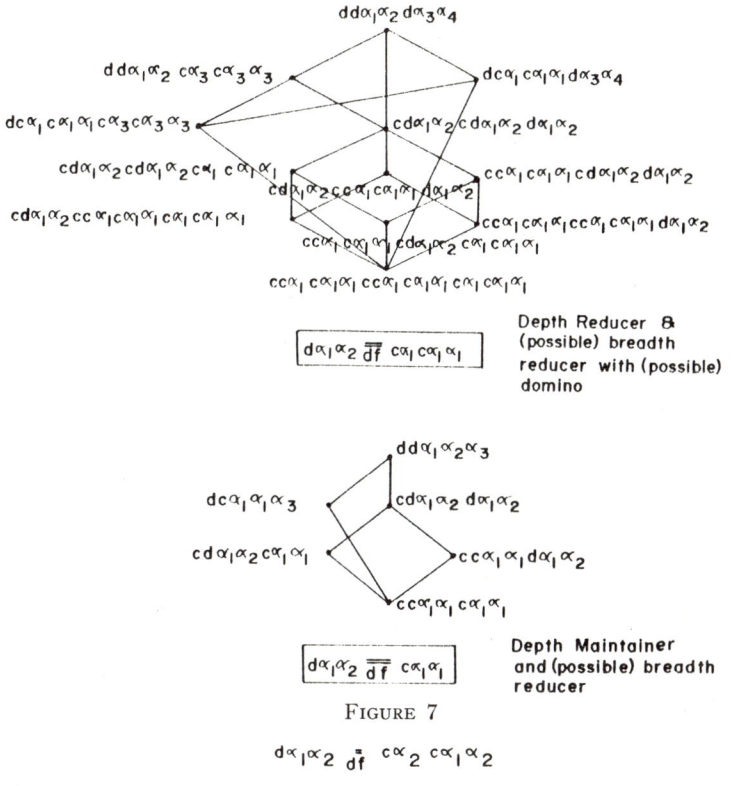

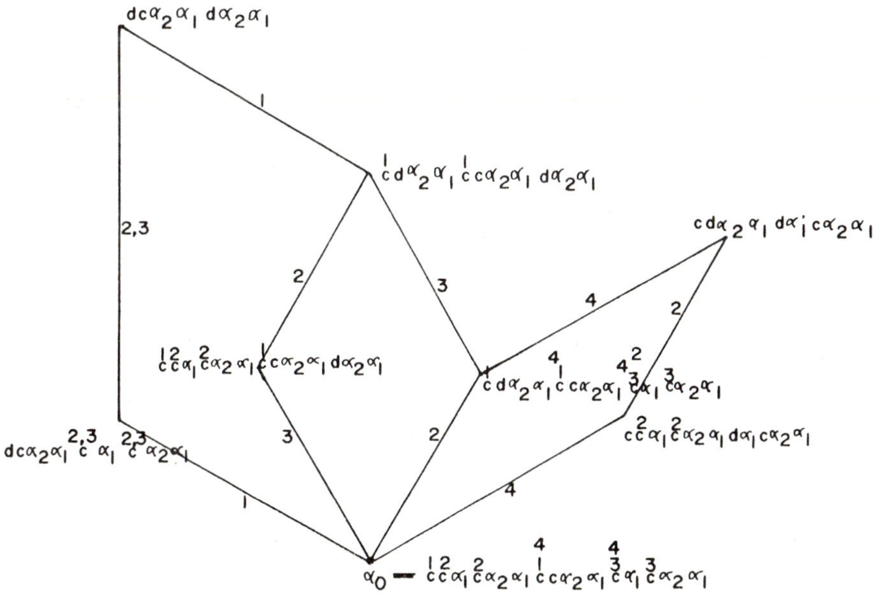

FIGURE 8. The Hasse diagram of $E(\alpha_0)$.

$d\alpha_1\alpha_2$	$\overline{\overline{\mathrm{df}}}$	$c\alpha_2 c\alpha_1\alpha_2$
$d_1\alpha_1\alpha_2\alpha_3$	$\overline{\overline{\mathrm{df}}}$	$c\alpha_1 c\alpha_2\alpha_3$
$d_2\alpha_1\alpha_2$	$\overline{\overline{\mathrm{df}}}$	$d_1\alpha_1\alpha_2\alpha_1$
$d\alpha_1\alpha_2$	$\overline{\overline{\mathrm{df}}}$	$d_2\alpha_2\alpha_1$

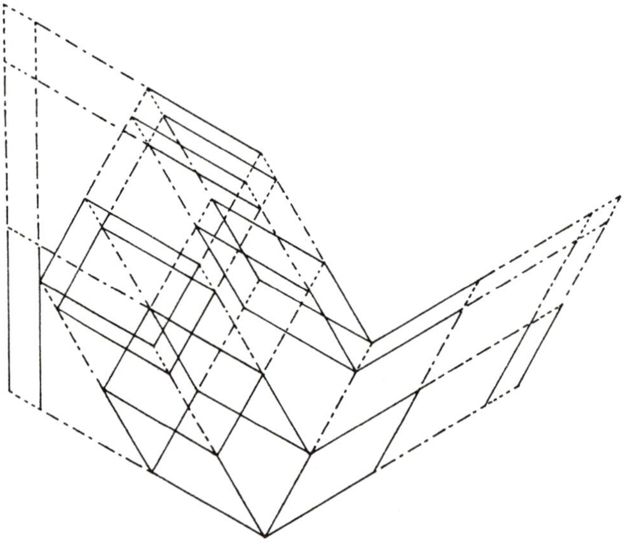

FIGURE 9. The Hasse diagram of $E(\alpha_0)$ when d is factored.

there is an infinite chain through β_0. It therefore suffices to show that $\beta_0 = \phi_d$ has an infinite chain if and only if the definition is depth expanding (assuming that it is not scope expanding). Now I_d is applicable to ϕ_d only at $*$, and $I_d * \phi_d = d\alpha_1 \cdots \alpha_n$; if I_d is to be applicable again it must be either at $*$ or at some, and hence any, i, $0 \leqslant i \leqslant n-1$; in either case, since d must not appear explicitly in ϕ_d, $\phi_d = \alpha_i$, and the definition must be a depth expander. The sufficiency is obvious; see, for example, Figure 6.

Let us now note some obvious conditions for domination and dominoes. From the definition, it is obvious that the tree domain D_1 dominates the tree domain D_2 with intermediary $D_0(D_0 \supset \{*\})$ if and only if $D_0 \subseteq D_2$ and there is an address a_2 such that $a_2 \cdot D_0 \subseteq D_1$. Hence D_3 contains a $D_1 D_2$-domino if and only if there is an address a_1 such that $a_1 \cdot D_1 \subseteq D_3$ and $a_1 \cdot a_2 \cdot D_2 \subseteq D_3$ (a_2 is the same address as in the domination definition above, i.e., the D_0-occurrence). It follows that D_3 contains a $D_1 D_2$-domino if and only if there is a D_1-occurrence in D_3, $a_1 : D_1 \subseteq Sa_1 D_3$, and a D_2-occurrence in D_3, $a_2 : D_2 \subseteq Sa_2 D_3$, such that $a_1 \cdot D_1 \cap a_2 \cdot D_2$ contains more than one element.

Correspondingly stronger statements apply to domination among forms, and form dominoes. In particular, β_0 contains a $\phi_d\phi_d$-domino if and only if there are two ϕ_d occurrences in β_0, $a_0 \in C^{-1}\phi_d\beta_0$ and $a_1 \in C^{-1}\phi_d\beta_0$, such that $a_0 \cdot D\phi_d \cap a_1 \cdot D\phi_d$ contains more than one address.

This will now permit us to prove a restricted converse of Theorem 3. In Theorem 3 we are given two d-maps, f_1 of β_1 into β_3, and f_2 of β_2 into β_3, and we are able to complete the following diagram under any conditions.

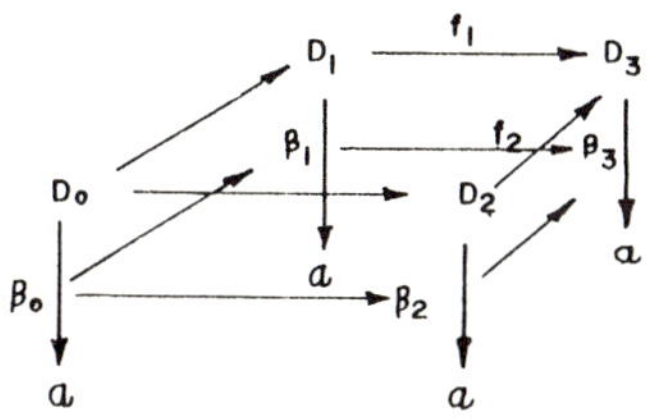

We are able to do this by use of d-eliminations alone, mainly because each d-elimination may multiply some other occurrences of d, or even cause them to disappear, but can never "hide" them.

If, however, we begin with β_0, β_1, and β_2, and try to complete this diagram by finding β_3, we must now work with d-introduction functors, and the disappearance of a ϕ_d occurrence with such a d-introduction may cause the disappearance of another when two ϕ_d occurrences "overlap." Occurrences of single characters cannot overlap, where occurrences of patterns can.

We could, at this point, introduce d-introduction functors, F^{-1}, which would operate on address sets which are unions of sets of the form $a \cdot D\phi_d$, where we restrict ourselves to the "interior" of ϕ_d. There would then be inverse rules to the rules 1 to 6 of the last section. However, Rule 7 now needs a restricted inverse, and the correspondents to Lemmas 5 and 6 are false without appropriate restriction.

Let us call the interior of $D\phi_d$ simply $D_i\,\phi_d$. If, then, $A_1 = \{a_1, \cdots, a_k\}$ is an address set such that $a_i \geqslant a_j \Rightarrow a_i \geqslant_d a_j$, it follows that

$$A_1 \cdot D_i\,\phi_d = \bigcup_{j=1}^{k} a_j \cdot D_i\,\phi_d$$

has no other partition into such a form, and it is only such address sets which possess unique complete d-introduction functors. For example, in Figure 8, $D_i\,\phi_d = \{*, 1\}$, and for $\beta_0 = cc\alpha_1\,c\alpha_2\,\alpha_1\,cc\alpha_2\,\alpha_1\,c\alpha_1\,c\alpha_2\,\alpha_1$ we have

$$C^{-1}\phi_d\,\beta_0 = \{*, 0, 1, 11\},$$

whence $D_i\,\phi_d \cdot C^{-1}\phi_d\,\beta_0 = \{*, 1\} \cup \{0, 01\} \cup \{1, 11\} \cup \{11, 111\}$ does not have the proper form because $a_{11} = \{10\}$, $a_{21} = \{0\}$, and $a_{22} = \{11\}$; in $C^{-1}\phi_d\,\beta_0$, $* \leqslant_d 0$, but $* \nleqslant_d 1$ and $1 \nleqslant_d 11$. Thus if $\beta_1 = dc\alpha_2\,\alpha_1\,c\alpha_1\,c\alpha_2\,\alpha_1$,

$$\beta_2 = cd\alpha_2\,\alpha_1\,cc\alpha_2\,\alpha_1\,c\alpha_1\,c\alpha_2\,\alpha_1,$$

and $\beta_3 = cc\alpha_1\,c\alpha_2\,\alpha_1\,d\alpha_1\,c\alpha_2\,\alpha_1$, then β_i has β_0 as complete d-eliminant with address sets $\{*\}$, $\{0\}$, and $\{1\}$ respectively, and $f_1^{-1}(*) = \{*, 1\}$, $f_2^{-1}(0) = \{0, 01\}$, and $f_3^{-1}(1) = \{1, 11\}$ where f_i is the d-map of β_0 into β_i. If we choose $A_1 = \{*, 1\}$, and $A_2 = \{0, 1\}$, then $A_1 \cdot D_i \phi_d = \{*, 1\} \cup \{1, 11\}$, where

$$\{*\} \cdot \{*, 1\} \cap \{1\} \cdot \{*, 1\} \neq \Lambda,$$

and hence there is no complete ϕ_d-eliminant above β_1 and β_3, while

$$A_2 \cdot D_i \phi_d = \{0, 01\} \cup \{1, 11\},$$

where $\{0\} \cdot \{*, 1\} \cap \{1\} \cdot \{*, 1\} = \Lambda$, and hence there is a complete ϕ_d-eliminant above β_2 and β_3, namely, $cd\alpha_2\alpha_1 d\alpha_1 c\alpha_2\alpha_1$.

LEMMA 8. *If β_0 contains no $\phi_d \phi_d$-domino, then there is a $\beta_3 \in \mathfrak{F}\mathfrak{A}_1$ such that $\beta_1 \leqslant_d \beta_3$ and $\beta_2 \leqslant_d \beta_3$.*

Sketch of proof. Let f_1 be the d-map of β_0 into β_1, and f_2 the d-map of β_0 into β_2. Also, let $A_1 \subseteq D\beta_1$ and $A_2 \subseteq D\beta_2$ be the occurrences of those d's whose d-multiplicity over β_0 in f_1 and f_2, respectively, is greater than zero. Finally, let $A_3 = f_1^{-1}(A_1) \cup f_1^{-1}(A_2)$. Because β_0 contains no $\phi_d \phi_d$-domino, the arguments of Lemma 3 (using height distributions of ϕ_d occurrences), Lemmas 5 and 6 can be adapted to yield a complete d-introduction functor for A_3, which is applicable to β_0; call the resulting tree β_3. The proof then proceeds in a way completely analogous to that of Theorem 3.

The following is now obvious.

THEOREM 6. *A necessary and sufficient condition that $E(\beta_1)$ be a lattice is that β_0, the normal form of β_1, contain no $\phi_d \phi_d$-domino.*

COROLLARY. *In order that every equivalence class be a lattice, it is necessary and sufficient that the definition be not self-dominating. If the definition is not a depth contraction, every equivalence class is a lattice.*

LEMMA 9. *If the definition of d is not a scope expander, then a necessary and sufficient condition that $E(\beta_1)$ be modular is that, for every pair of addresses a_1, $a_2 \in D_d \beta_1$ such that $a_2 \geqslant_d a_1$, say, $a_2 = a_1 \cdot a_{ij} \cdot a_1'$, the multiplicity of i is* 1.

Proof. Because the definition is not a scope expander, no multiplicities not already determined in $D_d \beta_1$ can be introduced. Hence, we can restrict our attention to $D_d \beta_1$. Suppose $E(\beta_1)$ is non-modular. Then there are two forms in $E(\beta_1)$, $\beta_3 \geqslant_d \beta_4$, such that there are two elimination chains of different length from β_3 to β_4. But this means that the reduction functor from the "active addresses" of β_3 into β_4, which is determined by the d-address map from β_4 into β_3, has two representations of different length. This is possible only if some commutation of the Rule 6 type is applicable in which the multiplicity $m_i > 1$; but all the addresses involved must be in $D_d \beta_1$.

Conversely, suppose we have addresses $a_2 \geqslant_d a_1$ such that $a_2 = a_1 \cdot a_{ij} \cdot a_1'$, and the multiplicity of i is greater than one; $a_1, a_2 \in D_d \beta_1$. Then

$$\langle a_1, a_1 \cdot i \cdot a_1' \rangle = \langle a_1 a_{i1} \cdot a_1', \ldots, a_1 \cdot a_{im_i} \cdot a_1', a_1 \rangle$$

by Rule 6, so that, because $m_i > 1$, the reduction functor is factored into the composition of two applications of g_d on the left and $m_i + 1 > 2$ on the right, call it F.

In either case, if β_0 is the normal form, and $F^{-1}\beta_0 = \beta_2$ is constructed by $\beta_2 = I_d\, a_1 \cdot i \cdot a_1' I_d\, a_1\, \beta_0$, there are two complete reduction chains from β_2 to β_0, one of length 2, and the other of length $m_i + 1 \neq 2$. Hence $E(\beta_1)$ is non-modular, and the theorem is proved.

THEOREM 7. *A necessary and sufficient condition that every d-equivalence class be modular is that the definition of d be not breadth expanding.*

For if every multiplicity in the definition is $\leqslant 1$, then the scope expansion condition of Lemma 9 becomes irrelevant, and the theorem is proved.

We shall not develop the structure theory any further here except to point to the following evidence that it is capable of further development.

We have already noted a classification of classes of explicit definitions: depth expanders, scope expanders, redundant, depth contractors, breadth contractors, etc. Many explicit definitions are combinations of these types. For example, the definition $d\alpha_1\, \alpha_2 =_{\text{df}} c_1\, \alpha_1\, c_2\, \alpha_1\, \alpha_1$ is simultaneously a scope expander, a depth contractor, and a breadth contractor (α_1 has multiplicity three, α_2 multiplicity zero).

There is clear evidence that explicit definitions may be "factored" into a sequence of definitions of "pure types," and that this "factorization" of the definitions is reflected in a factorization into an appropriate type of "product" of partially ordered sets, each of a corresponding "pure type." The relationship between Figures 8 and 9 is an example of such evidence.

Figure 8 shows a complete equivalence class having the following three properties:

(1) it fails to be a lattice,
(2) it fails to be modular,
(3) it contains Boolean subsets.

We can isolate these properties by introducing two "intermediate, ideal" definitions, each of a "pure type." The definition of d "caused" property 1 in this equivalence class because it was a self-dominating depth reducer, property 2 because it was a breadth reducer, and property 3 because there is in the definition, in addition to these two properties, a residual permutation effect. Let us therefore factor this out one at a time.

"$d_1\, \alpha_1\, \alpha_2\, \alpha_3 =_{\text{df}} c\alpha_1 c\alpha_2\alpha_3$" is a "pure depth reducer,"
"$d_2\, \alpha_1\, \alpha_2 =_{\text{df}} d_1\alpha_1\alpha_2\alpha_1$" is a "pure breadth reducer," and
"$d\alpha_1\, \alpha_2 =_{\text{df}} d_2\, \alpha_2\, \alpha_1$" is a "pure permuter."

If we "eliminated" d_1 and d_2, we would get the original "composite definition"

$$d\alpha_1\, \alpha_2 = d_2\, \alpha_2\, \alpha_1 = d_1\, \alpha_2\, \alpha_1\, \alpha_2 = c\alpha_2\, c\alpha_1\, \alpha_2.$$

If, now, instead of merely extending the original alphabet $\mathfrak{A}_0$ (which contains the element c) to $\mathfrak{A}_1 = \mathfrak{A}_0 \cup \{d\}$, we consider $\mathfrak{A}_1' = \mathfrak{A}_0 \cup \{d_1, d_2, d\}$ in which d_1 has stratification three, and d_2 and d have stratification two, then $E(\beta_0)$, where $\beta_0 = cc\alpha_1\, c\alpha_2\, \alpha_1\, cc\alpha_2\, \alpha_1\, c\alpha_1\, c\alpha_2\, \alpha_1$, changes from a partially ordered set of eight elements, as in Figure 8, to one of fifty-two elements in Figure 9. The original eight elements are easily detectable as those and only those which may be "approached" from the normal form along a path possessing the same number of solid segments (representing d_1-introductions), dashed segments (representing d_2-introductions), and dotted segments (representing d-introductions) in the same directions (representing the "equivalent" occurrences of the introductions).

The connected subsets composed of solid lines in Figure 9 are the only ones containing non-lattices. The basic component of this type is at the bottom centre of the diagram. The connected subsets composed of dashed lines in Figure 9 are all lattices, but are the only ones containing non-modular configurations. The basic component of this type can be seen toward the left and interior of the diagram. Finally, the connected subsets composed of dotted lines are all Boolean lattices, though the word used in illustration is too small to exhibit this effect dramatically.

If we consider the languages of the categories of $\mathfrak{P}\mathfrak{A}_0$ (including $\mathfrak{T}\mathfrak{A}_0$, $\mathfrak{P}_d\,\mathfrak{A}_0$, $\mathfrak{T}\mathfrak{A}_0$, etc.) and $\mathfrak{F}\mathfrak{A}_0$ to be "extended" by "extension of the alphabet" to $\mathfrak{A}_0' = \mathfrak{A}_0 \cup \{d_1\}$, $\mathfrak{A}_0'' = \mathfrak{A}_0' \cup \{d_2\}$, and $\mathfrak{A}_1' = \mathfrak{A}_0'' \cup \{d\}$, each successive "simple extension" has all the properties we have discussed. For example, every connected dashed substructure has a unique normal form in $\mathfrak{P}\mathfrak{A}_0'$; every connected dotted substructure has a unique normal form in $\mathfrak{P}\mathfrak{A}_0''$.

A further development of the present theory is therefore concerned with the properties of non-simple extensions, i.e., those involving more than one definition; and the normal form property is an example.

It would be wrong to think that the properties of equivalence by definition, which we have exhibited as an effect of language extension, apply at the object language level only and do not affect the meta-language beyond the requirements for form and equivalence recognizers. Exactly the same phenomena can occur simultaneously at both levels. For example, in Church's axiomatic system for the propositional calculus, the first axiom is

$$A_1\colon \vdash \alpha_1 \supset (\alpha_2 \supset \alpha_1).$$

The name, A_1, like the symbol "$\vdash$," and to my way of thinking, the control characters such as parentheses, all belong to the meta-language. But in the meta-language, we can consider that A_1 is an address in the theorem storage, or also, simultaneously, a meta-language operator with two operands in a programming portion of the meta-language: the command $A_1\alpha_1\alpha_2$ will cause (in prefix form) the production of the object "$\supset\alpha_1 \supset \alpha_2\, \alpha_1$." We could consider this to be a semantic interpretation of the explicit definition in the meta-language:

$$A_1\, \alpha_1\, \alpha_2 =_{\mathrm{df}} \supset \alpha_1 \supset \alpha_2\, \alpha_1.$$

Thus the syntactic relation

$$A_1\, A_1\, \alpha_1\, \alpha_2 \supset \alpha_2\, \alpha_1 \equiv \supset A_1\, \alpha_1\, \alpha_2\, A_1 \supset \alpha_2\, \alpha_1\, \alpha_1,$$

which is verified by the reduction of both terms to the same normal form

$$\beta_0 = \supset\supset \alpha_1 \supset \alpha_2\, \alpha_1 \supset\supset \alpha_2\, \alpha_1 \supset \alpha_1 \supset \alpha_2\, \alpha_1,$$

can be interpreted semantically as a statement that two derivation programmes will yield the same object language expression; furthermore, we know that there cannot exist a word β_1 in $\mathfrak{P}\{A_1, \supset, \alpha_1, \alpha_2\}$ from which both can be derived because of the $A_1\, A_1$-domino in β_0. The processing for theorem proving, and even for the derivation of meta-theorems, is no different from the processing at object-level.

It therefore follows that

(1) the introduction of explicitly defined symbols, as in $\supset \alpha_1\, \alpha_2 =_{\mathrm{df}} \vee \sim \alpha_1\, \alpha_2$,
(2) the introduction of the names of axioms, as in $A_1\, \alpha_1\, \alpha_2 =_{\mathrm{df}} \supset \alpha_1 \supset \alpha_2\, \alpha_1$,
or even
(3) the introduction of a system of names of theorems, as in $T_1\, \alpha_1 =_{\mathrm{df}} \supset \alpha_1\, \alpha_1$,
etc.,

all play exactly the same role in the meta-language as the introduction of new names of instructions, or of macro-instructions in a programming language. They can all be interpreted as the names of transformations on the object language, and therefore as operators in a meta-language over the object language. This is why we could identify the problems *a*, *b*, *c*, *d*, *e* mentioned in the introduction from information science, logic, and linguistics.

REFERENCES

[1] Boyer, M. Christine, "A Tree Structure Machine for Proving Theorems," Moore School Master's Thesis (Aug., 1964).
[2] Chroust, Gerhard, "A Heuristic Derivation Seeker for Uniform Prefix Languages," Moore School Master's Thesis (Aug., 1965).
[3] Curry, H. B., and Feys, R., *Combinatory Logic*, 1 (North Holland, 1958).
[4] Gorn, Saul, "Mechanical Pragmatics: A Time Motion Study of a Miniature Mechanical Linguistic System," *Communications of the ACM*, 5 (12) (Dec., 1962) 576–89.
[5] ——— "An Axiomatic Approach to Prefix Languages," *Proc. Symposium International Computation Centre; Symbolic Languages in Data Processing* (Gordon and Breach, 1962).
[6] ——— "Language Naming Languages in Prefix Form," presented at the IFIP Working Conference on Formal Language Description Languages (Vienna, Sept. 14–18, 1964).
[7] Newell, A., Shaw, J. C., and Simon, H. A., "Empirical Explorations of the Logic Theory Machine: A Case Study in Heuristics," *Proc. of the Western Joint Computer Conference* (1957), 218–30.
[8] ——— "Report on a General Problem-Solving Programme," *Information Processing*, (Paris: UNESCO, 1959), 256–64.
[9] Rasiowa, H., and Sikorski, R., "The Mathematics of Metamathematics" (Warsaw: Monografie Matematyczne, 1963).

Heuristic and Complete Processes in The Mechanization of Theorem Proving

J. A. Robinson*

Consider the theorem-proving problem: *to determine, given a statement B and a collection of statements A, whether or not B follows from A.*

The goal of developing an automatic, mechanical method for solving this problem is an old one. Leibniz (1646–1716) envisaged such a method. Much earlier, the ancient theory of the syllogism stemming from Aristotle (384–322 B.C.) was essentially the elaboration of a mechnical method of this kind for a specially simple type of statement. The familiar "Venn diagram" technique embodies the Aristotelian theory in a convenient and graphic computational form.

Old though it may be, this goal has only relatively recently been well defined and thoroughly investigated. One startling discovery, by Alonzo Church [1], is that the goal is *unattainable in principle.* There simply cannot be a fully automatic general method for deciding whether or not B follows from A, if no restrictions are placed on A and B.

Over against this negative result, however, we can set a positive result which is equally startling: there does exist a fully automatic general *complete* method with the property that, if B follows from A, this fact will certainly be detected and (in a suitable sense) "proved" merely by applying the method to A and B. If B does *not* follow from A, the method will in general leave its user in ignorance of this fact (as, by Church's result cited above, it must). The only condition attached is that B and the members of A must be expressed in the language of quantification theory. This condition is most liberal, since it appears that any statement likely to be encountered in a mathematical context can be so expressed.

The development of this complete method took place at several hands. Perhaps Löwenheim's 1915 paper [2] should be cited as the pioneering effort. This was followed by the work of Skolem in the nineteen twenties [3, 4] and culminated in classic papers by Gödel [5] and Herbrand [6] in 1930. In the mid-thirties Gentzen

*Rice University. This work was supported in part by National Science Foundation grant no. GP-2466.

[7] organized these ideas into a form later exploited by Beth [8], Wang [9], and Prawitz [10] in the nineteen fifties. Other logicians, notably Quine [11, 12] and Dreben [13], clarified and simplified the method without essentially changing its fundamental character.

Until about the mid-fifties interest in the method had been purely theoretical. The method called for large quantities of symbol-manipulation, in general so large as to be completely out of the question for performance by human labour. However, the advent, at about that time, of large-scale electronic computers as reasonably accessible facilities changed the situation very rapidly. Several logicians seized the opportunity to programme one or another version of the method on one or another computer, and to test its practicability, for the first time, on interesting examples. The reports of this work reached the open literature at the end of the fifties. Wang [9], Gilmore [14], Prawitz *et al.* [10], and Davis-Putnam [15] are the key papers.

Overall, the results reported by these workers pointed to the rather disappointing fact that, in the purely theoretical form in which they had implemented the method, it was impractical, even at electronic computing speeds, when applied to reasonably interesting examples.

Meanwhile, a completely different approach to the problem was being tried, and indeed, virtually a new discipline was being founded, by a number of people, notably Simon, Newell and Shaw [16], and Gelernter *et al.* [17, 18]. The leading principle of this approach was to isolate "heuristic" strategems employed by humans when seeking to solve problems of the theorem-proving sort, and to programme these for the computer. Another way of describing this "heuristic" approach would be to say that it consisted of simulating, on the computer, the problem-solving behaviour of intelligent human beings. More recently, the usage has grown up whereby a *heuristic* method is one which lacks (or is not known to possess) the *completeness* property of the complete method: it may, that is, fail to detect that B follows from A even in some cases where B *does* follow from A. On the other hand, a heuristic method, so characterized, could conceivably compensate for this possible loss of completeness by reaching its decisions with greater speed than that with which the complete method reaches them.

It cannot be said that the heuristic approach to the mechanization of theorem-proving, at least as far as these early results were concerned, was successful as an attempt to solve the theorem-proving problem—the theorems proved being quite trivial and the effort to prove them quite large. Nevertheless this work opened up an important new line of attack which might well be applicable to a much wider class of problems than the theorem-proving problem (this has proved to be so: Feigenbaum and Feldman's collection of papers [19] provides fascinating documentation). In any case, these investigations may also be viewed as studies, through simulation, of explanatory models of intelligent behaviour of humans.

Returning now to the other approach, which we may call the *complete* approach in order to distinguish it from the *heuristic* approach, it occurred to the early experimenters that there was much which could be done to improve the practical efficiency of the complete methods. In particular, Dag Prawitz detected a systematically removable source of inefficiency to which he devoted an important paper [21], whose obscurity and difficulty have prevented it from receiving the acclaim it deserves. It is possible to reorganize and develop Prawitz' idea into an extremely useful principle, which I have called the *resolution principle* [20].

It is interesting to note, as will be later discussed in more detail, that the underlying idea of the resolution principle, and the resolution principle itself, are very much like the sort of stratagem earlier characterized as *heuristic*—to the point of intuitively appearing, at first, to involve the "loss of completeness" now taken, as explained earlier, to be part of the essence of a heuristic method. In spite of its being heuristic in the sense of being the sort of thing an intelligent human problem-solver would do, the resolution principle turns out *not* to be heuristic in the "loss of completeness" sense.

This phenomenon of a device initially hit upon as a heuristic turning out, on analysis, not to involve a loss of completeness after all, occurred at least twice more in recent work. In the first of these, a powerful and elegant device originating with Wos [22], known as the *set of support principle*, was in effective use for quite some time as a heuristic before Wos and I were able to show, in the summer of 1964, that no loss of completeness was in fact involved. In the second of these, I myself hit upon a device, which I call the *hyper-resolution principle* [23], that initially presented itself in the form of a heuristic, and subsequently proved to be complete. These will be discussed later in somewhat more detail.

Now in order to explain some of the main ideas I shall next pose a purely abstract symbolic combinatorial problem. The intention is to show as directly as possible the nature of the device underlying the resolution principle. First, two definitions:

Expressions. An *expression* is either (*a*) an upper-case letter or (*b*) a lower-case letter, or (*c*) an upper-case letter followed by a parenthesized list of expressions. For example:

(1) A

(2) x

(3) $A(x, y, B(C(x, C(K, x)), y), z)$

are all expressions.

Instances. An expression β is an *instance* of an expression α when β can be obtained from α by replacing each lower-case letter in α with an expression (each occurrence of the *same* letter being replaced by the *same* expression). Thus the expression

(4) $A(K(y), x, B(C(K(y), C(K, K(y))), x), z$

is an instance of the expression (3) above, x being replaced by $K(y)$, y being replaced by x, and z by z.

Now the problem I pose is this: *given two expressions* α *and* β, *determine whether there is an expression* λ *such that* λ *is an instance both of* α *and of* β; *and, if there is, construct a most general such expression, i.e., an expression* λ *such that if* μ *is any expression which is an instance both of* α *and of* β, *then* μ *is an instance of* λ.

For example, given the two expressions

(5) $$A(x, B(x, y), z, C(x, y, z))$$

and

(6) $$A(D(u), v, E(u, v), w),$$

it can be seen that the expression

(7) $$A(D(K), B(D(K), J), E(K, B(D(K), J)), C(D(K), J, E(K, B(D(K), J))))$$

is an instance of both, while the expression

(8) $$A(D(u), B(D(u), y), E(u, B(D(u), y)), C(D(u), y, E(u, B(D(u), y))))$$

is a *most general* instance of both. To see that (8) is a most general common instance of (5) and (6), note that any common instance of (5) and (6) must also be a common instance of (6) and

(9) $$A(D(u), B(D(u), y), z, C(D(u), y, z),$$

which is the instance of (5) obtained by putting $D(u)$ for x. This is so because any instance of (6) will certainly have a subexpression of the *form* $D(u)$ in the place where (5) has its first occurrence of x; hence an instance of (5) which is also an instance of (6) will necessarily have, in place of x, a subexpression of this same form. Similarly, any common instance of (6) and (9) must also be a common instance of (9) and

(10) $$A(D(u), B(D(u), y), E(u, B(D(u), y)), w),$$

which is the instance of (6) obtained by putting $B(D(u), y)$ for v. In turn, any common instance of (9) and (10) must also be a common instance of (10) and

(11) $$A(D(u), B(D(u), y), E(u, B(D(u), y)), C(D(u), y, E(u, B(D(u), y)))),$$

which is the instance of (9) obtained by putting $E(u, B(D(u), y))$ for z. Finally, any common instance of (10) and (11) must also be a common instance of (11) and

(12) $$A(D(u), B(D(u), y), E(u, B(D(u), y)), C(D(u), y, E(u, B(D(u), y)))),$$

which is the instance of (10) obtained by putting $C(D(u), y, E(u, B(D(u), y)))$ for w. But (11) and (12) are the same expression as (8); thus any common instance of (5) and (6) must be an instance of (8).

The successive substitution process illustrated here always converges, for any two expressions α and β, on a most general common instance of α and β, if α and β have any common instance at all. If α and β do not have any common instance, this fact is detected in one of two ways: *either*, at some stage, the two current expressions first differ at two symbols neither of which is a lower-case letter, *or* the lower-case letter to be replaced occurs *within* the expression which is to replace it. These two conditions are very simple to recognize, and the overall process, given fully in [20], is entirely mechanical. It is called the *unification algorithm.*

The abstract combinatorial problem solved by the unification algorithm in fact arises most naturally within the complete proof procedures. If one construes expressions like (5) and (6) as statements, with lower-case letters playing the role of universally quantified variables and upper-case letters the role of relation- and function-symbols, then the search for common instances arises when we are investigating the mutual *consistency* of such statements. For example, suppose we were given (5) and the *negation* of (6) as a pair of statements to be examined for consistency; then the discovery that (7) is an instance of (5) as well as of (6) shows that (7) *would have to be both true and false* if both (5) and the negation of (6) were true; hence the given set is not consistent.

This symbolic combinatorial problem could also be solved by generating the set of *all* instances of the expression α, and the set of *all* instances of the expression β, in some systematic fashion, and determining whether the two sets have any common members. In the complete proof procedures before Prawitz, this in effect was the method used, and therein lay one major reason for their utter impracticality.

For the general theorem-proving problem a more general form of statement is required than that of what we are calling an expression. One such form is called a *clause.* Viewed purely symbolically, a clause is a finite non-empty sequence of *distinct* expressions; one of these expressions is required to be a special symbol, $\square$; the others (if any) are expressions of the sort we have defined above. The general appearance of a clause is thus

$$\alpha_1 \ldots \alpha_m \,\square\, \alpha_{m+1} \ldots \alpha_{m+n} \tag{13}$$

with $m \geqslant 0$ and $n \geqslant 0$. (The intended interpretation of the clause (13) as a statement is that it is *true* just in case, for each system of values of its variables, there is at least one of its components which is *correct*; the component α_i is correct just in case it is false and $i \leqslant m$ or it is true and $i > m$).

Now the resolution principle is a construction whereby, given two clauses, A and B, a finite set (possibly empty) of other clauses, called *resolvents* of A and B, is computed. The computation involves a systematic application of the unification algorithm to the right-hand side of A and the left-hand side of B. Specifically, one or more of the right-hand expressions of A are *marked*, and one or more of the left-hand expressions of B are *marked*; then the clauses A' and B' are computed (if they exist) as the most general instances of A and B respectively in which the marked expressions are all alike; finally, a clause C is formed from all the distinct *unmarked* expressions in A' and B', the left-hand expressions in C being the unmarked left-hand expressions of A' and B', the right-hand expressions of C being the unmarked right-hand expressions of A' and B'. C is then a *resolvent* of A and B; and for each of the finitely many ways of marking right-hand expressions of A and left-hand expressions of B, there is (possibly) a resolvent of A and B, depending on whether or not the marked expressions collectively have a most general common instance.

For our present purposes it is sufficient to know that (i) the computation of the resolvents (if any) of A and B is *purely mechanical*, and (ii) that every theorem-proving problem can be formulated as the problem to detect, given a set S of clauses, whether the *empty clause* $\square$ can be obtained as the last member of a sequence of clauses each member of which is either in S or is a resolvent of two earlier members of the sequence.

This being so, the process of seeking, for a particular S, a genesis of $\square$ if one exists, is then the fundamental process of any theorem-proving device based on the resolution principle. Now the basic theory of the resolution principle guarantees that the following *exhaustive* growth process will always provide a genesis of $\square$ from S if there is one:

Process 1. Let S_0 be S. Then, for $n \geqslant 0$, if $\square$ is in S_n, terminate successfully; otherwise, let S_{n+1} be $S_n \cup R_n$ where R_n is the set of all resolvents of pairs of clauses in S_n.

Process 1 may be likened to the "unrestrained growth" of a population whose initial generation is S; every pair of "parents" adds to the population all the "offspring" those parents are capable of having.

Not unexpectedly, Process 1 is a rather *explosive* process; the size of the sets S_n in general increases with n at an enormous exponential rate, and the result is that Process 1 is still not a practically feasible theorem-proving procedure, even though it is orders of magnitude more efficient than the older theorem-proving procedures which were based on the enumeration of instances.

The problem therefore presents itself of introducing some form of "birth control" or "selective breeding" into the fundamental growth process. This can be done in a great many ways, but the difficult part is to avoid losing the *completeness* property possessed by the fundamental process, whereby $\square$ will certainly turn up eventually if the process is continued long enough, provided only that the initial set S of clauses is unsatisfiable.

It is very natural to think of restrictions on the growth process which are *heuristic* in character, i.e., which reflect the principles of selection that an intelligent human, seeking to generate $\square$ from S, would employ.

Very little experience with Process 1 is needed before it is realized that, since, after all, *the point of the growth process is to try to generate* $\square$, there are some rather obvious heuristics which an intelligent human, given S and the job of generating $\square$ from S, would use. If such a heuristic is *empirically* successful, and *in fact* generates $\square$ in reasonable time from a variety of initial sets S, then one is strongly inclined not to worry very much whether the resulting process is theoretically complete. Indeed, one is inclined to think, intuitively, that the *more* successful a heuristic device is, the *less* likely it is to preserve the completeness property.

It is therefore all the more interesting that a particularly successful restrictive device (Wos [22]), which was originally hit upon as a most intelligent-seeming heuristic, should turn out after all to leave the completeness property undisturbed. Wos' idea is this: in most problems, S is made up of a set A of clauses which are the *axioms* or *basic assumptions* of the problem, and a set B of clauses which collectively express the *denial* of the theorem which is to be proved from A. This being so, it seems intelligent to do what most experienced mathematicians would instinctively do and to confine one's deductions of further clauses to those that in some sense can be *traced to* B. It is evident that elaborating consequences of A alone will not produce the desired contradiction since A is presumably consistent. Wos therefore defined, and successfully operated [22] the following process:

Process 2 (*set of support principle*). Let B be a non-empty subset of S; let S_0 be S and B_0 be B. Then, for $n \geqslant 0$, if $\square$ is in S_n, terminate successfully; otherwise, let B_{n+1} be the set of all resolvents of all pairs of clauses in S_n, *at least one member of which is in* B_n; and let S_{n+1} be $S_n \cup B_{n+1}$.

The set B is called the *set of support* of the growth process. It is, in the definition above, left as a parameter, but in accordance with the informal heuristic reasoning given immediately before, it is normally chosen by Wos in his computations in the particularly natural and compelling fashion there described.

Process 2 can be likened to a growth process in which the initial generation is segregated into an "in" group and an "out" group, and breeding is so restricted that two members of the population are allowed to breed only if at least one of them is descended from the "in" group.

Evidently, the rate of growth is much retarded under such a breeding restriction, and, moreover, it has the attraction of being "purposive" in the sense of being deliberately *steered* towards the production of $\square$ (at least when the set of support is chosen according to Wos' own prescription).

As it turned out, Process 2 *is* complete, provided only that B is so chosen that $(S - B)$ is satisfiable (which in particular is evidently the case when B is chosen in accordance with Wos' criterion). Thus what appeared to have been "merely" a heuristic later was found not to be one—and yet the fact that it preserves completeness in no way detracts from its "heuristic" attractiveness and efficiency-increasing effectiveness.

Most intrigued by this kind of situation, whereby heuristic principles can sometimes be successfully incorporated into the underlying fundamental machinery, I began to look for other ways of restricting the growth process which, though heuristic, would preserve completeness.

One line of attack was based on the observation that Process 1 tended to produce a great many distinct derivations of one and the same clause, and hence of $\square$. In some sense, it was thought, every derivation from S of some given clause C is equivalent to any other, and it ought to be possible to dispense with (at best) all but *one* of these—a sort of canonical form concept for derivations. Much of the redundancy observed seemed to originate in the permutability of steps, and what seemed to be needed was therefore a notion which was independent of the order of these more primitive operations. Such a notion was forthcoming as the basis of the following process:

Process 3. Let A be the set of clauses in S which have empty left-hand sides (call such clauses *positive* clauses); let B be $(S - A)$; let S_0 be S and A_0 be A. Then, for $n \geqslant 0$, if $\square$ is in S_n, terminate successfully; otherwise, construct S_{n+1} by adding to S_n all the positive clauses which can be obtained from S_n by repeated resolution, under the constraint that one of each pair of clauses resolved must be in A_n; and let A_{n+1} be the set of positive clauses in S_{n+1}.

It is easy to see that $(S_n - A_n)$ is B, for all n: i.e., Process 3 grows no clauses which are not positive. The sets $(S_{n+1} - S_n)$, i.e., the successive generations of clauses in the growth process, are each descended from S_n by a *series* of resolutions, not necessarily by only one, and in the birth of each new clause *many* clauses may have played a part, not necessarily only two. For this reason a new term was introduced to denote the construction of each new clause from its several immediate progenitors: *hyper-resolution.* This term is intended to suggest that the new construction is on a higher logical level than the old, as well as being inherently more powerful.

In [23] it is shown that Process 3 is in fact complete—a circumstance which is not trivial to prove and which indeed was psychologically later in being unearthed than the details of the heuristic hyper-resolution principle itself.

We are only just beginning, it seems to me, to get a *feel* for these growth processes based on resolution or indeed on any other analogous logical principles. I believe that there is still much to be discovered in the way of controlling the rate and direction of growth *intelligently yet automatically*, without disturbing the basic completeness property. I believe that there is nothing inherently conflicting in the two leading concepts—*heuristic control* of the process and *systematic, combinatorial control* of the process, and I have tried to illustrate how these two concepts overlap and merge into each other.

REFERENCES

[1] Church, A., "A Note on the Entscheidungs Problem," *The Journal of Symbolic Logic*, 1 (1936), 40–41. Correction, *ibid.*, 101–2.

[2] Löwenheim, L., "Über Möglichkeiten im Relativkalkül," *Mathematische Annalen*, 76 (1915), 447–70.

[3] Skolem, T., "Über die mathematische Logik," *Norsk matematisk tidskrift*, 10 (1928), 125–42.

[4] ——— "Über einige Grundlagenfragen der Mathematik," *Skrifter utgitt av Det Norske Videnskaps-Academi i Oslo, I. Matematisk-naturvidenskapelig klasse*, no. 4 (1929).

[5] Gödel, K., "Die Vollständigkeit der Axiome des logischen Functionenkalküls," *Monatshefte für Mathematik und Physik*, 37 (1930), 349–60.

[6] Herbrand, J., "Recherches sur la théorie de la démonstration," *Travaux de la Société des Sciences et des Lettres de Varsovie, Classe III sciences mathématiques et physiques*, no. 33 (1930).

[7] Gentzen, G., "Untersuchungen über das logische Schliessen," *Mathematische Zeitschrift*, 39 (1934–35), 176–210, 405–31.

[8] Beth, E., "Semantic Entailment and Formal Derivability," *Mededelinger der Kon. Nederl. Akad. van Wetensch., deel 18*, no. 13 (Amsterdam, 1955).

[9] Wang, H., "Towards Mechanical Mathematics," *IBM Journal of Research and Development*, 4 (1960), 2–22.

[10] Prawitz, D., *et al.*, "A Mechanical Proof Procedure and Its Realisation in an Electronic Computer," *Journal of the Association for Computing Machinery*, 7 (1960), 102–28.

[11] Quine, W. V., "A Proof Procedure for Quantification Theory," *The Journal of Symbolic Logic*, 20 (1955), 141–9.

[12] ——— *Methods of Logic* (rev. ed.), "Appendix" (New York: Henry Holt, 1959).

[13] Dreben, B., "On the Completeness of Quantification Theory," *Proceedings of the National Academy of Sciences*, 38 (1952), 1047–52.

[14] Gilmore, P. C., A Proof Method for Quantification Theory," *IBM Journal of Research and Development*, 4 (1960), 28–35.

[15] Davis, M., and Putnam, H., "A Computing Procedure for Quantification Theory, *Journal of the Association for Computing Machinery*, 7 (1960), 201–15.

[16] Newell, A., Shaw, J. C., and Simon, H., "Empirical Explorations with the Logic Theory Machine," *Proceedings of the Western Joint Computer Conference*, 15 (1957), 218–39. Reprinted in Feigenbaum and Feldman [19], 109–33.

[17] Gelernter, H., "Realization of a Geometry-Theorem Proving Machine," *Proceedings of an International Conference on Information Processing* (Paris: UNESCO House, 1959), 273–82. Reprinted in Feigenbaum and Feldman [19], 134–52.

[18] Gelernter, H., Hansen, J. R., and Loveland, D. W., "Empirical Explorations of the Geometry-Theorem Proving Machine," *Proceedings of the Western Joint Computer Conference*, 17 (1960), 143–7. Reprinted in Feigenbaum and Feldman [19], 153–63.

]19] Feigenbaum, E. A., and Feldman, J., eds., *Computers and Thought* (New York: McGraw-Hill, 1963).

[20] Robinson, J. A., "A Machine-Oriented Logic Based on the Resolution Principle," *Journal of the Association for Computing Machinery*, 12 (1965), 23–41.

[21] Prawitz, D., "An Improved Proof Procedure," *Theoria*, 26 (1960), 102–39.

[22] Wos, L., Robinson, G. A., and Carson, D. F., "Efficiency and Completeness of the Set of Support Strategy in Theorem-Proving," *Journal of the Association for Computing Machinery*, 12 (1965), 536–41.

[23] Robinson, J. A., "Automatic Deduction with Hyper-Resolution, *International Journal of Computer Mathematics*, 1, no. 3 (July, 1966), 227–34.

An Approach to Heuristic Problem Solving and Theorem Proving in the Propositional Calculus

Saul Amarel*

1. INTRODUCTION

One of the important questions in the area of heuristic problem solving by computer is that of transferability of results from specific exploratory problems to more general and real-life problems. Research in artificial intelligence has proceeded mainly by constructing complex systems for performing specific tasks such as elementary reasoning, question answering, scheduling, game playing, concept formation, etc. Are these complex systems isolated designs or can they be viewed as prototypes with wider usefulness? The answer to this question is not a clear-cut yes or no. In order to find general principles and transferable features in heuristic problem solvers, we must look below the surface of specific implementations and attempt to clarify the underlying logic and the common structural characteristics of these systems. Moreover, we must understand the process of describing a new problem area to a computer system. In what ways can a problem be represented to a machine? Are some ways better than others? Our general objective in this paper is to clarify these questions, to propose some answers for them, and to pinpoint areas of difficulty in the design of "truly general" problem solving systems.

We are considering in this paper problems of *derivation type*. Typical examples of such problems are theorem proving in formal systems and syntactic analysis of language. Advanced question answering systems (where a user asks a question—say from a remote console of a time sharing system—and the system generates an answer that is derived as a consequence of a body of stored information which is relevant to the question) are examples of complex systems that must include

*RCA Laboratories, Princeton, New Jersey. The research presented in this paper was sponsored in part by the Air Force Office of Scientific Research, of the Office of Aerospace Research, under contract no. AF49(638)-1184. The work was completed while the author was on a visiting appointment at the Computer Science Department of the Carnegie Institute of Technology, Pittsburgh, Pa. At the Carnegie Institute this research was partly sponsored by the Advanced Research Projects Agency of the Office of the Secretary of Defense under contract no. SD-146.

procedures for solving problems of derivation type. In section 2 we shall present a conceptual framework for the specification of a class of procedures that are used in the solution of derivation type problems; these procedures are called *reduction procedures*. In this framework, questions of *problem representation* are treated with the help of logical systems of natural inference, and questions of *solution strategy* are considered in the context of a feedback scheme that controls the search for solution. In subsequent sections, we shall use our framework in the specification of a set of procedures for the construction of simplest proofs to theorems in the propositional calculus. This will provide a detailed illustration of the design of reduction procedures along the lines suggested by the framework.* The proof construction problem which is formulated in section 3 was chosen because it is an ideal prototype for derivation problems and because of its intrinsic value for many future problem solving systems that are expected to require strong deductive capabilities.

We shall specify a sequence of three classes of procedures for solving the proof construction problem. This sequence is intended to represent a hypothetical evolution of a proof construction system where problem solving power increases as a procedure moves up in the evolutionary scale. This evolution follows closely the actual evolution of our own work in the area of proof construction procedures. The three stages of the evolution are presented in detail in sections 4, 5, and 6. At the third (highest) stage of development we have a powerful procedure for constructing simplest proofs. This procedure utilizes a new representation of logical proofs which is in the form of a special type of directed graphs.

The transitions between the main stages of the evolution are characterized by a change in problem representation. One of our main reasons for considering the hypothetical evolution of procedures is to examine in detail some aspects of the question of problem representation, and specifically the effect of changes in representation on problem solving power. Our work shows that changes in problem representation have a profound effect on the power of a problem solving system. By examining the factors that are involved in the evolutionary transitions, we can make some preliminary assessments about the possibility of mechanizing the process of changing a problem representation in "the right direction." If the evolutionary transitions are mechanizable, then we can attain problem solving systems that are substantially more general and powerful than the ones that we have at present. However, there are major difficulties in mechanizing the process of choosing an "appropriate" problem representation—the right way to look at a problem. Indeed, they seem to be related to the problem of mechanizing certain creative processes (see [1]). It is our hope that the hypothetical evolution of procedures presented in this paper will provide both a coherent view of major problems that we face in the design of systems that can choose their own representations, and also some clues as to how they can be approached. A discussion of the main ideas that enter in the evolution of our proof construction procedures

*In addition to the theorem proving problem presented in this paper, we have applied our framework successfully to other problems—among them the syntactic analysis problem [1.1].

is given in section 7. Conclusions that are relevant to the questions posed at the beginning of this introduction are given in section 8.

2. A FRAMEWORK FOR REDUCTION PROCEDURES

Most heuristic procedures for the solution of derivation problems can be regarded as procedures of reduction type. In reduction procedures, a problem undergoes a sequence of nested transformations that result in a set of simpler (reduced) subordinate problems whose solution implies the solution of the original problem. It is useful for the design and the comparative study of reduction procedures to have a unifying conceptual framework in which procedures from this class can be expressed. We shall introduce in this section such a framework. It will consist of two main parts: *logical systems of natural inference** for treating the logical-linguistic aspects of the procedures, and a *feedback scheme* for considering the dynamic aspects of decision making and control in them. A reduction procedure $\Pi_r(U)$ is defined in terms of a search space and a search strategy. U denotes the specific universe of discourse from which the problem to be solved by the procedure originates. We associate to $\Pi_r(U)$ a logical system of natural inference, which we call $N(U)$. In our present framework, the problem solving activity in $\Pi_r(U)$ will be interpreted as proof construction activity in the corresponding system $N(U)$. The specification of $N(U)$ amounts to the specification of the search space of $\Pi_r(U)$.

The main elements that enter in the specification of the *search space* of $\Pi_r(U)$ are the following.

(i) *States*. States are descriptions of the essential elements of problem situations. They have the form $S = (x_1, \ldots, x_n \Rightarrow x_{n+1})$, where $x_1, \ldots, x_{n+1}$ are expressions that have interpretations in the universe U; the double arrow, $\Rightarrow$, symbolizes the gap between the left-side expressions that usually describe initial positions (sources, premises) and the right-side expression that describes a terminal position (a destination, a consequence). In the logic system $N(U)$, the double arrow $\Rightarrow$ is a relation to be interpreted as "x_{n+1} is derivable from $x_1, \ldots, x_n$," and S is a proposition about derivability, whose truth value has to be established via the proof construction process (this process corresponds to the problem solving activity which attempts to close the problematic gap in S).

(ii) *Moves*. There are two types of moves, M_r and M_t. The moves M_t are terminal moves. They represent procedures that can be applied directly on a state in order to solve the problem given in that state. A common type of terminal move is a recognition operator that establishes (by table lookup) that a state represents an instance of a solved problem. The moves M_r are non-terminal moves or reduction moves that transform a state S_0 into a set of other states $S_1, \ldots, S_m$ with the intention of reducing the problem given in S_0 to a set of

*These are Gentzen type systems [10]. They have been used by Wang for machine theorem proving [4]. We shall describe such a system in detail in section 4.

problems that can eventually be solved directly by terminal moves. Often, there are several alternative moves that are applicable on S; these are the *relevant moves* of S.

In $N(U)$, terminal moves correspond to applications of axioms of the system, and non-terminal moves to applications of rules of inference. Thus, a move M_r has the logical interpretation of an implication $S_1 \& \ldots \& S_m \rightarrow S_0$ in the system $N(U)$.

(iii) *Solutions.* A solution for a state S is a tree rooted at S (sometimes it has the degenerate form of a chain) that traces the successive reductions of S via moves of type M_r—one reduction move per state—to a set of states all of which are conclusively handled by terminal moves. In the logic interpretation, a solution for a state S corresponds to a proof of the proposition of derivability expressed in S.

(iv) *Initial states.* An initial state is a state that describes the problem at the outset of the problem solving activity in a form acceptable for handling by $\Pi_r(U)$.

(v) *A search tree (also called problem solving tree) for a state S.* This is a (not necessarily finite) tree which is rooted at S that shows the reductions of S via all the relevant moves of S, and then the reductions of all the descendant states of S via their sets of relevant moves, and so on. During the problem solving activity, $\Pi_r(U)$ generates *part* of the search tree for an initial state S_0. A solution is a subtree of the search tree. Consider as an example, the search tree shown in (2.1).

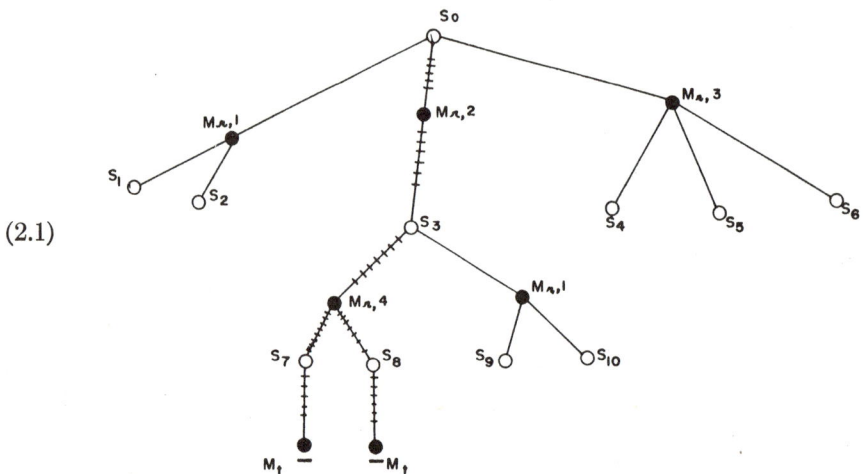

(2.1)

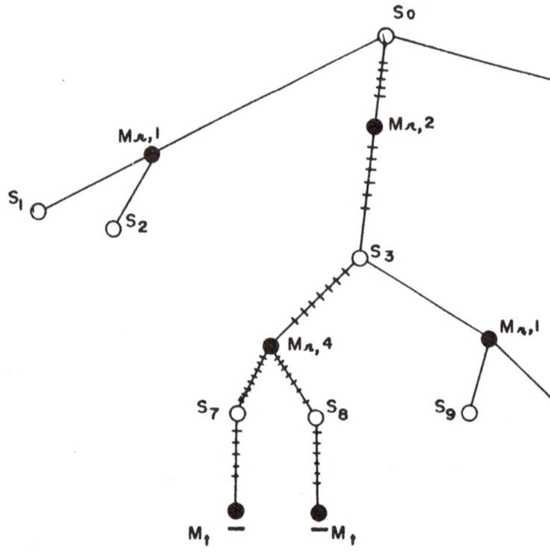

S_0 is the initial state, with three relevant moves $M_{r,1}$, $M_{r,2}$, $M_{r,3}$: The application of a reduction move on a state produces a characteristic number of descendant states (for the move) in a way that satisfies the logic of reduction. Thus, the logic interpretation of the state transition that is established by $M_{r,3}$ is given by the implicational formula $S_4 \& S_5 \& S_6 \rightarrow S_0$, (here, the S are interpreted as propositions). There is a solution tree in the search tree; it is shown by ⧺⧺ lines.

A specification of the search space for a procedure $\Pi_r(U)$ amounts to the assignment of a specific interpretation to its states and moves. Given a class of problems that originates in the universe U, a specific *choice of representation* for the solution of these problems by reduction procedures is equivalent to the choice of a search space for a procedure $\Pi_r(U)$. The problem of choosing a correct representation for a class of problems (i.e., interpretations of states and moves that satisfy the requirements of the logic of reduction) is far from easy—especially when the initial, raw, formulation of the problems is not within a formal system. The problem of choosing an "appropriate" representation (in the intuitive sense that it will make more efficient the problem solving task) is much more difficult. We shall discuss this question further in subsequent sections.

Given an initial state S_0, the goal of the reduction procedure is to find a solution tree for S_0. To this end, the procedure selectively grows parts of the search tree until it finds a solution tree that is embedded in the growth. The problem is to control the growth process in such a manner that a solution is found with a small amount of search-growth effort. This suggests that a measure of *efficiency* for a procedure Π_r can be defined in terms of the size of the search tree that is grown until a solution is found. The efficiency is higher if the number of extra states created (those that are not parts of the solution tree) is smaller. The advantage of this notion of problem solving efficiency is that it is independent of specific mechanizations, and it can be used without difficulty for comparing procedures with different search spaces. This latter property is significant for studies of the effect of different choices of representations on problem solving efficiency.

The dynamic aspects of a reduction procedure are considered in our framework in terms of a feedback scheme, which is schematically shown in (2.2).

(2.2)

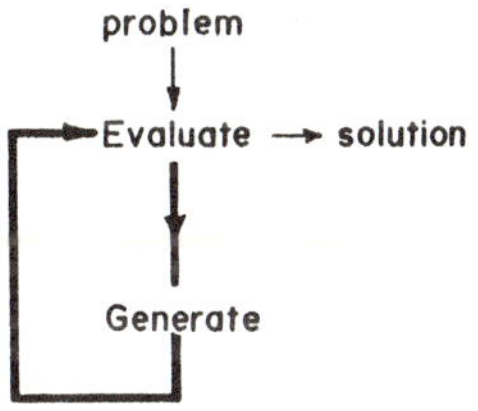

The operation of a procedure proceeds in cycles. Each cycle corresponds to a traversal of the feedback loop and it consists of a *generation phase* and an *evaluation phase*. During the generation phase, decisions on the growth of the search tree are made, and bursts of growth take place in accordance with these decisions. During the evaluation phase, the new growth is appraised and various performance measures are re-adjusted over the search tree. These new performance measures guide the decisions in the next cycle. The process continues until (i) a solution is found, (ii) it is discovered that a solution does not exist (this can be determined in some cases—as in our procedures of sections 5 and 6), or (iii) an upper limit of computational effort (which is imposed by the user) has been reached. The feedback structure of reduction procedures, their mode of operation, and some simple mechanizations are discussed in [2].

The *search strategy* used by a procedure determines certain essential features of its generation and evaluation subprocedures. These features are:

(i) *Move selection.* The restrictive rules for selecting relevant moves from a state, and the rules for ordering relevant moves by preference.

(ii) *Evaluation.* The choice of features for characterizing performance, and the methods of estimation and inference that are used to determine the values of these features over the search tree.

(iii) *Attention control.* The rules for selective direction of attention that essentially determine "where to go next" in the process of search.

Rules for restricting relevant moves are usually *heuristic rules.* Their creation and evaluation depends on the knowledge that is available to the designer about the class of problems under consideration. Choosing a good basis for estimating performance is a difficult problem; its satisfactory solution strongly depends on the specific problem area. Problems of search strategy that can be approached on more general grounds include decision theoretical problems of drawing best inferences from uncertain estimates of performance, and control theoretical problems of finding a "path of attention" that will lead to a solution with the least amount of spurious movement in the search space.

The specification of a search strategy for a procedure $\Pi_r(U)$ strongly depends on the choice of a representation. The questions of move relevance and estimation of progress are especially affected by such a choice. While it is possible to improve the performance of a procedure that has a given search space through refinements in the search strategy, the substantial improvements come when the search is transferred to a "more appropriate" search space.

In the remaining part of this paper, we shall consider specifically a proof construction problem of a certain type, and we shall propose for its solution a sequence of heuristic procedures. Each procedure in this sequence has at its core a reduction procedure or is itself a reduction procedure. The specification of these reduction procedures will be carried out in terms of the general framework introduced in this section.

3. THE PROOF CONSTRUCTION PROBLEM

A. BACKGROUND AND OBJECTIVES

In recent years, considerable effort was directed to the study of methods for mechanizing proof construction in various systems of symbolic logic. Both the propositional calculus and the predicate calculus have received attention by several investigators ([3], [4], [5], [6]). There are two main motivations underlying work in this area. The first is concerned with *logic* itself and with the potential of computers for strengthening and developing the foundations of mathematics. One of the hopes that has been expressed in this connection ([4], [7]) is that progress in the mechanization of inferential procedures may bring logic to the present state of arithmetic. The second motivation is concerned with computer realizations of *problem solving* processes in general. One of the current promising approaches of research in machine problem solving is the study of procedures for reasoning by machine. A major justification for this approach is that it provides a useful model for the exploration of ideas in an important class of problems, namely, problems of derivation type. In these problems, the main objective is to find a sequence of steps, chosen from a given set of allowable steps, for bridging the gap between a given initial "position" and a desired terminal "position." Furthermore, an efficient system for reasoning by computer has a significant intrinsic value for machine problem solving. Such a system is likely to be required as a component in many different types of complex problem solving systems (not limited to the solution of derivation type problems) that computers will realize in the future.

Our main motivation for working with the propositional calculus is problem-solving-oriented. Within this general context, we are interested to find efficient mechanizable procedures for logical deduction. But this is also an important goal of the logic-oriented research in this area. In general, it appears that as more progress is being made in machine theorem proving—and as attention is focusing on the more difficult issues—the distinction between the logic-oriented and the problem-solving-oriented approaches is gradually disappearing.

The propositional calculus, in its conventional formulation as an "axiom system," has already provided a fruitful vehicle problem for the development of the early concepts and methods in heuristic problem solving procedures [3]. In our work, we are using a "rule system," called *SU*, for the formulation of the propositional calculus. Proofs in *SU* are constructed by the *method of suppositions*. The system *SU* is similar to the systems of supposition or of "subordinate proofs" of Fitch [8] and Nidditch [9], that in turn derive from earlier systems of "natural inference" of Gentzen [10] and Jaskowski [11]. A system which is close to our system, *SU*, has been used by Guard and Mott [12] in their study of theorem proving by man-machine interaction. In order to cast the proof construction problem in the general problem solving framework that we have presented in the previous section, we associate with *SU* a system *N*(*SU*) of natural inference which has the form of Gentzen's system of *sequents*. An extension of the system *N*(*SU*) (that we shall introduce for formulating advanced procedures) is closely related to the formulation of the propositional calculus that was used by Wang [4] in his work on the realization of decision procedures by computers. However, unlike Wang, our objectives in this area go beyond the formulation of decision procedures. *Our problem is to develop and study efficient machine procedures for the construction of the simplest proof of a theorem in the system SU of the propositional calculus.* Our notion of proof simplicity will be clarified subsequently in section 4.B.

One of the reasons for choosing this problem comes from our interest in the nature of a "minimal" or "most elegant" logical argument. This is a subject in which logic *per se* is little interested, but it is of significance (at least) for the design of efficient reasoning machines. Another reason is that the problem constitutes a useful elementary model for situations where a solution is to be found that satisfies certain metric conditions of quality in addition to logical conditions of feasibility. Most problems of planning and design are of this latter type. Also, there is no efficient algorithm for finding simplest proofs; therefore, it is natural to approach this problem with heuristic procedures.

Our main emphasis in this research is to clarify the relationship between different representations of the proof construction problem and the efficiency of problem solving procedures that can be formulated for each of those representations. We shall outline three main stages of development in a hypothetical *evolution* of a proof construction system, each based on a different representation, and each featuring an improvement in performance with respect to the previous stages. We emphasize here the dynamic nature of the concept of evolution because we can already identify some of the main elements that enter in the inter-stage transitions, and in view of this, we feel that it may be possible to conceive a mechanization of such an evolution—which would add a new dimension of power to problem solving systems. Thus, in addition to the specification of each of the stages, we shall make observations, when possible, about notions that are relevant to the inter-stage transitions.

In earlier work at RCA Laboratories ([13], [14]), an effective, but moderately efficient, procedure was developed for constructing a simplest proof of a theorem in *SU*. This procedure is part of the first stage of evolution of our problem, and it will be outlined below during our discussion of the first stage. The developments in the subsequent two stages grew out of this earlier work.

B. INTRODUCTION TO PROOFS BY THE METHOD OF SUPPOSITIONS

The method of suppositions reflects a form of human reasoning which is commonly (naturally) used in the construction and testing of mathematical proofs. We shall illustrate this type of reasoning by tracing two verbal arguments; one leading to the *construction* of a proof for Peirce's law*,

$$T = (((p \rightarrow q) \rightarrow p) \rightarrow p), \tag{3.1}$$

and another for *establishing the validity* of this law via the presentation of a given proof.

(α) *The proof construction argument.* This argument runs as follows: (a) to prove T, it suffices to show that p is a consequence of the supposition $((p \rightarrow q) \rightarrow p)$; ($b$) this can be shown to hold if we can obtain a contradiction as a consequence of the supposition $\sim p$ and the previous supposition $((p \rightarrow q) \rightarrow p)$ (here, a *reductio ad absurdum* argument [in reverse] is attempted); (c) let $\sim p$ and p be the formulas that would establish the contradiction; thus, the theorem will be proved if we show that both p and $\sim p$ are consequences of the suppositions made so far; (d) $\sim p$ is an obvious direct consequence of the supposition $\sim p$

*For the purposes of this example, we shall use conventional infix logical notation and the symbols "$\rightarrow$" for implication and "$\sim$" for negation. This is not our regular notation in SU, where, as we shall see shortly, we are using Polish prefix notation.

(here, a link closing part of the argument is established); (*e*) p can be shown to be a consequence of the existing suppositions if both $((p \rightarrow q) \rightarrow p)$ and $(p \rightarrow q)$ are shown to be themselves consequences of the same suppositions (here, a *modus ponens* argument [in reverse] is attempted); (*f*) the formula $((p \rightarrow q) \rightarrow p)$ is an obvious direct consequence of the supposition $((p \rightarrow q) \rightarrow p)$ (here, another part of the argument is closed); (*g*) to show that $(p \rightarrow q)$ is a consequence of the existing suppositions, it suffices to show that q is a consequence of a new supposition p and the previous suppositions; (*h*) but this is indeed the case, since from the suppositions p and $\sim p$, any formula follows and specifically the formula q (here, a version of the *reductio* argument is used [in reverse] and it closes the last open part in the total argument); hence, the theorem T is proved.

(β) *The argument for establishing validity.* This argument runs as follows: (i) suppose $((p \rightarrow q) \rightarrow p)$; (ii) given this first supposition, suppose $\sim p$; (iii) because of the first supposition, the formula $((p \rightarrow q) \rightarrow p)$ can be introduced as a consequence after the second supposition; (iv) given the first two suppositions, suppose p; (v) from the suppositions p and $\sim p$ we get a contradiction, thus we can infer q; (vi) hence, on the first two suppositions, we can obtain as a consequence $(p \rightarrow q)$; (vii) since the formulas $((p \rightarrow q) \rightarrow p)$ and $(p \rightarrow q)$ are both deduced under the first two suppositions, then we can also obtain (via *modus ponens*) p as a consequence under these suppositions; (viii) from this p and the second supposition, $\sim p$, we have a contradiction; (ix) hence, we obtain (via *reductio*) that p is a consequence under the first supposition $((p \rightarrow q) \rightarrow p)$; (x) from this we can deduce that the formula $(((p \rightarrow q) \rightarrow p) \rightarrow p)$ is logically valid, and the theorem is proved.

The system SU for the propositional calculus is devised to capture and formalize the essence of the reasoning modes that we have just presented. We show next the proof of Pierce's law* in the system SU.

	1	$((p \rightarrow q) \rightarrow p)$	supposition
	2	$\sim p$	supposition
	3	$((p \rightarrow q) \rightarrow p)$	reiteration of 1
	4	p	supposition
	5	$\sim p$	reiteration of 2
	6	p	reiteration of 4
(3.2)	7	X	X introduction by 5, 6
	8	q	X elimination by 7
	9	$(p \rightarrow q)$	$\rightarrow$ introduction by 4, 8
	10	p	$\rightarrow$ elimination by 3, 9
	11	$\sim p$	reiteration of 2
	12	X	X introduction by 10, 11
	13	p	$\sim$ elimination by 2, 12
	14	$(((p \rightarrow q) \rightarrow p) \rightarrow p)$	$\rightarrow$ introduction by 1, 13

*In this paper we shall use Pierce's law as a running example for the illustration of the relationships between various proof construction methods.

This proof has a *sequential form.* It establishes the validity of Peirce's law, T, via a sequence of 14 steps, each of which contains a formula and associated information that "justifies" the presence of the formula in that step. Each formula in the sequence is either a supposition or it is obtained from previous formulas according to the rules of SU. The final formula in the sequence is the theorem T.

Even without a detailed specification of the system SU, we can note at this point the correspondence between the previously given verbal arguments and the proof in SU, which we have just presented. Consider first the relationship between the argument (β) which establishes the validity of Peirce's law) and the proof sequence in (3.2). The steps 1, 2, 3, 4 of the proof in SU correspond to the steps (i), (ii), (iii), and (iv) respectively of the verbal argument; the steps 5, 6, 7, 8 of the proof in SU correspond to the step (v) of the verbal argument; the step 9 corresponds to the step (vi); the step 10 corresponds to the step (vii); the steps 11, 12 correspond to (viii); the step 13 corresponds to (ix), and finally the step 14 of the proof in SU corresponds to the step (x) of the verbal argument. In general, the order and the nature of the verbal argument that establishes the validity of a theorem by the method of suppositions are reflected precisely in a sequential proof of the theorem as formulated in SU.

While the validation argument is captured directly by tracing the proof sequence in SU, the proof construction argument (α) is reflected less directly in the SU-proof. The reason for this is that the steps in the development of the verbal argument for proof construction correspond to stages of formation of the SU-proof, where formulas appear in different parts of the final sequence in an order which depends on the specific proof construction strategy employed by the problem solver. It can be easily verified that the relationship between the verbal proof construction argument for Peirce's law and the SU-proof given in (3.2) is as follows.

Assume that the proof construction argument starts with a statement of the theorem to prove. This corresponds to a partial specification of step 14 in the SU-proof, where the formula appears without its associated justification (we will denote here by n_f the formula part of step n, by n_j the justification part, and by n the complete step); thus, the argument starts with step 14_f already taken. The step (a) in the verbal argument corresponds to the steps 1, 13_f, 14_j of the SU-proof; the step (b) corresponds to 2, 12_f, 13_j; the step (c) corresponds to 10_f, 11_f, 12_j; the step (d) corresponds to 11_j; the step (e) corresponds to 3_f, 9_f, 10_j; the step (f) corresponds to 3_j; the step (g) corresponds to 4, 8_f, 9_j; the final step (h) corresponds to 5, 6, 7, 8_j in the SU-proof; now the proof is completely formed (both the structure of formulas and their associated justifications are specified), and, in view of the validation argument, the theorem is proved valid. Note that the proof is formed by starting from the end of the proof sequence, and then working from both ends of the sequence inwards. One step of the SU-proof is created in two steps (usually successive) of the proof construction argument, except for supposition steps in the SU-proof that are specified within a single proof construction step. The formula part of a step in the SU-proof is formed first and it is then followed by the justification part of that step. Although the proof construction process can be represented and developed in the notation of *SU*-proofs, we will show that it can be handled more directly and naturally in a system of natural inference *N*(*SU*) that we shall associate with *SU*. However, before moving to another system, we shall specify formally the system *SU* in which we are assuming that our proof construction problem is initially formulated.

C. THE SYSTEM *SU*

The *language* of *SU* has the following *vocabulary*:

p, q, r, etc.	denoting propositional variables
C, D, I, N	denoting conjunction, disjunction, implication and negation respectively
X	denoting contradiction (this special symbol stands for the conjunction of any formula and its negation)

The *formation rules* for the language of *SU* are as follows: (i) a propositional variable is a formula; (ii) if x is a formula, then Nx is a formula; (iii) if x and y are formulas, then Cxy, Dxy, Ixy are formulas; (iv) X is a formula. The formation rules were chosen so that formulas take the parenthesis free, or prefix form due to Lukasiewicz and the Polish school; here, propositional connectives precede the symbols on which they operate. This form is especially convenient for computer processing and typing. Also, it provides a representation of a formula's structure which we find especially appropriate in the development of advanced proof procedures (see section 6). To illustrate this notation, Peirce's law, which appears in conventional infix notation as $(((p \rightarrow q) \rightarrow p) \rightarrow p)$, is represented in *SU* by $IIIpqpp$.

The notion of a *proof in SU* is as follows. The proof consists of a finite sequence of steps, each of which includes a formula and its justification. The justification indicates that the formula is a supposition or it is a consequence from a rule of deduction in *SU* which is applied to previous formulas in the proof sequence. The last formula in the proof sequence is the theorem to which the proof applies. The proof sequence has an internal structure in the form of nested segments (see, for example, the proof in (3.2)). The outer segment, referred to as the *main proof*, has no supposition, and its last step contains the theorem. Each segment which is nested within the outer segment is referred to as a *subordinate proof*, and it has a single supposition which appears at its first step. Thus, in the *SU*-proof shown in (3.2), there are three subordinate proofs, and the subordinate proof with supposition p is *included in* the subordinate proof with supposition $\sim p$.

Formally, a *supposition* is a formula which has the status of an *ad hoc* hypothesis in an *SU*-proof. However, in the construction of proofs, and especially of simplest proofs, supposition formulas are chosen (by the proof construction procedure) with the purpose of effecting certain desired consequences (we shall return to this point later).

The ten *rules of deduction* in SU are as follows:*

1. *Conjunction introduction* (C_{in}). If x_1 and x_2 are in a given proof segment, then Cx_1x_2 is a consequence in the same segment. In the special case where x_1, x_2 are a formula and its negation, then the consequence is X, and the rule is labeled X_{in}.

2. *Conjunction elimination* (C_{el}). If Cx_1x_2 is in a given proof segment, then x_1 (alternatively, x_2) is a consequence in the same segment.

3. *Disjunction introduction* (D_{in}). If x_1 (alternatively, x_2) is in a given proof segment, then $Dx_1\,x_2$ is a consequence in the same segment.

4. *Disjunction elimination* (D_{el}). If Dx_1x_2 is in proof segment j, and if there are two subordinate proofs that are immediately included in this segment, one with supposition x_1 and final formula x_3 and the other with supposition x_2 and final formula x_3, then x_3 is a consequence in the proof segment j.

5. *Implication introduction* (I_{in}). If x_1 and x_2 are respectively the supposition and the final formula of a given subordinate proof, then Ix_1x_2 is a consequence in the proof segment that immediately includes the subordinate proof.

6. *Implication elimination* (I_{el}). If x_1 and Ix_1x_2 are in a given proof segment, then x_2 is a consequence in the same segment (this rule corresponds to a *modus ponens* argument).

7. *Negation introduction* (N_{in}). If x is the supposition of a given subordinate proof whose final formula is X (the symbol for contradiction), then Nx is a consequence in the proof segment that immediately includes the subordinate proof.

8. *Negation elimination* (N_{el}). This is the same as 7, except x is replaced by Nx and Nx by x. (This rule, and the previous rule 7, correspond to *reductio ad absurdum* arguments.)

9. *Contradiction elimination* (X_{el}). If X is in a given proof segment, then any formula x is a consequence in the same segment.

10. *Reiteration* (Reit). If x is in a given proof segment, then x is a consequence in the same segment or in any proof segment that is included in the given proof segment.

A re-examination of the illustrative proof given in (3.2) may help at this point in familiarization with the deduction rules that we have just presented and with the method of their use for establishing the validity of a proposed SU-proof.

The system SU is a *consistent and complete* characterization of the propositional calculus. Consistency is easily established. A proof of completeness can be obtained by showing that the axioms of well-known formulations of the propositional calculus are provable as theorems in SU†.

It is important to note that the "rule system" SU for the propositional calculus does *not* provide rules for constructing an SU-proof to a theorem. Its rules offer

*In these rules, each of the symbols x, x_1, x_2, x_3 stands for any formula in SU.
†Such a proof is given in [12].

a direct method for *checking* the well-formedness of SU-proofs. To enable a straightforward formulation of procedures for constructing SU-proofs, we must choose now a more appropriate system in which our problem can be represented. The system is chosen so that it can be conveniently used in the formulation of reduction procedures (of the type discussed in section 2) for proof construction. This leads us to the system of natural inference $N(SU)$. The transition $SU \rightarrow N(SU)$ takes us from the "initial" representation of the problem to a representation where known methods of problem solving (the search-reduction methods) can be directly used for its solution. The system $N(SU)$ provides the basis for the first phase of development of our proof construction procedures.

4. PROOF CONSTRUCTION PROCEDURES: THE FIRST STAGE

A. THE NATURAL INFERENCE SYSTEM $N(SU)$

The formulas in $N(SU)$ are expressions of the form

(4.1) $$(\eta \Rightarrow x),$$

and they are called *sequents*; η denotes a finite sequence (possibly empty) of formulas in SU where some of the formulas are especially designated (we mark them with asterisks); x is a formula in SU; the double arrow, $\Rightarrow$, denotes a reflexive and transitive relation that gives the following interpretation to a sequent: "x is deducible in SU from the sequence of formulas η," i.e., there exists a subsequence of a proof sequence in SU which includes the formulas of η, in the order of their appearance in the sequence η, and which is followed by the formula x; the formulas of η that are marked with asterisks are supposition formulas in the proof sequence. We call the formulas in η the *information formulas* or *source formulas* of the sequent, and the formula x the *goal formula* or the *destination formula* of the sequent. It is clear from the definition of a sequent and from the completeness of the system SU that the problem of establishing whether a formula T is logically valid is equivalent to establishing that the sequent $(\Rightarrow T)$ is valid in $N(SU)$.

The system $N(SU)$ has a single axiom schema and a set of rules of inference, which corresponds in a straightforward manner to the rules of deduction in SU. The axiom schema, $\mathfrak{A}_1$, of $N(SU)$ is as follows: A sequent $(\eta \Rightarrow x)$ is *valid* if there exists a formula in the sequence η which is identical with x. The assertion that a sequent is logically valid by $\mathfrak{A}_1$ (such a sequent is called a *conclusive sequent*) will be represented graphically by an *axiom link* as follows:

(4.2) $$\begin{array}{c} (\eta \Rightarrow x) \\ \uparrow \\ \bullet \;\; \mathfrak{A}_1[1] \end{array}$$

The number in brackets is the *weight* of the axiom link. This notion will be discussed later (in 4.B).

We give in Table 4.1 the *rules of inference* of $N(SU)$ in the form of transition schemata between sequents. These schemata are structured as trees (some of them are degenerate trees, i.e., chains); we call them *inference trees*. The arrows in each tree show the direction of the inference. The root sequent in each tree (an arrow points into it) is called the *consequent sequent*. The sequents at the tree terminals (arrows point away from them) are called the *antecedent sequents*. The logical property of an inference tree is as follows:

(4.3)   The consequent sequent is valid if *all* the antedecent sequents are valid.

TABLE 4.1
TRANSITION SCHEMATA FOR THE RULES OF INFERENCE OF $N(SU)$

(The sequents shown below are sequent forms, each standing for a [infinite] set of sequents; the variables x, x_1, x_2, x_3 stand for any formulas in SU; η stands for an arbitrary [possibly empty] sequence of formulas in SU, some of which may be suppositions.)

$(\eta \Rightarrow Cx_1x_2)$ $C_{in,a}^N[2]$ $(\eta \Rightarrow x_1)$ $(\eta, x_1 \Rightarrow x_2)$

$(\eta \Rightarrow Cx_1x_2)$ $C_{in,b}^N[2]$ $(\eta, x_2 \Rightarrow x_1)$ $(\eta \Rightarrow x_2)$

$(\eta \Rightarrow x_1)$ $C_{el,a}^N[1]$ $(\eta \Rightarrow Cx_1x_2)$

$(\eta \Rightarrow x_1)$ $C_{el,b}^N[1]$ $(\eta \Rightarrow Cx_2x_1)$

$(\eta \Rightarrow Dx_1x_2)$ $D_{in,a}^N[1]$ $(\eta \Rightarrow x_1)$

$(\eta \Rightarrow Dx_1x_2)$ $D_{in,b}^N[1]$ $(\eta \Rightarrow x_2)$

$(\eta \Rightarrow x_3)$ $D_{el}^N[7]$ $(\eta \Rightarrow Dx_1x_2)(\eta, Dx_1x_2, x_1^* \Rightarrow x_3)(\eta, Dx_1x_2, x_2^* \Rightarrow x_3)$

$(\eta \Rightarrow Ix_1x_2)$ $I_{in}^N[3]$ $(\eta, x_1^* \Rightarrow x_2)$

$(\eta \Rightarrow x_2)$ $I_{el,a}^N[2]$ $(\eta \Rightarrow Ix_1x_2)(\eta, Ix_1x_2 \Rightarrow x_1)$

$(\eta \Rightarrow x_2)$ $I_{el,b}^N[2]$ $(\eta, x_1 \Rightarrow Ix_1x_2)(\eta \Rightarrow x_1)$

$(\eta \Rightarrow Nx)$ $N_{in}^N[3]$ $(\eta, x^* \Rightarrow X)$

$(\eta \Rightarrow x)$ $N_{el}^N[3]$ $(\eta, Nx^* \Rightarrow X)$

$(\eta \Rightarrow X)$ $X_{in,a}^N[2]$ $(\eta \Rightarrow x)$ $(\eta, x \Rightarrow Nx)$

$(\eta \Rightarrow X)$ $X_{in,b}^N[2]$ $(\eta, Nx \Rightarrow x)(\eta \Rightarrow Nx)$

$(\eta \Rightarrow x)$ $X_{el}^N[1]$ $(\eta \Rightarrow X)$

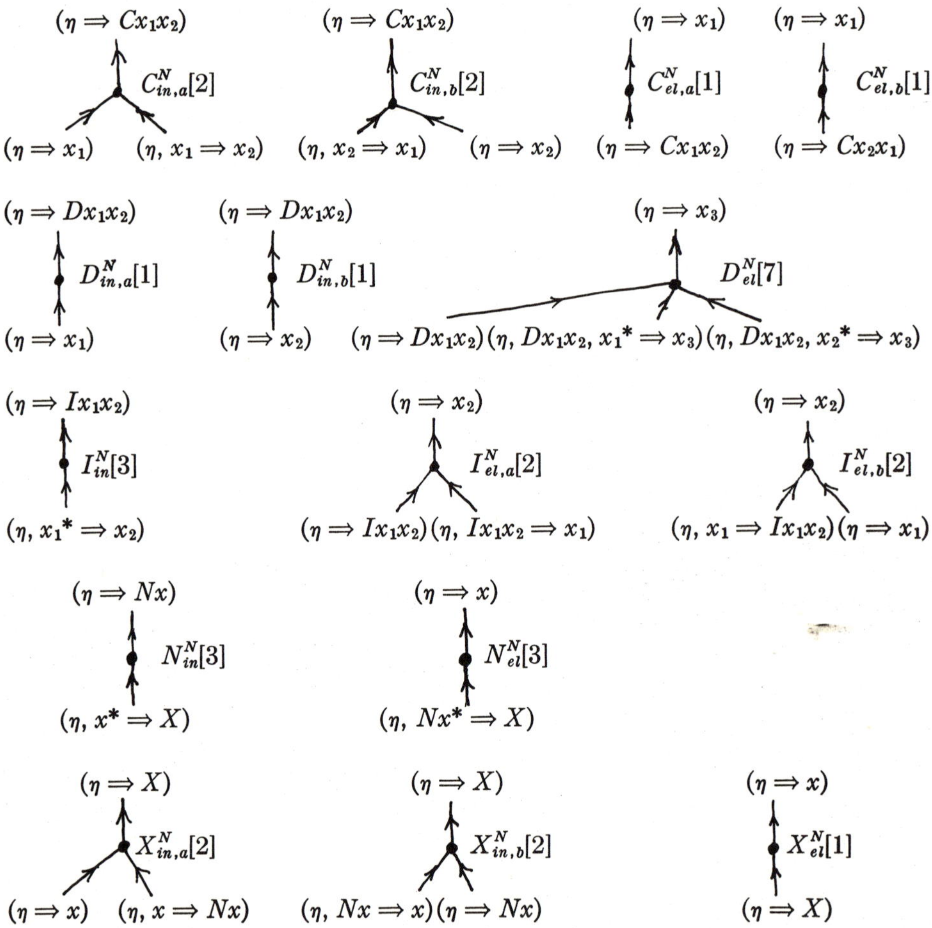

The label associated with the node of an inference tree is the name of the rule of inference, and it clearly indicates the rule of deduction of SU to which the inference corresponds. The number in brackets that is associated with the name of an inference denotes the *weight of the inference*; the latter will be discussed shortly (in 4.B).

Note that for each rule of introduction or elimination in SU, there are one or two inference trees in $N(SU)$; also, the axioms $\mathfrak{A}_1$ correspond to the rule of reiteration.

When the variables in the sequents of an inference tree assume specific values, then the tree will be called a *specified inference tree.*

A *proof in tree form* in the system $N(SU)$ is a labeled tree that is made of specified inference trees, and it terminates exclusively with axiom links to $\mathfrak{A}_1$. More specifically, the proof tree is rooted at a sequent which is the consequent sequent of a specified inference tree. The antecedent sequents of this specified inference tree are themselves either consequent sequents of other specified inference trees or conclusive sequents (linked to $\mathfrak{A}_1$), and so on, until all the tree terminals are linked to $\mathfrak{A}_1$.

A sequent is a *theorem of* $N(SU)$ if it is at a root node of a tree proof in $N(SU)$. If a theorem of $N(SU)$ has the form $(\Rightarrow T)$, then T is a theorem in logic and the tree proof in $N(SU)$ is its proof. This statement is easily verified on the basis of the logical properties of the system SU, and the correspondences between the systems SU and $N(SU)$.

As an example, we show next in (4.4) the *proof in tree form of Peirce's law* (using now our prefix notation, this theorem reads $IIIpqpp$). (The sequential proof of this theorem in the system SU is given in [3.2].)

It is relatively easy to transform a tree proof in $N(SU)$ to an equivalent sequential proof in SU. To each specified inference tree in the tree proof, there corresponds a distinct step in the equivalent sequential proof that is justified by a specific rule of deduction of SU; to each axiom link in the tree proof, there corresponds a reiteration step in the SU proof; in addition to the steps that correspond to specified inference trees and axiom links, the SU proof has one step for each supposition. Thus, in our example, the tree proof in (4.4) has seven specified inference trees and four axiom links, and the equivalent sequential proof in (3.2) has 14 steps, 11 of which are in direct correspondence to the $7 + 4$ steps in the tree proof and three are supposition steps.

To "read" the tree proof of (4.4) with a view to establishing the validity of the theorem, we have to trace the information in the direction of the arrows (from the tree terminals to the root). Each conclusive sequent is valid by $\mathfrak{A}_1$. Each consequent sequent that has all its antecedent sequents valid is itself valid. Propagating the validity arguments from the tree terminals upward, we can finally show that the root sequent is valid.

If we reverse the direction of attention and we trace the tree proof against the arrows (from root to terminals), then we capture precisely the verbal proof construction argument (α) that we have presented previously. This can be checked

[w = 20]

(4.4)

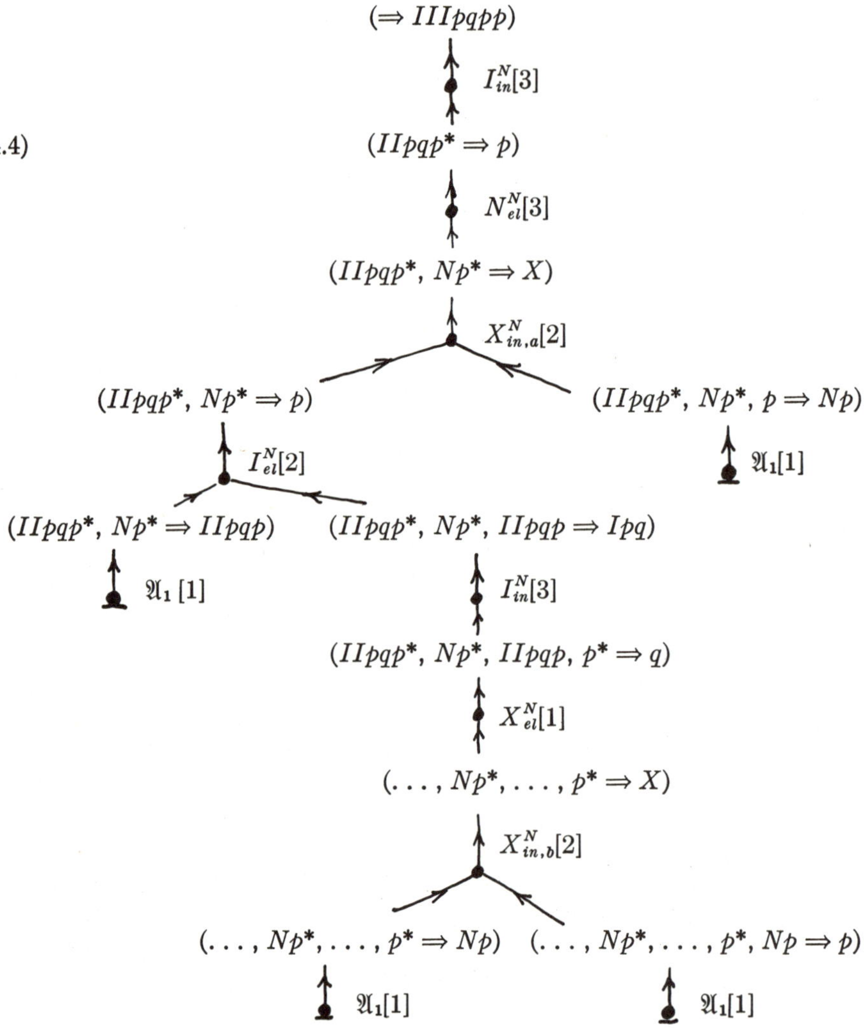

in detail by referring to the argument in question and also to the correspondence that we have established between this argument and the sequential *SU*-proof. In general, a proof construction argument of type (α) would start by considering a state of affairs corresponding to a sequent ($\Rightarrow T$), where T is the theorem to be proved. It would then attempt to apply in reverse an inference tree on ($\Rightarrow T$) (i.e., the inference tree would have ($\Rightarrow T$) as its consequent sequent). Subsequently, it would focus attention on the antecedent sequents of the inference tree and treat each of them recursively in the same manner as the initial segment ($\Rightarrow T$), until conclusive sequents are finally found for all the branches of the argument.

The "backward reasoning" scheme that we have just outlined is expressable in a natural manner within the problem solving framework that we have introduced in section 2. The following correspondences can be established. Problem solving states can be interpreted as sequents in $N(SU)$, non-terminal moves can be interpreted as reverse applications of rules of inference, terminal moves can be interpreted as applications of the axiom schema $\mathfrak{A}_1$, and a solution tree can be considered as a proof tree in $N(SU)$. Our main objective in introducing the system $N(SU)$ was to establish this type of correspondence between proof construction procedures in the propositional calculus and the more general class of problem solving procedures of reduction type. Furthermore, the treatment of the proof problem in $N(SU)$ provides a *knowledge basis* from which properties of improved proof construction procedures can be suggested and in which their relevance/usefulness can be eventually demonstrated. We are using here the concept of knowledge basis to cover the ensemble of problem-specific information that is essential for the evolution of a problem solving system in the problem area of interest. We shall return to this concept later.

It is important to note that the rules of inference that we have specified for $N(SU)$ reflect a specific *mode of using* the rules of deduction of SU in the process of proof construction. Specifically, our approach in the present system is a pure "backward reasoning" approach; every rule of inference focuses on the goal formula of a state (see Table 4.1). This enables the proof construction activity to move systematically, from a goal formula (a destination) to information formulas (sources). It is possible to formulate other natural inference systems on the basis of SU, with rules of inference that reflect other modes of applying the rules of deduction of SU in the process of proof construction. For example, a set of rules of inference can be formulated for a combined approach to proof construction, where activity proceeds simultaneously from sources and destinations. Such a system will be introduced later, in the third stage of development of our problem solving procedures (see section 6).

B. MINIMAL PROOFS

The notion of proof complexity (or simplicity) that we are using in this work is based on the assignment of numerical *weights*, w, to the rules of inference and to the axiom schema of $N(SU)$. The weights that we have chosen are shown in square brackets in (4.2) for axiom links and in Table (4.1) for inference trees. The sum of weights of the inference trees and the axiom links that constitute the building blocks of a given tree proof gives the *weight of the proof*. Thus, it can be verified that the weight of the proof of Peirce's law, which is shown in (4.4), is 20; incidentally, this is, according to our definitions, the simplest proof of Peirce's law in $N(SU)$, i.e., there is no tree proof of this theorem which has a smaller weight.

Our approach to weight assignment is mainly aimed to capture a measure of structural complexity in the proof and also a measure of interaction between the formulas that participate in the proof development. We arbitrarily assign a weight of one to an axiom application. Such an application amounts to establishing a simple connection, a direct link, between two identical formulas. When a rule of inference is applied, a certain number of formulas in the antecedent sequents, we call them *input formulas*, participate in a *reasoning interaction* that results in the consequent sequent. The input formulas are the formulas that enter in the antecedent of the deduction rule of SU which corresponds to the inference rule in question. We assign to each rule of inference a weight which equals the number of its input formulas, where supposition formulas are counted twice. We justify the heavier participation of a supposition formula in the weight count on the grounds of its special structural role as a determinant of a distinct subordinate proof in the system SU. To illustrate: In the two rules of inference $C^N_{in,a}$, $C^N_{in,b}$, the input formulas are x_1, x_2 and the weight of the inference is two; in the rule of inference D^N_{el}, the input formulas are Dx_1x_2, x_1^*, x_3, x_2^*, x_3 and the weight is seven.

Consider a sequential proof in SU which is equivalent to a given tree proof in $N(SU)$. It is natural to assign identical weights to both proof forms. The proof weight is related to the form of the sequential proof as follows: the proof weight equals the number of formulas in the sequential proof (except the formula in the final step, i.e., the theorem), augmented by the number of subordinate proofs in the sequential proof and by the number of reiteration steps. Alternatively, the proof weight equals the total number of references to previous steps in the justification part of all the steps, augmented by the number of suppositions. (As a check of these relationships, see the sequential proof shown in (3.2).)

The appropriateness of any definition of proof complexity is always open to debate. The main virtue of our definition is that it provides us with a well-defined problem of derivation type that has "reasonable" metric constraints, and for which no efficient decision procedure is known. The reasonableness of our formulation rests on a considerable amount of empirical evidence which indicates that minimum weight proof (in the sense that we have just defined) coincide with intuitively simplest proofs. Furthermore, it appears at present that our definition may be theoretically fruitful, as it provides a measure of the total flow of formulas in a proof. We shall return to this point later (in section 6) in our discussion of graph representations of proofs.

C. PROCEDURES $\Pi_r(PC, 1)$

As stated previously, in section 3.A, our problem as designers is to develop efficient machine procedures for the construction of the simplest proof of a theorem in the system SU of the propositional calculus. In view of the transition to the system $N(SU)$, and the definition of proof simplicity that we have just given, we can restate as follows the problem that our procedures are required to solve: given a theorem T in the propositional calculus, construct a proof tree of minimal weight for T in the system $N(SU)$. Moreover, carry out the construction process in the most efficient manner (in the sense of efficiency discussed in section 2).

The formulation of the proof problem in the system $N(SU)$ enables us to specify the overall structure of our procedures in the general form that we have outlined in section 2, i.e., as procedures of *reduction type*, Π_r. We have already shown in 4.A how the notions of state, root (or initial) state, non-terminal move, terminal move, and solution can be interpreted in terms of concepts in the system $N(SU)$ in which the proof problem has been formulated. We shall adopt this interpretation for a class of procedures that we call $\Pi_r(PC, 1)$. These procedures are characteristic of the first stage of development of our proof construction system. The interpretation of states and moves is an essential part of the specification of $\Pi_r(PC, 1)$. In order to complete the specification, it remains to consider the generation and evaluation sub-procedures of $\Pi_r(PC, 1)$.

A first step in the specification of a generation procedure is to examine the set of relevant moves.

(a) Non-Terminal Moves

Let us consider first the set, μ, of applicable non-terminal moves, i.e., the set of all moves that can be applied on *any* non-conclusive state $(\eta \Rightarrow x)$ in accordance with the correspondence that we have just established between the system $N(SU)$ and the problem solving system. We shall use here the definitions of inference rules of $N(SU)$ that were presented in Table (4.1). Given an inference rule, R, we denote by $\bar{R}$ the move that corresponds to the reverse application of R. We will present the moves in μ in terms of the characteristics of the states on which they apply.

(4.5) *The set μ of applicable non-terminal moves.*

(i) Given a non-conclusive state $(\eta \Rightarrow x)$, for any goal formula x which is not a contradiction (i.e., $x \neq X$), there exist

(1) a set of six moves, whose application to the given state depends on the form of x (these are $\bar{C}^N_{in,a}$, $\bar{C}^N_{in,b}$, $\bar{D}^N_{in,a}$, $\bar{D}^N_{in,b}$, $\bar{I}^N_{in}$, $\bar{N}^N_{in}$),

(2) two moves $\bar{N}^N_{el}$, $\bar{X}^N_{el}$ that are applicable independently of the form of x,

(3) two (infinite) sets of moves $\{\bar{C}^N_{el,a,b}\}$, $\{\bar{I}^N_{el,a,b}\}$ that are applicable independently of the form x. (The set $\{\bar{C}^N_{el,a,b}\}$ is generated by all reverse applications of the rules of inference $C^N_{el,a}$ or $C^N_{el,b}$ on any sequent $(\eta \Rightarrow x)$ where in the antecedent sequent of the inference (i.e., in $(\eta \Rightarrow Cx_1x)$ or in $(\eta \Rightarrow Cx\,x_1)$) the variable x_1 ranges over all possible formulas in SU. The set $\{\bar{I}^N_{el,a,b}\}$ is generated by all reverse applications of the inference $I^N_{el,a}$ or $I^N_{el,b}$ on any sequent $(\eta \Rightarrow x)$, where in the two antecedent sequents of the inference (i.e., in $(\eta \Rightarrow Ix_1x)$ and $(\eta, Ix_1x \Rightarrow x_1)$, or in $(\eta, x_1 \Rightarrow Ix_1x)$ and $(\eta \Rightarrow x_1)$) the variable x_1 (which is the antecedent formula in the *modus ponens* argument characterized by this inference) ranges over all possible formulas.)

(ii) For any goal formula x, including $x = X$, there is an infinite set of moves $\{\bar{D}_{el}^N\}$ that are applicable independently of the form of x. (This set is generated by all reverse explications of the inference D_{el}^N on any sequent $(\eta \Rightarrow x)$, where in the three antecedent sequents $(\eta \Rightarrow Dx_1x_2)$, $(\eta, Dx_1x_2, x_1^* \Rightarrow x)$, $(\eta, Dx_1x_2, x_2^* \Rightarrow x)$ of the inference the variables x_1, x_2 range over all possible formulas.)

(iii) If the goal formula x is X, there is an infinite set $\{\bar{X}_{in,a,b}^N\}$ of applicable moves. (This set is generated by all reverse application of $X_{in,a}^N$ or $X_{in,b}^N$ on $(\eta \Rightarrow X)$, where in the two antecedent sequents (i.e., in $(\eta \Rightarrow x_1)$ and $(\eta, x_1 \Rightarrow Nx_1)$, or in $(\eta, Nx_1 \Rightarrow x_1)$ and $(\eta \Rightarrow Nx_1)$), the variable x_1 ranges over all possible formulas.)

In order to formulate an effective proof procedure, we must restrict the infinite set of applicable moves that we have just described to a finite set. We achieve this restriction by limiting the set of formulas that can appear anywhere in a tree proof of a theorem T to *subformulas or negations of subformulas* that occur in T. We call this set of formulas $\sigma(T)$.

It can be proved (and we shall do so subsequently) that the restriction that we have just imposed on the form of constructible proof does not entail loss of *completeness* for the proof procedures $\Pi_r(PC, 1)$. In an early stage of development of these procedures, it is possible to assume that no rigorous proof of completeness is known to the designer, and the completeness property of the proposed procedures has the status of a conjecture. The strength of this conjecture increases with the accumulation of supporting empirical evidence—via the application of the procedure to many theorems. Given such a state of knowledge about completeness, the first stage procedures $\Pi_r(PC, 1)$ have the status of *heuristic proof procedures.* The same procedures assume the status of *proof algorithms* at a later time (in the second stage of development of the proof construction system) when all the elements for a proof of completeness exist, and therefore we assume that the proof is known.

It is important to note at this point—because this situation is typical of the processes underlying the development of many problem solving procedures—that a conjecture has been formulated about the relationship between a given property of the overall search space (to be searched under the procedure's direction) and a desired functional property of the procedure, with the general intention to decrease the cost of using the procedure. In the case just discussed, the intention is to obtain a practically feasible effective procedure. In most cases, however, increased efficiency of a procedure is the objective (we shall discuss such cases later). As the knowledge of properties of the search space increases, it may be possible to prove the validity of the conjecture. Also, the new knowledge may support a second round of changes in procedure that will further increase their efficiency.

Completeness is one of the desired functional properties of our proof construction procedures; minimality is the other (the constructed proof should have minimal weight). We now extend the previous conjecture to cover both completeness and minimality, i.e.,

CONJECTURE $\mathfrak{C}_\sigma$. *For any theorem T in SU, the formulas that appear anywhere in the minimal proof of T are in* $\sigma(T)$.

The empirical evidence supporting this conjecture is considerably strong from the early stages of development of the procedures $\Pi_r(PC, 1)$. However, even after the conjecture of completeness is proved during the development of the second-stage procedures, the conjecture of minimality retains its status as an unproved but strong conjecture. It seems plausible at present that the validity of this conjecture will be demonstrated during the third stage of development of the procedures.

The restriction on possible formulas that we have just discussed entails a restriction on the set of applicable non-terminal moves; we call this reduced set the *relevant set of non-terminal moves under* $\mathfrak{C}_\sigma$, and we denote it by μ^*.

The reduced set μ^* is related to the set μ as follows: the moves of μ that are given in (i)(1) and (i)(2) of (4.5) are also in μ^*. The infinite sets of moves $\{\bar{C}^N_{el,a,b}\}\{\bar{I}^N_{el,a,b}\}$ in (i)(3) of (4.5) are reduced here to finite (possibly empty) sets of moves; a $\bar{C}^N_{el,a}$ (or $\bar{C}^N_{el,b}$) move applies on a state $(\eta \Rightarrow x)$ if a formula Cxy (or Cyx) is among the formulas or subformulas of η, where y can be any formula; a $\bar{I}^N_{el,a}$ or $\bar{I}^N_{el,b}$ move applies on $(\eta \Rightarrow x)$ if a formula Iyx is among the formulas or subformulas of η, when y can be any formula. The infinite set of moves $\{\bar{D}^N_{el}\}$ in (ii) of (4.5) is restricted here to a finite (possibly empty) set, and a member of this set applies on $(\eta \Rightarrow x)$ if a formula Dyz is among the formulas or subformulas of η, where y, z can be any pair of formulas; each such disjunction Dyz is a candidate for a disjunction elimination argument. The infinite set of moves $\{\bar{X}^N_{in,a,b}\}$ in (iii) of (4.5) is restricted to a finite set which is generated as follows: each move corresponds to the reverse application of the inference $X^N_{in,a}$ or $X^N_{in,b}$ on $(\eta \Rightarrow X)$, where in the two antecedent sequents (i.e., in $(\eta \Rightarrow x_1)$ and $(\eta, x_1 \Rightarrow Nx_1)$ or in $(\eta, Nx_1 \Rightarrow x_1)$ and $(\eta \Rightarrow Nx_1)$), the pair of contradictory formulas Nx_1, x_1 ranges over all the distinct pairs of formulas or subformulas of η and their contradictions. This later set of relevant moves, whose function is to specify a pair of contradictory formulas as a goal for a *reductio ad absurdum* argument, is the least restricted by $\mathfrak{C}_\sigma$. The wealth of alternatives that it provides at each state is responsible for a major part of the problem solving (search) effort that is associated with the procedures $\Pi_r(PC, 1)$.

The specification of the set μ^* of relevant moves is equivalent to a specification of the original set μ of applicable moves together with a set of *restrictive principles* relevant to these moves; the restrictive principles derived from the property $\mathfrak{C}_\sigma$ of the search space. As an example, the following restrictive principle is relevant to $\bar{C}^N_{el,a}$ moves: for all y, an inference tree where $(\eta \Rightarrow x)$ is the consequent sequent and $(\eta \Rightarrow Cxy)$ the antecedent sequent is not on a minimal proof tree if Cxy is not among the formulas or subformulas of y. The logical status (truth, plausibility) of such a principle depends, of course, on the logical status of the conjecture $\mathfrak{C}_\sigma$. Its effect in the problem solving process is to inhibit the application of a set of moves from a state, hence, to inhibit search activity in a part of state space. Its value is directly related to the size of the "barren" territory avoided (this, of course, is related to useless search effort). In general, the value of a reasonable restrictive principle, which is relevant to a set of moves (reasonable in the sense that there exists, at best, a demonstration, otherwise, good evidence that its use doesn't cut out fruitful territory), is quite appreciable, and the discovery of such a principle is an important objective in the development of improved problem solving procedures.

(b) Terminal Moves

In the procedures $\Pi_r(PC, 1)$, we associate with each state S a value $v_t(S)$, $(t = 0, 1, 2, \ldots)$ which is intended to reflect the state of certainty of the problem solving system at time t about whether "S has a solution" (i.e., the sequent corresponding to S is at the root of a minimal tree proof in $N(SU)$). We assume that $v_t(S)$ can take two possible numerical values, 1 or 1/2. The value $v_t(S) = 1$ corresponds to *certainty* that S has a solution, and $v_t(S) = 1/2$ corresponds to *uncertainty* about whether S has a solution.

Any procedure in the class $\Pi_r(PC, 1)$ has $\bar{\mathfrak{A}}_1$ as a terminal move. This corresponds to a reverse application of the axiom $\mathfrak{A}_1$ on a state. A state that is "recognized" by a terminal move $\bar{\mathfrak{A}}_1$ is a valid conclusive state, and its value is 1.

By increasing the number of terminal moves that establish that a state is conclusive, or "solved," we can often increase the efficiency of a problem solving procedure. A certain amount of search time is eliminated, at the expense of some increase in storage required for the additional terminal moves and time required for attempted matchings at each state. The specific choice of number of terminal moves depends on the storage/timing properties of the machine under consideration.

In advanced procedures of $\Pi_r(PC, 1)$, additional terminal moves are introduced to establish that states having certain forms are conclusive, "solved," states (i.e., their value is 1). While the terminal move $\bar{\mathfrak{A}}_1$ has a weight of 1 (this is also the weight of the "lightest" possible proof tree schema that consists of a conclusive state followed by $\bar{\mathfrak{A}}_1$), the additional terminal moves have weights 2, 3, 4, and they correspond to certain proof tree schemas of weights 2, 3, and 4 respectively. The state forms on which terminal moves of weights 2, 3, and 4 can be applied are as follows*.

$$
\begin{aligned}
w = 2&: (\eta[x] \Rightarrow Dxy),\ (\eta[Cxy] \Rightarrow x)\\
w = 3&: (\eta[x] \Rightarrow DDxx_1x_2),\ (\eta[Cxy] \Rightarrow Dxx_1)\\
&\quad (\eta[Cx_1Cxy] \Rightarrow x)\\
w = 4&: (\eta[x] \Rightarrow DDx_1Dxy\, x_2),\ (\eta[Cxy] \Rightarrow DDxx_1x_2)\\
&\quad (\eta[CCxyx_1] \Rightarrow Dxx_2),\ (\eta[x, y] \Rightarrow Cxy)\\
&\quad (\eta[Cx_1Cx_2Cxy] \Rightarrow x),\ (\eta[x] \Rightarrow Iyx)\\
&\quad (\eta[Iyx, y] \Rightarrow x),\ (\eta[x, Nx] \Rightarrow X).
\end{aligned}
\tag{4.6}
$$

*The notation $\eta[x, y]$ used here indicates that the formulas x, y are among the formulas of the sequence of formulas η; also, x, x_1, x_2, y are any formulas in SU. Commutativity of both D and C should be assumed in reading these forms.

Clearly, each terminal move that applies on one of these state forms can be defined as a tree proof schema of given weight. If $\bar{\mathfrak{A}}_1$ doesn't apply on a given state, S, and if S has one of the two forms that correspond to a terminal move of weight 2, then the tree proof schema that defines the move is minimal for S. If terminal moves of weight 2 do not apply on S, but one of the moves of weight 3 applies, then its defining tree proof schema is minimal for S. Similarly, for terminal moves of weight 4. If none of the terminal moves of weights up to 4 apply on a state, then the minimal weight of a proof of S can be at least 5.

The additional terminal moves just presented cover most of the elementary arguments that are used in closing reasoning gaps during the construction of proofs by the method of suppositions. The trend towards an increase in the number of such terminal moves is a natural line of evolution *within* the class $\Pi_r(PC, 1)$. In deciding whether additional and (more complex) terminal moves are desirable, we face the problem of finding a balance between the "recognition effort" that goes into the direct evaluation of a terminal state on the basis of its form and "search effort" that goes into growing a search tree below the state. This optimization problem deserves serious study.

Another important type of terminal move, which can reduce in a significant manner the search effort for a given problem, amounts to recognizing that the (already established) solution of a subproblem (state) in the search tree is applicable to the subproblem which is now under consideration, and therefore it can be transferred to it. In the procedures $\Pi_r(PC, 1)$ such a terminal move develops early and it can be defined as follows.

"Similarity Recognition" terminal move. For a state $(\eta_1 \Rightarrow x_1)$ which is under consideration in a given search tree, if there is in the search tree a state $(\eta_2 \Rightarrow x_2)$ which is at the root of a minimal proof tree, and is such that (i) $x_2 = x_1$, (ii) all the formulas of η_1 are in η_2, and (iii) the formulas of η_2 that are used as "input formulas" in the tree proof of $(\eta_2 \Rightarrow x_2)$ are in η_1, then the tree proof of $(\eta_2 \Rightarrow x_2)$ can be transferred to (or "hung" below) $(\eta_1 \Rightarrow x_1)$.

The utilization of the known solution of a given problem for a second problem, by recognizing that there exists a certain relationship between the two problems (which is known to justify such a transfer), is one of the fundamental stratagems for efficient problem solving. It is an important objective in the development of improved problem solving procedures to discover relationships between problems that permit the application of this stratagem. The recognition of syntactic symmetries between element of states has already proven greatly useful in the transfer of solutions between states (or, viewed differently, in avoiding repetitions of identical solutions within certain symmetries) in the Geometry Theorem Proving machine [15]. The problem of recognizing symmetries is discussed further in section 6 and also in example 3 of Appendix II.

If no terminal move applies on a terminal state, S, of a search tree at time t, then we assign a value of 1/2 to that state, i.e., $v_t(S) = 1/2$.

(c) Evaluation Procedure

While the notion of weight was introduced in connection with inference rules and proofs (correspondingly with moves and solutions), it is also useful to associate a notion of weight, $w_t(S)$, with a state S at time t. If S is a conclusive state (and $v_{t_1}(S) = 1$), then the weight of S equals the weight of its terminal move for all $t \geqslant t_1$. A state $S = (\eta \Rightarrow x)$ is called *redundant* in a given search tree if there exists a state $S_1 = (\eta_1 \Rightarrow x_1)$ which dominates S in that tree (there exists a tree chain on which both S and S_1 lie, and which is such that S_1 is closer to the tree root than S) and (i) $x = x_1$, (ii) the formulas in η are included in η_1.

If S is a terminal non-redundant state and $v_t(S) = 1/2$, then we associate with S the *minimal possible weight* (at time t) of a proof tree for S; thus, if all terminal moves up to $w = 4$ were considered at time t and they were found not applicable on S, then $w_t(S) = 5$.

If S is a terminal redundant state, then we associate with S an *upper bound of possible weight* of a proof tree for S (say, a large number W which exceeds any conceivable weight of the proof trees under consideration).

The last two weight assignments were chosen with the purpose of formulating simple decision functions (where state weights enter as arguments) that would control the order in which the problem solving activity could be efficiently carried out in the search tree. Before discussing these decision functions, we shall first present the basis for assigning values and weights to non-terminal states.

Consider a non-terminal state S, on which the relevant moves $M_1, M_2, \ldots, M_n$ apply; and further, consider the set of descendant states from each move M_i ($i = 1, \ldots, n$); this set contains 1, 2, or 3 states. Suppose that all the descendant states reached from S via the application of the relevant moves have values and weights associated with them. If a descendant state is terminal, then its v and w are assigned in accordance with our previous discussion. If it is non-terminal, then we assume that it has been assigned v, w according to the scheme now under discussion. The *conditional weight* of S, given a relevant move M_i, is as follows:

$$w_t(S/M_i) = w(M_i) + \sum_{j=1}^{b_i} w_t(S_j), \tag{4.7}$$

where S_j denotes a descendant state resulting from the application of M_i on S and b_i denotes the number of such states. From the conditional weights, the weight of S is determined as follows:

$$w_t(S) = \operatorname*{Min}_{i=1}^{n} w_t(S/M_i). \tag{4.8}$$

The *conditional value* of S, given a relevant move M_i, is as follows:

$$v_t(S/M_i) = \operatorname*{Min}_{j=1}^{b_i} v_t(S_j). \tag{4.9}$$

Consider the subset of relevant moves of S such that $w_t(S) = w_t(S/M_i)$. Let us denote the moves in this subset by $M_1, M_2, \ldots, M_m$.

We can now define the value of the non-terminal state S as follows:

$$v_t(S) = \operatorname*{Max}_{i=1}^{m} v_t(S/M_i). \tag{4.10}$$

Thus, the assignment of w and v to a non-terminal state at a given time reflects the current state of knowledge about the maximum value of proof trees of minimum possible weight that "hang" below S. We now have completed the specification of the main elements in the evaluation subprocedures of $\Pi_r(PC, 1)$.

(d) Generation Procedure: Attention Control

Suppose that a burst of growth takes place at time t from a set of terminal states of a search tree that is rooted at a state S_0. At each state of this set, the relevant moves are applied and new descendant states are created. Clearly, just before the growth takes place, we have $v_t\,(S_0) = 1/2$; otherwise, the problem would have been solved, no further growth would have been necessary, and the procedure would have terminated. The new states created by the burst of growth are evaluated as new terminal states, and these evaluations initiate a wave of re-evaluations (in accordance with the rules (4.8) and (4.10)) which moves backwards over the search tree up to S_0. If $v_{t+1}(S_0) = 1$, then the proof construction problem is solved; a minimal proof of weight $w_{t+1}(S_0)$ exists, and it can be readily extracted from the search tree. If $v_{t+1}(S_0) = 1/2$, then further search (tree growth) is necessary. The efficiency of the proof construction procedure depends on the appropriate sequencing of the decisions about *where to grow next.* In view of our basis for evaluations, it is possible to form at the start of each generation cycle (by local decisions in the search tree) an *attention path* from S_0 to the set of terminal states of the search tree from which the next burst of growth is to take place. The formation of the attention path (which has the form of a subtree of the search tree and is rooted at S_0—sometimes it is just a chain—) is controlled as follows.

If the attention path has reached a non-terminal state, S, that has a value 1/2 and a certain weight $w_t(S)$, then attention is directed to a move M among those applied at S, for which $w_t(S) = w_t(S/M)$. If more than one move exists that satisfies this condition, then the choice is random among these moves. Suppose now that the attention path has reached a non-terminal move node in the search tree. If the move has a single descendant state, then attention moves to that state. If the move has more-than-one descendant states, then attention is directed to *each one* of these descendant states that have value 1/2. This is the source of branching for the attention path; it results, in general, in a burst of growth distributed over more-than-one (sometimes quite remote) terminal states.

The procedure just described has the effect of growing the search tree in width, layer by layer, scanning in a controlled manner the space of potential proof trees from the simplest to the more complex. It is easy to verify that this systematic procedure will find a minimal proof, if one exists, without spending any appreciable effort in searching for proofs of higher complexity than the minimal. Since there is no guarantee that this procedure will terminate if a candidate formula is not a theorem, then an upper limit on search effort has to be specified in order to stop the machine in reasonable time. This limit can be based on an estimate of the complexity of the expected proof, or on an arbitrary bound on proof complexity.

D. IMPROVEMENTS AND VARIATIONS OF $\Pi_r(PC, 1)$

One way of further augmenting the efficiency of the procedures in $\Pi_r(PC, 1)$ is to utilize the experience with proof construction in order to formulate additional restrictive principles on the relevant moves. If such principles are introduced without proof that they do not affect the completeness and minimality of the proof procedures, i.e., the problem solving capacity of the procedures, then the heuristic status of the procedures is further accentuated. However, in view of the great labour of search needed by a procedure from $\Pi_r(PC, 1)$, the gain in expected efficiency obtained by introducing reasonable move restrictions justifies, in many cases, the possible loss of capacity. This type of judgment, where a capacity *versus* efficiency balance is attempted, occurs quite frequently in the development of problem solving procedures. It is not uncommon that, in view of the necessity to increase problem solving efficiency, initial goals for problem solving capacity are re-adjusted (usually restricted). In such cases, we are confronted with the interesting theoretical question of characterizing the problem solving capacity that corresponds to a given procedure (which is proposed on the basis of efficiency ideas).

A useful restrictive principle on relevant moves at the root state $(\Rightarrow T)$ is suggested from our experimentation, and it is as follows.

> *Move Restriction.* A $\bar{N}_{el}^{N}$ move is not relevant for a root state $(\Rightarrow T)$, where $T = Ixy$ and x, y are arbitrary formulas of SU.

It should be noted that the usefulness of this restriction derives from the fact that many interesting theorems in logic are in the form of implications, and the restriction allows us to start all proofs for such theorems with a single opening move, i.e., $\bar{I}_{in}^{N}$. The move restriction just introduced does not affect proof completeness. This can be seen from the discussion of completeness in the second stage of development of the procedures. A proof that this restriction does not affect proof minimality has to proceed by analysis of *proof forms*, as in the case of the move restrictions induced by the conjecture $\mathfrak{C}_\sigma$. This is a difficult task that is still to be performed. However, it seems highly plausible that it can be carried out within the framework of the third stage of development of the proof procedures. We denote by μ^{**} the set of relevant moves which is restricted both by $\mathfrak{C}_\sigma$ and by the principle just introduced.

In order to achieve flexible control over both the capacity and the efficiency features of the procedures, we can try to establish a *simple order* (which can be determined empirically) among the relevant moves, and we can then use it to formulate a modification of the generation subprocedure discussed previously, in such a way that subsets of relevant moves will be used in different stages of growth. The subsets are chosen so that high order relevant moves are considered first (and the tree growth proceeds rapidly in depth on basis of them), and they are subsequently followed by lower order relevant moves, under the overall control of a search strategy which is dependent on an allocation of computational effort to the different stages of growth. The desired effect of the strategy on the growth dynamics of the search tree is: (i) to first grow one tree (or a small number of trees) in depth in such a way that, at each decision point (state), one or two strong moves are taken until a proof is obtained or a given maximum proof weight is exceeded; (ii) to go back to a state closest to the root state (in the beginning, to the root state itself) to consider there the next one or two untried strongest moves and to proceed by a selective deep growth until a better proof is obtained (i.e., a proof of lesser weight than the current best proof) or until a weight is reached which exceeds that of the previous best proof or the given maximum weight; (iii) to repeat the cycles of deep growth by considering in succession the untried relevant moves of depth 1 followed by their strongest continuations, then by considering all the untried relevant moves of depth 2, etc. The desired objective of such a strategy is to maximize the quality of the proof obtained for any amount of search effort spent. First, a reasonably quasi-minimal proof is sought rapidly and then by successive investments of effort the solution space is selectively searched in order either to improve the quality of the current best proof or eventually to demonstrate (if we are willing to pay the price of the total search) that the current best proof is indeed minimal. Such a demonstration is conditioned, of course, by the restrictive assumptions on the relevant moves. The strategy which we have just outlined is readily implemented within the general framework of reduction procedures. An *available effort* variable can be associated with each state (together with v and w), and it can be used within the attention control mechanism to determine the extent of growth from a state that lies on the attention path. Such an implementation is described in [2]. More work remains to be done in order to establish a rationale for effort allocation that produces an optimal search pattern in a *strategy of successive improvements* of the type discussed here. Clearly, the success of this strategy depends primarily on the choice of an appropriate ordering relation for the relevant moves. In a machine-aided evolution of procedures $\Pi_r(PC, 1)$, we expect a specific order assignment to evolve via statistical learning techniques (such as discussed in [16]). Since our primary objective in this paper is to examine a sequence of stages in the development of problem solving procedures, with the intention of making only preliminary remarks about the processes of transitions between stages, we will not discuss further here the problem of evolving an order assignment for the relevant moves. During our experimentation with the present procedures, the following ordering scheme for move types from the set μ^{**} has emerged:

(4.11) $$\langle \bar{C}^N_{el}, \bar{I}^N_{el}, \bar{D}^N_{el}, \bar{C}^N_{in}, \bar{D}^N_{in}, \bar{I}^N_{in}, \bar{X}^N_{in}, \bar{N}^N_{in}, \bar{X}^N_{el}, \bar{N}^N_{el} \rangle.$$

Within each move type, it is possible to impose further order according to the characteristics of the specific formulas that participate in the move application. For example, given a state ($\eta \Rightarrow x$), the move $\bar{D}^N_{el}$ is first tried for each disjunction Dyz (for any y, z) which is among the formulas of η, and then for each disjunction in subformulas of η. In each case, first are tried disjunctions where one of the disjuncts is x or Cxy (or Cyx), then those with a disjunct Iyx (for any y), then the rest. The $\bar{X}_{in}^{N}$ moves, where two contradictory formulas have to be chosen as goals of a *reductio* argument, are the most difficult to order at this stage of development of the procedures. The order suggested by our experiments is as follows: given a state $(\eta \Rightarrow X)$, consider first, as one of the contradictory formulas (this, of course, fixes a candidate pair), any formula of the form Nx in η (for an x that does not start with a N), and among them consider first those with longer x; then consider subformulas of the form Nx in η, and here again consider those with longer x first; then consider the remaining formulas and subformulas in η.

It becomes evident in the course of attempting to formulate a good ordering relation among relevant moves, that no fixed ordering is completely appropriate; one needs also to consider the *stage of the overall argument.* It appears that in each proof there are "molecular arguments" that are usually composed from a number of atomic arguments of the type that each single move represents. This has implications on the move variety to be considered from each state. Therefore, it is important to search for such molecular arguments in the solution space and to establish accordingly move orderings and search strategies. Also, it appears (from experimentation) that there exist some conditions that control the utilization of information formulas in moves on basis of the history of previous utilization of these formulas. By finding such conditions, we can hope to further reduce the number of relevant moves that apply on successive states of search trees.

The search for "molecular arguments" and for restrictive conditions on the utilization of information formulas, and also the desire to formulate better successive improvements strategies lead to the second stage of development of the proof procedures, i.e., to the procedures $\Pi_r(PC, 2)$. We shall discuss these procedures next.

5. PROOF CONSTRUCTION PROCEDURES: THE SECOND STAGE

A. THE NATURAL DECISION SYSTEM $\mathfrak{N}(SU)$

We shall now extend the natural inference system $N(SU)$ into a system $\mathfrak{N}(SU)$ that is better suited for the formulation of the new features in the procedures $\Pi_r(PC, 2)$; we call $\mathfrak{N}(SU)$ a *natural decision system.*

The formulas in $\mathfrak{N}(SU)$ are sequents of the form $(\eta \Rightarrow x)$, as in $N(SU)$. However, in $\mathfrak{N}(SU)$ we distinguish in each sequent a set of formulas that we call *logically irredundant formulas.* The set of logically irredundant formulas of a sequent $(\eta \Rightarrow x)$ includes

(i) the goal formulas x of the sequent, and
(ii) a subset of the formulas η of the sequent, that we call *logically irredundant information formulas.*

In the sequents of $\mathfrak{N}(SU)$, a logically irredundant information formula is especially tagged (in this paper, we mark such formulas by a lower bar, e.g., if x is a logically irredundant information formula we write $\underline{x}$).

(a) Logically Irredundant Information Formulas

We shall characterize the set of logically irredundant information formulas (henceforth l.i. information formulas for brevity) of a sequent in terms of certain logical properties of sequents.

By our interpretation of the double arrows in a sequent $(\eta \Rightarrow x)$, i.e., "x is deducible from (or is a logical consequence of) the formulas in η," and by the deduction theorem,* we can make the following assertion:

(5.1) A sequent $(\eta \Rightarrow x)$ is valid if and only if $[C(\eta_t) \rightarrow x]$ is a theorem in logic.

In (5.1), η_t denotes the total (non-ordered) set of information formulas that are contained in the sequence η, $C(\eta_t)$ denotes the conjunction of all the formulas in η_t, and the arrow, $\rightarrow$, denotes implication.†

Let η_e, $\eta_{\bar{e}}$ be two non-intersecting subsets of η_t such that

$$\eta_e \cup \eta_{\bar{e}} = \eta_t. \tag{5.2}$$

Clearly, we have

$$C(\eta_t) \equiv C(\eta_e) \mathbin{\&} C(\eta_{\bar{e}}). \tag{5.3}$$

A set η_e is called a set of *eliminable information formulas* when the following equivalence holds:

$$(C(\eta_t) \rightarrow x) \equiv (C(\eta_{\bar{e}}) \rightarrow x). \tag{5.4}$$

This means that it is possible to eliminate the information formulas in η_e from the sequent without disturbing its validity. In other words, the formulas of η_e are not necessary for deciding about the validity of the sequent.

Any subset η_e of η_t, to which no new member can be added without violating the equivalence (5.4), defines a complementary set $\eta_{\bar{e}}$ which is a minimal set of non-eliminable formulas. The set of l.i. information formulas that we are using in $\mathfrak{N}(SU)$ is one of these minimal sets of non-eliminable formulas. It is chosen so that it excludes longer formulas in favour of shorter formulas, whenever possible. This choice is clarified below in our discussion of the rules of decision.

The following two theorems are useful for testing the eliminability of information formulas, and thus for identifying the l.i. information formulas of a sequent.

*For a discussion of this theorem see Rosenbloom [17], chap. II, sec. 3.

†In order to distinguish between the object language of SU and the meta-language in which we are discussing proofs of theorems in SU and in its associated systems $N(SU)$, $\mathfrak{N}(SU)$, we are using in the meta-language conventional infix logical notation and the symbols $\rightarrow$, $\vee$, &, $\equiv$ for implication, disjunction, conjunction, and equivalence respectively; also we use $C(\alpha)$ for the conjunction of the formulas that are members of a set α.

THEOREM 5.1. *Given a sequent* $(\eta \Rightarrow x)$, *a subset* η_e *of its information formulas is eliminable if and only if the following implication holds*:

$$(C(\eta_e) \rightarrow x) \rightarrow (C(\eta_{\bar{e}}) \rightarrow x). \tag{5.5}$$

Proof. From (5.3) we have

$$[C(\eta_t) \rightarrow x] \equiv [(C(\eta_e) \;\&\; C(\eta_{\bar{e}})) \rightarrow x]. \tag{5.6}$$

Hence we can state the eliminability condition (5.4) as follows:

$$[(C(\eta_e) \;\&\; C(\eta_{\bar{e}})) \rightarrow x] \equiv [C(\eta_{\bar{e}}) \rightarrow x]. \tag{5.7}$$

To prove the theorem, we have to prove that the propositional formula (5.7) which expresses the eliminability condition is equivalent to the propositional formula (5.5) which is stated in the theorem. This proof can be obtained in the system $N(SU)$ (via one of our procedures $\Pi_r(PC, 1)$ if we introduce the definition for logical equivalence, i.e.,

$$(p \equiv q) \overset{\text{def}}{=} (p \rightarrow q) \;\&\; (q \rightarrow p); \tag{5.8}$$

however, for convenience, we pursue here a conventional form of proof. In view of the theorem:

$$[(p \;\&\; q) \rightarrow x] \equiv [(p \rightarrow x) \vee (q \rightarrow x)], \tag{5.9}$$

the equivalence in (5.7) is equivalent to the following equivalence:

$$[(C(\eta_e) \rightarrow x) \vee (C(\eta_{\bar{e}}) \rightarrow x)] \equiv [C(\eta_{\bar{e}}) \rightarrow x]. \tag{5.10}$$

Now, in view of the theorem:

$$[(p \vee q) \equiv q] \equiv (p \rightarrow q), \tag{5.11}$$

the equivalence in (5.10) is equivalent to the following implication:

$$(C(\eta_e) \rightarrow x) \rightarrow (C(\eta_{\bar{e}}) \rightarrow x),$$

which proves the theorem.

Before proceeding to the next theorem, we shall introduce the notion of a *set of sequents* and the validity of such a set. A set Σ of n sequents,

$$\Sigma = \{(\eta_1 \Rightarrow x_1), (\eta_2 \Rightarrow x_2), \ldots, (\eta_n \Rightarrow x_n)\}, \tag{5.12}$$

is valid when x_1 is deducible from the formulas in η_1, *and* x_2 is deducible from the formulas in η_2, *and* . . . , for all the n sequents in the set. By following the same argument as in the case of a single sequent (see (5.1)), we can make the following assertion for a set of sequents.

> A set of sequents Σ (as given in (5.12) above) is valid if and only if the propositional formula
>
> $$[C(\eta_{1,t}) \rightarrow x_1] \;\&\; [C(\eta_{2,t}) \rightarrow x_2] \;\&\; \ldots \;\&\; [C(\eta_{n,t}) \rightarrow x_n] \tag{5.13}$$
>
> (which we can also denote by $\underset{i=1}{\overset{n}{\&}} [C(\eta_{i,t}) \rightarrow x_i]$)
>
> is a theorem in logic.

Consider the sequent $(\eta_1 \Rightarrow x_1)$ in Σ. Using similar notation and definitions as in the case of a single sequent, we can assert that a subset $\eta_{1,e}$ of $\eta_{1,t}$ is eliminable when the following equivalence holds:

$$\underset{i=1}{\overset{n}{\&}} [C(\eta_{i,t}) \rightarrow x_i] \equiv \left\{ [C(\eta_{1,\bar{e}}) \rightarrow x_1] \,\&\left[\underset{i=2}{\overset{n}{\&}} [C(\eta_{i,t}) \rightarrow x_i] \right]\right\}. \tag{5.14}$$

THEOREM 5.2. *Given the set of sequents* $\Sigma = \{(\eta_i \Rightarrow x_i)\}$, $1 \leqslant i \leqslant n$, *the subset* $\eta_{1,e}$ *of information formulas in the sequent* $(\eta_1 \Rightarrow x_1)$ *of* Σ *is eliminable if and only if the following implication holds*:

$$\left\{ [C(\eta_{1,e}) \rightarrow x_1] \,\&\left[\underset{i=2}{\overset{n}{\&}} [C(\eta_{i,t}) \rightarrow x_i] \right]\right\} \rightarrow [C(\eta_{1,\bar{e}}) \rightarrow x_1]. \tag{5.15}$$

Proof. By proceeding as in the proof of theorem 5.1, we can write the eliminability condition (5.14) as follows:

$$\left\{ [C(\eta_{1,e}) \rightarrow x_1] \,\&\left[\underset{i=2}{\overset{n}{\&}} C(\eta_{i,t}) \rightarrow x_i \right]\right\} \vee \left\{ [C(\eta_{1,\bar{e}}) \rightarrow x_1] \,\& \right.$$
$$\left.\left[\underset{i=2}{\overset{n}{\&}} C(\eta_{i,t}) \rightarrow x_i \right]\right\} \equiv \left\{ [C(\eta_{1,\bar{e}}) \rightarrow x_1] \,\&\left[\underset{i=2}{\overset{n}{\&}} [C(\eta_{i,t}) \rightarrow x_i] \right]\right\}. \tag{5.16}$$

Now, in view of the previously used theorem (5.11), and also the theorem

$$(p \,\&\, q) \rightarrow (r \,\&\, q) \equiv (p \,\&\, q) \rightarrow r, \tag{5.17}$$

we can write the eliminability condition (5.16) as stated in (5.15), which proves the theorem. Note that here, also, as in theorem (5.1), the theorem can be proved by our procedures $\Pi_r(PC, 1)$ after it is appropriately stated.

The notion of l.i. information formulas provides a basis for a more restricted, selective, utilization of information formulas in the process of proof construction. Furthermore, it enables the formulation of an axiom schema for refutation in $\mathfrak{N}(SU)$, which in turn suggests the development of a set of "molecular arguments" and a decision procedure for the propositional calculus.

(b) Axiom Schemata

The axiom schemata of $\mathfrak{N}(SU)$ are as follows.

For validation

$\mathfrak{A}_1$: Same as $\mathfrak{A}_1$ in $N(SU)$. The weight associated with $\mathfrak{A}_1$ is 1.

$\mathfrak{A}_{1,1}$: A sequent $(\eta \Rightarrow x)$ is valid if two contradictory formulas y, Ny are among the formulas of η. The weight associated with $\mathfrak{A}_{1,1}$ is 5; this is the weight of a tree proof of $(\eta \Rightarrow x)$ in $N(SU)$, via a X_{el}^N inference followed by a X_{in}^N inference (that takes y, Ny as contradictory goal formulas), and terminated by two applications of $\mathfrak{A}_1$.

For refutation

$\mathfrak{A}_0$: A sequent $(\eta \Rightarrow x)$ is not valid if its l.i. formulas are all atomic,* and if

*An atomic formula is a propositional variable or the negation of a propositional variable.

neither of the axiom schemata $\mathfrak{A}_1$ or $\mathfrak{A}_{1,1}$ apply on its l.i. formulas. In other words, x is not logically deducible from the formulas in η if both x and the l.i. information formulas in η are atomic, and if x is not identical with any of the l.i. information formulas or if no contradictory formulas are included among the l.i. information formulas.

(c) Rules of Decision

The axiom schema for refutation suggests that a decision procedure for the propositional calculus can be built on the basis of a method for reducing a sequent $(\Rightarrow x)$, where x is any formula in SU, to a logically equivalent set of sequents in each of which the l.i. information formulas are all atomic. This idea leads to the formulation of the following general requirement for the rules of decision of $\mathfrak{R}(SU)$: the rules of decision should provide the links for the establishment of the desired overall transition between a sequent $(\Rightarrow x)$ and a logically equivalent set of sequents, the l.i. formulas of which are all atomic.

There is no unique way in which this general requirement can be satisfied by a set of rules of decision. Our specific approach to the formulation of rules of decision for $\mathfrak{R}(SU)$ is guided by a preference for *functional homogeneity* in the specification of the rules. This means that we seek a specification of identical functional requirements for each rule such that the general requirement which is imposed on the set of rules is satisfied. This approach has the virtue of making the process of rule formulation relatively simple. We are imposing two conditions on each rule of decision; one is logical and the other is morphological.

Our notion of a rule of decision in $\mathfrak{R}(SU)$ is an extension of the notion of a rule of inference. A rule of decision establishes the logical validity (valid or not) of a consequent sequent on basis of the validity of a set of antecedent sequents. Thus, the logical condition imposed on every rule of decision is the following:

(5.18) For any rule of decision that has a consequent sequent S and a set of antecedent sequent Σ, the sequent S is valid if and only if Σ is valid.

This logical property is in accordance with the desired property of logical equivalence in the overall transition between sequents. We now turn to the general morphological property of our rules of decision.

Consider a rule of decision D which effects a transition between a set of antecedent sequents $\Sigma = \{(\eta_i \Rightarrow x_i)\}, 1 \leqslant i \leqslant n$, and a consequent sequent $S = (\eta \Rightarrow x)$ (in our rules of decision, n is 1 or 2). We assume that the applicability of D is determined by one of the formulas of S, which we call the *key formula* of D. Only a non-atomic l.i. formula of S can assume the role of a key formula. We can represent the general form of a key formula of D as Kyz or Ky, where $K \epsilon \{C, D, I, NC, ND, NI\}$, or $K = NN$ respectively, and y, z are arbitrary formulas in SU. We call K the *characteristic operator* of the key formula. We define the *set of l.i. formulas of* Σ to be the union of sets of l.i. formulas of its member sequents. We can now specify the morphological condition that we impose on our rules of decision.

(5.19) For any rule of decision D that has a consequent sequent S, a key formula in one of the forms Kyz or Ky in S, and a set of antecedent sequents Σ, the set of l.i. formulas of Σ equals the set of l.i. formulas of S with the following exception: if the key formula is Kyz, then Kyz does not appear in the set of l.i. formulas of Σ, but it is replaced by formulas y, z, negated or unnegated, and similarly if the key formula is Ky, it is replaced by y.

Thus the gross morphological effect of a rule of decision that carries out a transition from a sequent to a logically equivalent set of sequents is *to strip out one main connective* in the set of l.i. formulas. This way, a step is taken toward the attainment of a desired set of l.i. formulas that is free of propositional connectives and, hence, it can be conclusively handled via the refutation axiom. The more specific effect of a rule of decision (the precise distribution of the fragments of the key formula over the antecedent sequents) depends both on the general *location* of its key formula in the given sequent (i.e., "information" or "goal" side) and on the general *form* of the key formula (i.e., its characteristic operator). Therefore, we classify rules of decision by the characteristic operators and by the general location of their key formulas. This leads to the specification of the rules of decision in terms of 14 main transition schemata between sequents (since we have 7 characteristic operators and 2 general locations for key formulas).

A transition schema for a rule of decision in $\mathfrak{N}(SU)$ is constructed with transition schemata of rules of inference of $N(SU)$ (these are given in Table (4.1)). The overall structure of the transition has the form of a tree where the consequent sequent is at the root, and the antecedent sequents are at tree terminals. The weight of a rule of decision is the sum of weights of its constituent rules of inference. Thus, a rule of decision assumes the function of a "molecular argument" that has a specific atomic structure in the form of a configuration of rules of inference from $N(SU)$. Furthermore, the application of such a molecular argument in reverse (from the consequent sequent to the antecedent sequents) always guarantees to advance the overall argument towards a decisive resolution.

In Appendix I, we give the detailed structural definitions of the rules of decision of $\mathfrak{N}(SU)$ in terms of the rules of inference of $N(SU)$. We have chosen these definitions so that they satisfy the required logical and morphological properties described above, with the least possible weight.

A significant step in the transition from the first stage to the second stage of development of our proof procedures is the formation of such a set of "molecular arguments" that can function as rules of decision. It would be a challenging task to develop procedures that can automatically carry out such a formation process. It appears that this task would be feasible if the language of formation is specified (tree combinations of rules of inference), if the logical and morphological conditions to be satisfied by each rule are clearly defined, if a basis for identifying sets of l.i. information formulas exist (the theorems 5.1 and 5.2, for example), and if a good proof procedure for the propositional calculus is available (the best versions of $\Pi_r(PC, 1)$ can be used for this bootstrapping activity). The puzzling question for artificial intelligence is how to arrive at a *formulation* of this formation problem without the explicit aid of a human designer, especially how to evolve by machine the concepts of l.i. formulas and the overall decision schema that led to the specification of the logical and morphological conditions for the rules of decision! Another point is worth noting here: if we assume that somehow the formation problem outlined above is appropriately specified, then a problem solving system designed to cope with it will require an efficient proof procedure in logic as one of its main components. As we have mentioned before in section 3.A, we feel that such a requirement will appear in an increasing number of future problem solving systems.

(d) Decision trees

A *decision tree* in the system $\mathfrak{N}(SU)$ is a labeled tree that is made of specified transition schemata for decision rules (these are trees as shown in Appendix I, with the sequents specified explicitly in terms of formulas), and it terminates exclusively with axiom links (to validation or refutation). More specifically, the root sequent of a decision tree is the consequent sequent of a single specified transition tree for a decision rule; each antecedent sequent of this transition tree is either the consequent sequent of a single specified transition tree for a decision rule, or it is linked to an axiom; and so on, until all the antecedent sequents of transition trees are linked to axioms. If all the terminating axioms of a decision tree are validating axioms, then the decision tree is a *proof tree* in $\mathfrak{N}(SU)$, and its root sequent is a *theorem of* $\mathfrak{N}(SU)$. If at least one of the terminating axioms of a decision tree is a refutation axiom, then the decision tree is a *refutation tree* relative to the root sequent. If a theorem of $\mathfrak{N}(SU)$ has the form $(\Rightarrow x)$, then the propositional formula x is a theorem in logic, and the proof tree in $\mathfrak{N}(SU)$ is its proof. This *consistency property* is based on our definitions of the rules of decision in terms of the rules of inference of $N(SU)$ (which are given in Appendix I) and on the consistency of the system $N(SU)$. Similarly, it is easy to verify from the properties of the decision rules that if $(\Rightarrow x)$ is at the root of a refutation tree, then x is not a theorem in logic. Note that a proof tree in $\mathfrak{N}(SU)$ is also a proof tree in $N(SU)$, but the converse is not always true. This means that, in general, we cannot build in $\mathfrak{N}(SU)$ all the different proofs of a theorem that can be built in $N(SU)$. Specifically, in most cases, we cannot build in $\mathfrak{N}(SU)$ the minimal proof of a theorem (in the sense of minimality that we have discussed in 4.B).

As an example, we show in (5.20) the proof of Pierce's law in the system $\mathfrak{N}(SU)$. The double parentheses indicate *boundary sequents* of decision rules (antecedent or consequent sequents of such rules). Also, using the same conventions as in Appendix I, we have not specified completely all the sequents that are not boundary sequents of decision rules, and we have merged in several places N^N_{el} and X^N_{in} inferences that occur consecutively. (The minimal proof of this theorem was given previously in sequential form in (3.2), and in tree form—in the system $N(SU)$—in (4.4).)

(5.20)

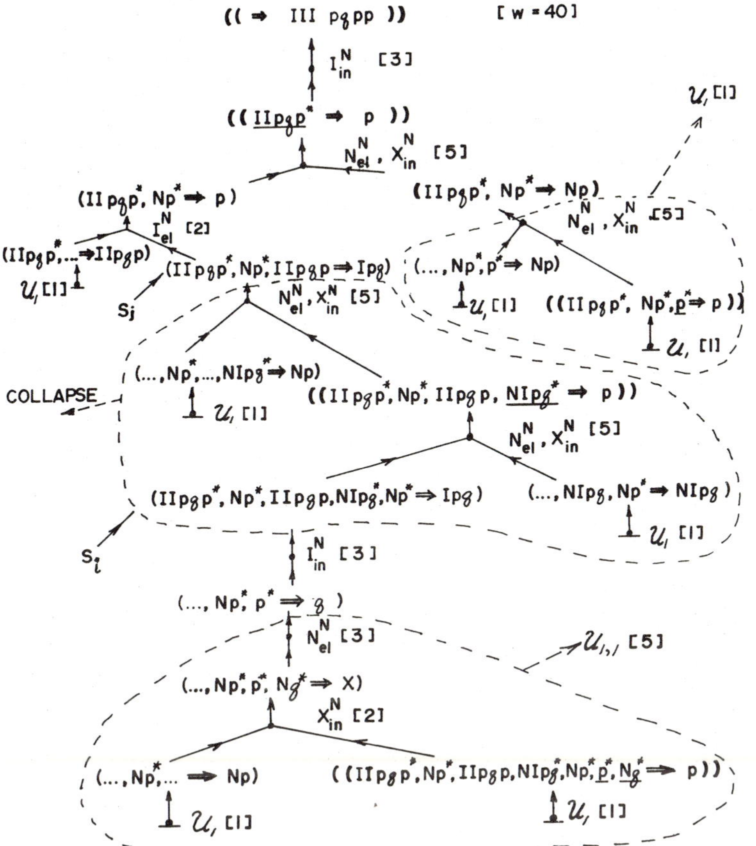

Note that the weight of this proof is 40, which is double the weight of the minimal proof. An examination of the proof will show that it includes a number of obviously circuitous and unnecessary arguments. This is the result of the generality of treatment provided by the "molecular arguments." The situation is analogous to many design situations that are based on the flexible utilization of a set of large *modules*, where a certain amount of redundancy at the fine level cannot be avoided. Following the same analogy, the advantages of modular design are also the principal advantages of proof construction in $\mathfrak{N}(SU)$. Once an appropriate set of modules (in our case, the decision rules) is designed, then the number of design decisions in any proof construction situation is strongly reduced. This is where we gain sharply in the efficiency of the proof procedure.

We will come back to this point later. As for the minimality of a proof, our approach will be to take a proof in $\mathfrak{N}(SU)$ as a point of departure for a simplification procedure which will attempt to eliminate redundant arguments by considering the detailed structure of the proposed proof. This approach will be presented following a discussion of proof construction in the system $\mathfrak{N}(SU)$.

B. THE SYSTEM $\mathfrak{N}^*(SU)$

It is useful to have abbreviated, functional, definitions of the rules of decisions that reflect the transformations that these rules effect on relevant formulas of sequents. To this end, we can use abbreviated representations of sequent forms, where only sets of l.i. formulas and their location (in the "information" or "goal" side) are indicated. We call such representations *characteristic representations* of sequent forms. A rule of decision can be now specified as a transition schema between the characteristic representation of a consequent sequent form and the set of characteristic representations of antecedent sequent forms. We call such a transition schema, a *characteristic transition schema* for a rule of decision. A characteristic transition schema can be represented as a tree (in a similar way to the one used for the representation of a rule of inference in Table 4.1) or in a linear form. In Table 5.1, we give the linear form representation of the characteristic transition schemata for the rules of decision of $\mathfrak{N}(SU)$. In this table, the leftmost string in parentheses is a name for a rule of decision. The name consists of a pair, where the first element identifies the characteristic operator of the key formula of the rule, and the second element indicates the location of the key formula (right or left of the double arrow in the consequent sequent). As an example, the name of the first rule in Table 5.1 is (C, R) and it indicates that the key formula is in the right side of the consequent sequent and it is a conjunction C. In the table, a name is followed in some cases by a specification of a version, (a) or (b), of the rule. Each version is associated with a different structure of the transition schema for a rule (see Appendix I), and in the case of the rules (D, R), (NC, R), it is associated with different characteristic transition schemas. The left element of each characteristic transition schema in the table corresponds to the consequent sequent and the right element to the set of antecedent sequents. The number associated with the transition arrow denotes the weight of the rule of decision. Since characteristic representations of sequents contain only l.i. formulas, we do not underline l.i. information formulas in such representations.

It is clear from the above discussion of abbreviated representations that we can project the proof problem from $\mathfrak{N}(SU)$ into a more abstract decision system with the following properties: the formulas of the system are characteristic representations of sequents, its rules are given by the characteristic transition schemata of Table 5.1, and its axiom schemata are identical† to those of $\mathfrak{N}(SU)$.

†Since all the formulas in characteristic representations of sequents are l.i., then all the formulas enter in the test of applicability for $\mathfrak{A}_0$. The validation axioms $\mathfrak{A}_1$, $\mathfrak{A}_{1,1}$ apply over all formulas, as in $\mathfrak{N}(SU)$, but note that their application is implicitly limited here over the l.i. formulas only.

TABLE 5.1

CHARACTERISTIC TRANSITION SCHEMATA FOR THE RULES OF DECISION OF $\mathfrak{N}(SU)$

(Each transition is between characteristic representations of sequent forms; the variables x, x_1, x_2, x_3 stand for any formula in SU; η_r, θ_r stand for sets [possibly empty] of relevant information formulas.)

(C, R)	(a), (b)	$:(\eta_r \Rightarrow Cx_1x_2) \overset{[2]}{\longleftrightarrow} \{(\eta_r \Rightarrow x_1), (\eta_r \Rightarrow x_2)\}$
(C, L)		$:(\theta_r, Cx_1x_2 \Rightarrow x_3) \overset{[20]}{\longleftrightarrow} \{(\theta_r, x_1, x_2 \Rightarrow x_3)\}$
(D, R)	(a)	$:(\eta_r \Rightarrow Dx_1x_2) \overset{[14]}{\longleftrightarrow} \{(\eta_r, Nx_1 \Rightarrow x_2)\}$
	(b)	$:(\eta_r \Rightarrow Dx_1x_2) \overset{[14]}{\longleftrightarrow} \{(\eta_r, Nx_2 \Rightarrow x_1)\}$
(D, L)		$:(\theta_r, Dx_1x_2 \Rightarrow x_3) \overset{[8]}{\longleftrightarrow} \{(\theta_r, x_1 \Rightarrow x_3), (\theta_r, x_2 \Rightarrow x_3)\}$
(I, R)		$:(\eta_r \Rightarrow Ix_1x_2) \overset{[3]}{\longleftrightarrow} \{(\eta_r, x_1 \Rightarrow x_2)\}$
(I, L)		$:(\theta_r, Ix_1x_2 \Rightarrow x_3) \overset{[20]}{\longleftrightarrow} \{(\theta_r, Nx_1 \Rightarrow x_3), (\theta_r, x_2 \Rightarrow x_3)\}$
(NC, R)	(a)	$:(\eta_r \Rightarrow NCx_1x_2) \overset{[14]}{\longleftrightarrow} \{(\eta_r, x_1 \Rightarrow Nx_2)\}$
	(b)	$:(\eta_r \Rightarrow NCx_1x_2) \overset{[14]}{\longleftrightarrow} \{(\eta_r, x_2 \Rightarrow Nx_1)\}$
(NC, L)	(a), (b)	$:(\theta_r, NCx_1x_2 \Rightarrow x_3) \overset{[20]}{\longleftrightarrow} \{(\theta_r, Nx_1 \Rightarrow x_3), (\theta_r, Nx_2 \Rightarrow x_3)\}$
(ND, R)		$:(\eta_r \Rightarrow NDx_1x_2) \overset{[17]}{\longleftrightarrow} \{(\eta_r \Rightarrow Nx_1), (\eta_r \Rightarrow Nx_2)\}$
(ND, L)	(a), (b)	$:(\theta_r, NDx_1x_2 \Rightarrow x_3) \overset{[20]}{\longleftrightarrow} \{(\theta_r, Nx_1, Nx_2 \Rightarrow x_3)\}$
(NI, R)		$:(\eta_r \Rightarrow NIx_1x_2) \overset{[8]}{\longleftrightarrow} \{(\eta_r \Rightarrow x_1), (\eta_r \Rightarrow Nx_2)\}$
(NI, L)		$:(\theta_r, NIx_1x_2 \Rightarrow x_3) \overset{[15]}{\longleftrightarrow} \{(\theta_r, x_1, Nx_2 \Rightarrow x_3)\}$
(NN, R)		$:(\eta_r \Rightarrow NNx) \overset{[6]}{\longleftrightarrow} \{(\eta_r \Rightarrow x)\}$
(NN, L)		$:(\theta_r, NNx_1 \Rightarrow x_2) \overset{[12]}{\longleftrightarrow} \{(\theta_r, x_1 \Rightarrow x_2)\}$

A decision tree in $\mathfrak{N}^*(SU)$ has a unique decision tree in $\mathfrak{N}(SU)$ that corresponds to it. The correspondence is readily established via the definitions of rules of decision given in Appendix I. As an example, we show in (5.21) the proof tree for Pierce's law in the system $\mathfrak{N}^*(SU)$. It is easy to verify that the transition between the proof in (5.21) and the proof in (5.20) is straightforward.

(5.21)

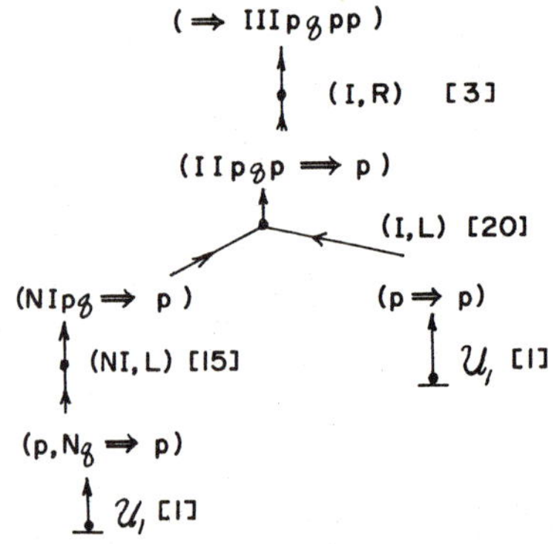

The system $\mathfrak{N}^*(SU)$ provides an appropriate space for handling the decision problem, as it abstracts only the essential elements that are required for making choices in the course of constructing a decision tree.

Note that our system $\mathfrak{N}^*(SU)$ is very similar to the decision system that was used by Wang [4] for machine theorem proving in the propositional calculus. We have sketched here a possible sequence of evolutionary steps that would take us from a system of suppositions to a system that includes a Wang-type decision system. It is certainly an "evolutionary advantage"—relative to a system of suppositions or to an axiomatic system that has a small number of rules of inference—to represent a proof construction problem in a Wang-type system. This is the case even if (as in our problem, where proof minimality is an objective) the Wang-type system is a higher order system which is used to construct the broad lines, the plan, of the proof. The challenge for problem solving research is to find ways of mechanizing the process of acquiring such an "evolutionary advantage."

C. PROCEDURES $\Pi_r(PC, 2a)$

As we have done previously in the case of $N(SU)$, we shall use the formulation of the problem in $\mathfrak{N}^*(SU)$ as a basis for the specification of solution procedures of the reduction type. In these procedures, the construction of a decision tree proceeds from "top to bottom," against the arrows, and it reflects a "backward reasoning" approach which moves by "reasoning leaps," rather than by steps, one "molecular argument" at a time. These procedures form a class which we call $\Pi_r(PC, 2a)$. The interpretation of the main elements in these procedures is as follows: states correspond to formulas of $\mathfrak{N}^*(SU)$, non-terminal moves correspond to reverse application of rules of decision in the abstract system, terminal moves correspond to applications of the axiom schemata, and a solution tree corresponds to a decision tree in $\mathfrak{N}^*(SU)$. Since the non-terminal moves of $\Pi_r(PC, 2a)$ are defined, in general, as compositions of moves from $\Pi_r(PC, 1)$, we call them *macromoves*. If the name of a rule of decision is D, we denote the macromove that corresponds to it by $\bar{D}$. Given a state with n non-atomic formulas (note that all the formulas in a state are l.i. formulas by definition), there are n relevant macromoves at that state, one for each non-atomic formula. As an example, in our illustrative proof of (5.21), there is a single relevant macromove for each of the three states in which a non-atomic formula occurs. The macromove $\overline{(I, R)}$ at the root state $(\Rightarrow IIIpqpp)$, the macromove $\overline{(I, L)}$ at the next boundary state $(IIpqp \Rightarrow p)$, and $\overline{(NI, L)}$ at $(NIpq \Rightarrow p)$.

In the procedures $\Pi_r(PC, 2a)$, we associate with each state S a value $v_t(S)$, $(t = 0, 1, 2, \ldots)$, which is to be interpreted as in our previous procedures. In the present case, however, $v_t(S)$ can take *three* possible numerical values, 1, 1/2, and 0. The values 1, 1/2 have their previous interpretation; the value $v_t(S) = 0$ corresponds to a *certainty that S has no solution.* A terminal state that is "recognized" as valid by $\bar{\mathfrak{A}}_1$ or $\bar{\mathfrak{A}}_{1,1}$ (these are terminal moves that correspond to reverse applications of the axiom schemata for validation) has the value 1. A terminal state that is "recognized" as non-valid by the terminal move $\bar{\mathfrak{A}}_0$ has the value 0. If none of the terminal moves $\bar{\mathfrak{A}}_1$, $\bar{\mathfrak{A}}_{1,1}$, $\bar{\mathfrak{A}}_0$ applies on a terminal state S at time t, then $v_t(S) = 1/2$.

Consider a non-terminal state S and the set $\{M\}$ of macromoves that are relevant to S. The application of any M_j from this set on S produces either a single descendant state S_j or two descendant states S_{j_1} and S_{j_2}. The value of S is related to the values of the descendant states as follows:

$$v_t(S) = \left\{ \begin{array}{l} v_t(S_j), \text{ if } M_j \text{ has a single descendant} \\ \text{Min}\, [v_t(S_{j,1}), v_t(S_{j,2})], \text{ if } M_j \text{ has two descendants} \end{array} \right\} \tag{5.22}$$

for any $M_j \in \{M\}$.

This relationship provides the basis for propagating values from terminals of the search tree to its root. Let S_0 be the root state. The procedure starts at t_1 with $v_{t_1}(S_0) = 1/2$, it grows a search tree by successively applying macromoves to non-conclusive terminal states of the search tree, and it stops when $v_{t_2}(S_0)$ is 1 or 0, after a finite interval of time $\Delta t = t_2 - t_1$.

From the relationship (5.22), it is clear that it is sufficient to apply at each state any (single) macromove from the set of relevant macromoves in order for the procedure to effectively produce a decision about the validity of S_0. It is possible, however, to arrive at a decision by a variety of paths, according to the order in which successive macromove applications are chosen. Our objective for the procedures $\Pi_r(PC, 2a)$ is both to maximize the speed with which a proof tree is constructed, and also to attain one or more proof trees that provide "good" starting points for subsequent simplification procedures. These starting points should enable us to reach a minimal proof with a post-processing effort which is as small as possible. With this objective in mind, we introduce the following method for deciding which among the relevant macromoves to apply at a state.

We assign to each formula that occurs in a state a numerical *complexity index*,

c, which provides a measure of deviation of the formula from atomic form. The assignment is as follows.

For arbitrary formulas x, y, z,

(5.23)
(i) if x is atomic, then $c(x) = 0$,
(ii) if x has the form Kyz, where $K \in \{C, D, I, NC, ND, NI\}$, then $c(x) = 1 + \text{Max}\,[c(y), c(z)]$,
(iii) if x has the form NNy, then $c(x) = 1 + c(y)$.

Note that the complexity index of a formula corresponds to the maximal depth of a tree representing the formula. Given a state $(x_1, x_2, \ldots, x_n \Rightarrow x_{n+1})$, with associated complexity indices $c(x_1), \ldots, c(x_{n+1})$, the macromove to be applied on that state is determined by choosing its key formula as follows:

(5.24) Consider the subset of formulas whose complexity index equals $\text{Max}[c(x_i)]$, for $i = 1, \ldots, n + 1$. If this subset has a single formula, choose it as key formula; if it has more-than-one formulas, choose according to the following priorities:
(i) a goal formula precedes an information formula,
(ii) candidate information formulas are rank-ordered by their characteristic operator as follows: $\langle D, NN, NI, C, I, ND, NC \rangle$.

The rationale for our choice of this decision mechanism for macromove selection is roughly as follows: a conclusive state is reached if the goal formula (not necessarily atomic) is identical with an information formula, or if two information formulas are contradictory. Identical or negated formulas have the same index of complexity. The gradual reduction of c-indices has the effect of permitting a systematic test, one complexity level at a time, of all the potential pairs of contradictory information formulas, and of all the potential identities between the goal formula and information formulas of the same complexity. The order of preference among information formulas of the same complexity is based on weight; "lighter" macromoves are tried first.

We assume that in a procedure $\Pi_r(PC, 2a)$, a single decision tree is in the process of growth at any one time. If alternative decision trees are desired, then they are grown one after another in sequence. Under these conditions, no attention control mechanism is required in the generation subprocedure. All the terminal non-conclusive states of a decision tree in the process of growth need attention before the tree is completed, and the order of distributing attention to them can be chosen only on the basis of machine processing convenience.

D. LOGICAL PROPERTIES OF PROCEDURES

(COMPLETENESS) THEOREM 5.3. *For any formula x in SU, if x is a theorem, then the sequent $(\Rightarrow x)$ is shown to be valid in the system $\mathfrak{N}^*(SU)$, via the procedure* $\Pi_r(PC, 2a)$.

Proof. The procedure $\Pi_r(PC, 2a)$ has macromoves for reducing any initial state $(\Rightarrow x)$ to a logically equivalent set of states, each of which contain only atomic formulas. The validity of the states in this set can be directly tested (via the terminal moves that correspond to axiom schemata of $\mathfrak{N}^*(SU)$), and it can be used directly to establish the validity of $(\Rightarrow x)$. The reduction process terminates after a finite number of macromove applications since the complexity index of x is finite, and each application of a macromove on a key formula monotonically decreases the complexity indices in the "progeny" of that formula. Hence, the system $\mathfrak{N}^*(SU)$ is complete, and $\Pi_r(PC, 2a)$ is a decision procedure for provability which yields a proof for each theorem in SU.

COROLLARY 5.4. *For any formula x in SU, if x is a theorem, then its proof in $N(SU)$ is obtainable via a procedure from $\Pi_r(PC, 1)$ that has a set μ^{**} of relevant moves.*

Proof. This follows from the previous completeness theorem and from the fact that (i) each macromove is expressable in terms of moves from μ^{**} (see Appendix I), (ii) the terminal move $\bar{\mathfrak{A}}_1$ is identical in $\Pi_r(PC, 1)$ and $\Pi_r(PC, 2a)$, and (iii) the terminal move $\bar{\mathfrak{A}}_{1,1}$ is defined in terms of moves from μ^{**} and the terminal move $\bar{\mathfrak{A}}_1$. Consider the proof of x constructed by $\Pi_r(PC, 2a)$, and suppose that this proof is converted to a proof in $N(SU)$ via the appropriate definitions of macromoves and terminal moves. Call this latter proof P_m, and let its weight be w_m. Since an elementary procedure in $\Pi_r(PC, 1)$ proceeds by growing systematically a search tree based on all the relevant moves from μ^{**}, and in such a manner as to check all potential proofs of a certain weight before checking those of a higher weight, it will either find a proof of weight w, where $w \leqslant w_m$, or it will find the proof P_m. Hence, the elementary procedure $\Pi_r(PC, 1)$ is always guaranteed to find a proof of weight $w \leqslant w_m$ to x, if x is a theorem.

E. PROOF CONVERSION AND SIMPLIFICATION

A proof $P^*_m(x)$ of a theorem x in $\mathfrak{N}^*(SU)$ (obtained via a procedure $\Pi_r(PC, 2a)$) can be regarded as an abbreviated representation of a corresponding proof $P_m(x)$ of x in the system $N(SU)$. The proof $P^*_m(x)$ can be converted to a unique corresponding proof in $\mathfrak{N}(SU)$, via the definitions of the macromoves in terms of moves, and this proof can be converted to the proof $P_m(x)$ by simply eliminating the distinction between l.i. and non-l.i. formulas. As an example, the proof of Peirce's law given in (5.20), with the lower bars removed from the relevant formulas, is the corresponding proof in $N(SU)$ of the proof in $\mathfrak{N}^*(SU)$ that is given in (5.21). In general, such a proof $P_m(x)$ is not a minimal proof (as we can see with the examples of proofs for Pierce's law), however, it has the advantage of usually requiring a vastly lesser processing effort for its construction than a proof obtained via $\Pi_r(PC, 1)$.

Given a proof $P_m(x)$ which is the projection in $N(SU)$ of a proof produced by a procedure $\Pi_r(PC, 2a)$, it is possible to use it as an input to a simplification procedure which attempts to remove some of the reasoning redundancies that occur in $P_m(x)$, with the aim of eventually attaining a minimal proof.

We outline next an *elementary simplification procedure*, which consists of a truncation procedure and a condensation procedure in cascade.

The *truncation procedure* is as follows. The input proof tree $P_m(x)$ is systematically scanned from the root sequent down. We assume that with each sequent S in the proof tree, there is an associated weight $w(S)$, which is the weight of the subtree "hanging" below S. At each sequent S encountered during the scan, the application of the terminal moves of weights 1, 2, 3, and 4 (given in (4.6)) is attempted in sequence, in such a manner that only terminal moves not exceeding in weight $w(S)$ are tried. If $w(S) > 5$, then, following the previous attempts to apply terminations, the $\mathfrak{A}_{1,1}$ terminal move is also attempted. If any of the attempts is successful, then the corresponding truncating termination is applied, and it substitutes the "heavier" subtree that originally hung below S. Then the scan advances to the next sequent of the proof tree (skipping sequents that are newly introduced with the truncating terminations). If no truncating termination applies to a sequent, then the scan advances to the next sequent. This process is repeated until the proof tree $P_m(x)$ is completely scanned. Let $P_{m,t}(x)$ denote the proof tree which is the output of the truncated procedure. The weight of $P_{m,t}(x)$ is at worst equal to that of $P_m(x)$, and in many cases it is smaller.

The *condensation procedure* is as follows. A test is made of whether the same goal formula appears in at least two sequents that are located on the same tree chain of $P_{m,t}(x)$. If the result of the test is negative, no condensation is possible. Suppose that the result is positive and there are two such sequents S_i, S_j where S_j is closer to the tree root. Let $\eta_{t,i}$, $\eta_{t,j}$ denote the sets of information formulas of S_i and S_j respectively. If the condition $\eta_{t,j} \subset \eta_{t,i}$ holds, and if no formula from the set $\eta_{t,i} - \eta_{t,j}$ has participated in the specification of any move below S_i in the tree $P_{m,t}(x)$, then a condensation is possible and the segment of $P_{m,t}(x)$ between S_j and S_i is collapsed. The subtree that initially hung below S_i (with the formulas from the set $\eta_{t,i} - \eta_{t,j}$ deleted) becomes the new continuation of S_j. If these conditions are not satisfied, then a condensation between S_i and S_j is not possible. The attempt to collapse a proof segment between a pair of sequents is tried in succession for all pairs located on the same tree chain and that have identical goal formulas. Pairs that are separated by the "heaviest" proof segments are tried before pairs with "lighter" separations. We denote by $P_{m,t,c}(x)$ the proof tree which is the output of the condensation procedure. Usually, the weight of $P_{m,t,c}(x)$ is smaller than the weight of $P_{m,t}(x)$. The major effect of the condensation procedure is to eliminate circuitous and redundant arguments by identifying such arguments in the two sides of boundary lines of macromoves.

In essence, the condensation procedure just described is a restricted variant of the "similarity recognition" terminal move which was previously discussed in connection with the procedures $\Pi_r(PC, 1)$. In general, it is possible to formulate a spectrum of simplification procedures, each of which is based on a systematic utilization of different classes of terminal moves that are expressed either as general schemata or as "similarity recognition" procedures. Each case should be judged by the relationship between its simplification power and the required processing effort. The understanding of this relationship is not satisfactory at present. Taking a broader view, a simplification procedure should be able to perform transformations that would take an input proof to a logically equivalent proof that is simpler. To this end, an exploration of the *set of logically equivalent proofs* is important, and it would constitute a reasonable path of development in the evolution of the proof procedures. In our work, however, we have not pursued this path to any appreciable extent.

We will now illustrate the performance of our elementary simplification procedure by applying it to the proof P_m of Peirce's law which is identical to the proof shown in (5.20), except for the tagging of l.i. formulas. The truncation procedure applies in two places of the proof P_m; once at the sequent $(IIpqp^*, Np^* \Rightarrow Np)$, where a simple $\mathfrak{A}_1$ truncation of weight one is substituted for the cumbersome termination of weight seven, and once at the sequent $(IIpqp^*, Np^*, IIpqp, NIpq^*, Np^*, p^* \Rightarrow q)$, where a $\mathfrak{A}_{1,1}$ truncation of weight five is substituted for a termination of weight seven. These truncations are indicated by dashed lines in (5.20). Thus, the truncation procedure has reduced the initial proof weight of 40 to 32. The condensation procedure applies to a single pair of sequents, which we have denoted by S_j, S_i in the proof of (5.20). Here we have $\eta_{t,j} = \{IIpqp, Np\}$ and $\eta_{t,i} = \{IIpqp, Np, NIpq\}$. Clearly, the condition $\eta_{t,j} \subset \eta_{t,i}$ holds and the single formula $NIpq$ in the set difference $\eta_{t,i} - \eta_{t,j}$ does not participate in the development of the proof below S_i. Hence, the proof segment between S_j and S_i can be collapsed (we show this segment within dashed lines in (5.20)). The collapsed segment, which has a weight of 12, is clearly redundant and it occurs characteristically in the two sides of the interface of two macromoves (here $\overline{(I, L)}$ and $\overline{(NI, L)}$). Thus, the condensation procedure has effected a further weight reduction of 12, which gives us a proof $P_{m,t,c}$ of weight 20. This proof is precisely the minimal proof of Peirce's law which is shown in (4.4).

F. PROCEDURES $\Pi_r(PC, 2)$: IMPROVEMENTS

The procedures $\Pi_r(PC, 2)$ are composed of a procedure $\Pi_r(PC, 2a)$ that constructs one-or-more proof trees in $\mathfrak{N}^*(SU)$ if a proof exists (otherwise, a refutation tree in $\mathfrak{N}^*(SU)$ is produced and the procedure stops), of a "projection" procedure that converts the candidate proof(s) in $\mathfrak{N}^*(SU)$ to proofs in $N(SU)$, and of a simplification procedure that attempts to simplify the candidate proofs in $N(SU)$ and that produces as output the simplest proof attained. We have considerable experimental evidence that the procedures $\Pi_r(PC, 2)$ produce minimal or quasi-minimal proofs. Their main advantage is *reduction in proof construction effort* relative to the more thorough search procedure in $\Pi_r(PC, 1)$. The fact that

$\Pi_r(PC, 2)$ includes a decision procedure for the propositional calculus, which also presupposes considerable progress in the status of knowledge about the proof problem, is another important evolutionary advantage with respect to the previous procedures. While the completeness condition for proof construction is demonstrably satisfied by $\Pi_r(PC, 2)$, there exist only informal arguments, supported by empirical evidence, that the minimality condition for proofs is satisfied by some procedures in the class $\Pi_r(PC, 2)$. Therefore, these procedures are, at this stage, reasonably efficient heuristic procedures.

An important idea in the procedures $\Pi_r(PC, 2)$ is that a *main avenue* to solution is surely and rapidly created, and then simplification resources are "run over it" to obtain a more direct way to solution by trying to find appropriate bypasses at various points on the avenue. In essence, this idea is a special case of the idea of *successive improvements* that we have discussed in connection with some advanced procedures in $\Pi_r(PC, 1)$. However, in $\Pi_r(PC, 2)$, the decision procedure is a much more efficient and sure provider of initial candidate proofs than the selective deep exploratory growth of $\Pi_r(PC, 1)$.

It is possible to combine the main features of $\Pi_r(PC, 2)$ with the successive improvement mode of the $\Pi_r(PC, 1)$ procedures, so that a more systematic approach to the search for a minimal proof can be obtained. It should be remembered that the evolutionary trigger for the development of the notion of l.i. information formulas and "molecular arguments" came from a desire to formulate more efficient "successive improvement" strategies in the procedures of the previous stage of development. We shall outline next an improved procedure of "successive improvements" that utilizes features of $\Pi_r(PC, 2)$.

First, a proof tree $P_m^{(1)}$ is generated via $\Pi_r(PC, 2)$, where the macromove selection in the subprocedure $\Pi_r(PC, 2a)$ is carried out in accordance with the preference rules given in (5.24). Subsequently, proofs $P_m^{(2)}, P_m^{(3)}, \ldots, P_m^{(n)}$ are generated in succession, where each proof reflects different orders of macromove applications in $\Pi_r(PC, 2a)$. The choice of different macromove orderings is carried out as follows. In the state closest to the root of the proof tree under construction, where a choice of key formulas exists, try in succession the key formulas in the order of preference given in (5.24); i.e., goal formula of highest c, information formulas of highest c in the order indicated in (5.24), then formulas of lower c in a similar order, and so on. After a macromove corresponding to the chosen key formula has been applied, continue the tree growth in accordance with the rules (5.24). Repeat this process in a systematic way for states at each level of tree depth where choice of key formulas exist until all possibilities are exhausted. The set of proof trees $P_m^{(1)}, \ldots, P_m^{(n)}$ (that can be regarded as a subset of the search tree in $N(SU)$) provide now a starting point for a further process of "successive improvements" of the type described in 4.D for advanced procedures of type $\Pi_r(PC, 1)$.

If no effort limitation exists, then this overall procedure will always reach a minimal proof (of course, under the assumption that the restrictions on the relevant set of moves μ^{**} do not affect minimality). In general, a proof which is quite close to minimal is obtained early in the process, and further investment of processing effort either improves the current best proof or it strengthens its claim to minimality.

Work with such improved "successive improvement" procedures reveals that there are some stable patterns in proofs that seem to be determined by the occurrence of certain configurations of subformulas in the candidate theorem. The "molecular arguments" of $\mathfrak{N}(SU)$ do not seem to relate well with these patterns, and slight reformulations of these "molecular arguments" do not change the situation appreciably. It appears that a different type of "molecular argument" has to be formulated that can capture the characteristic patterns in proofs. Furthermore, it appears that the flow of formulas in a proof must play an essential role in these new formulations. The search for such new *global* properties in the space of proofs, and also for methods of using these properties in a manner that would increase the power of proof procedures, leads to the third stage of development in the treatment of the proof construction problem. We shall discuss next the main features of this stage.

6. PROOF CONSTRUCTION PROCEDURES: THE THIRD PHASE

A. THE GRAPH GENERATING SYSTEM $\Gamma(SU)$

We shall now consider proofs as graphs of a special type that represent a connecting pattern between formulas in SU. A proof graph will be regarded as a word in a proof language, and such a language will be given in terms of a graph generating system $\Gamma(SU)$. We start by specifying the system $\Gamma(SU)$ which is equivalent (in the sense of producing the same proofs—within isomorphism) with the proof construction system $\Pi_r(PC, 1)$. The system $\Gamma(SU)$ is determined by (*a*) a vocabulary of nodes and branches, (*b*) a set of generating rules, and (*c*) an especially designated set of nodes called starting nodes.

(*a*) *Vocabulary of* $\Gamma(SU)$

(i) *Nodes.* The system has auxiliary, unlabelled nodes, denoted by a small dot, and two types of nodes that are labelled by formulas of SU; the two types are as follows.

(i1) *Source nodes.* This node type is denoted by $\oslash x$, and it indicates that x is an information formula. Furthermore, the notation $\circledast x$ indicates that x is a supposition formula, and $\oplus x$ designates a deduced formula. Thus, $\oslash$ is a variable with values $\circledast$, $\oplus$.

(i2) *Destination nodes.* This node type is denoted by $\bigcirc x$, and it indicates that x is a goal formula.

(ii) *Directed branches.* The system has auxiliary, unlabelled, directed branches, and also labelled directed branches, denoted by $\overset{w}{\rightarrow}$, where w stands for the *weight* of the branch. Branches indicate, in general, inferential couplings between formulas (in the logical interpretation of moves in $\Pi_r(PC, 1)$).

(b) Rules of Generation of $\Gamma(SU)$

There is one rule of generation for each relevant move of $\Pi_r(PC, 1)$ (non-terminal and terminal). To express the rules of generation, we introduce the following definition: a destination node of a graph is called a *processed destination node* if it has at least one arrow pointing into it; otherwise, it is called an *un-processed destination node.* The rules of generation are shown below in Table 6.1

TABLE 6.1

$C_{tn\ (a\ \text{or}\ b)}$:

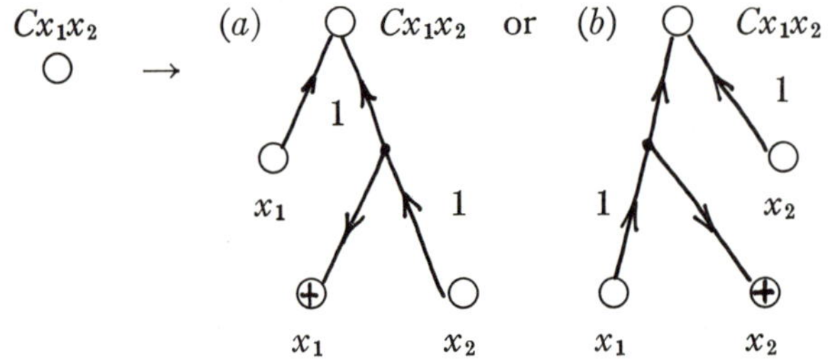

$D_{tn\ (a\ \text{or}\ b)}$:

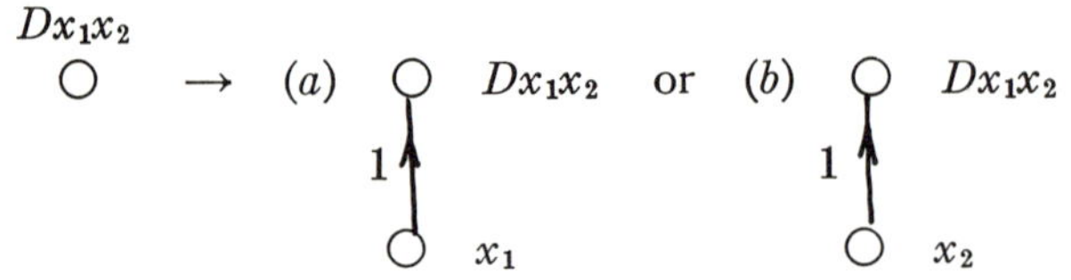

I_{tn}:

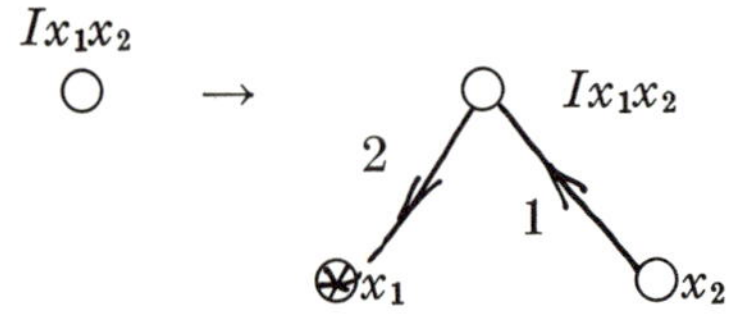

N_{tn}:

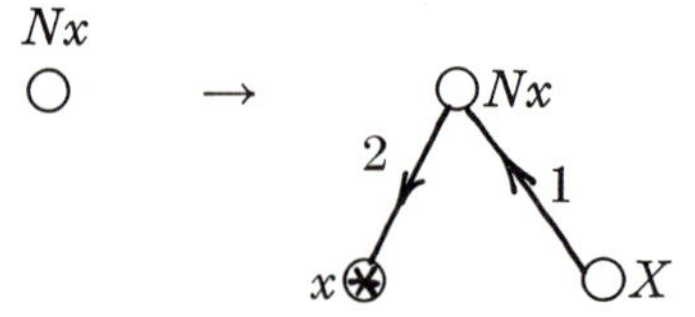

$C_{el\ (a\ \text{or}\ b)}$:

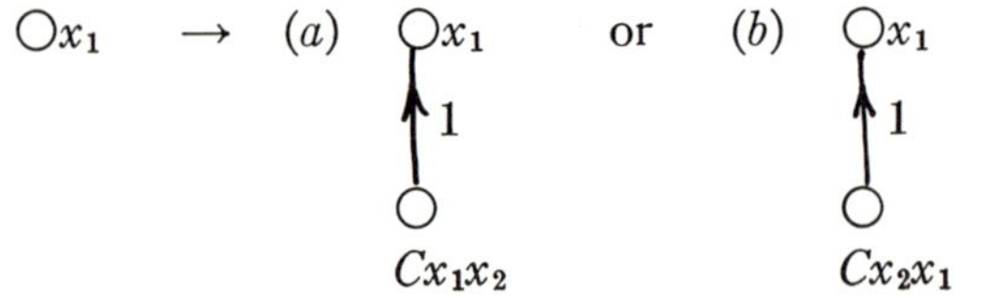

; Cx_1x_2 (or Cx_2x_1) must be in $\gamma(x_1)$.

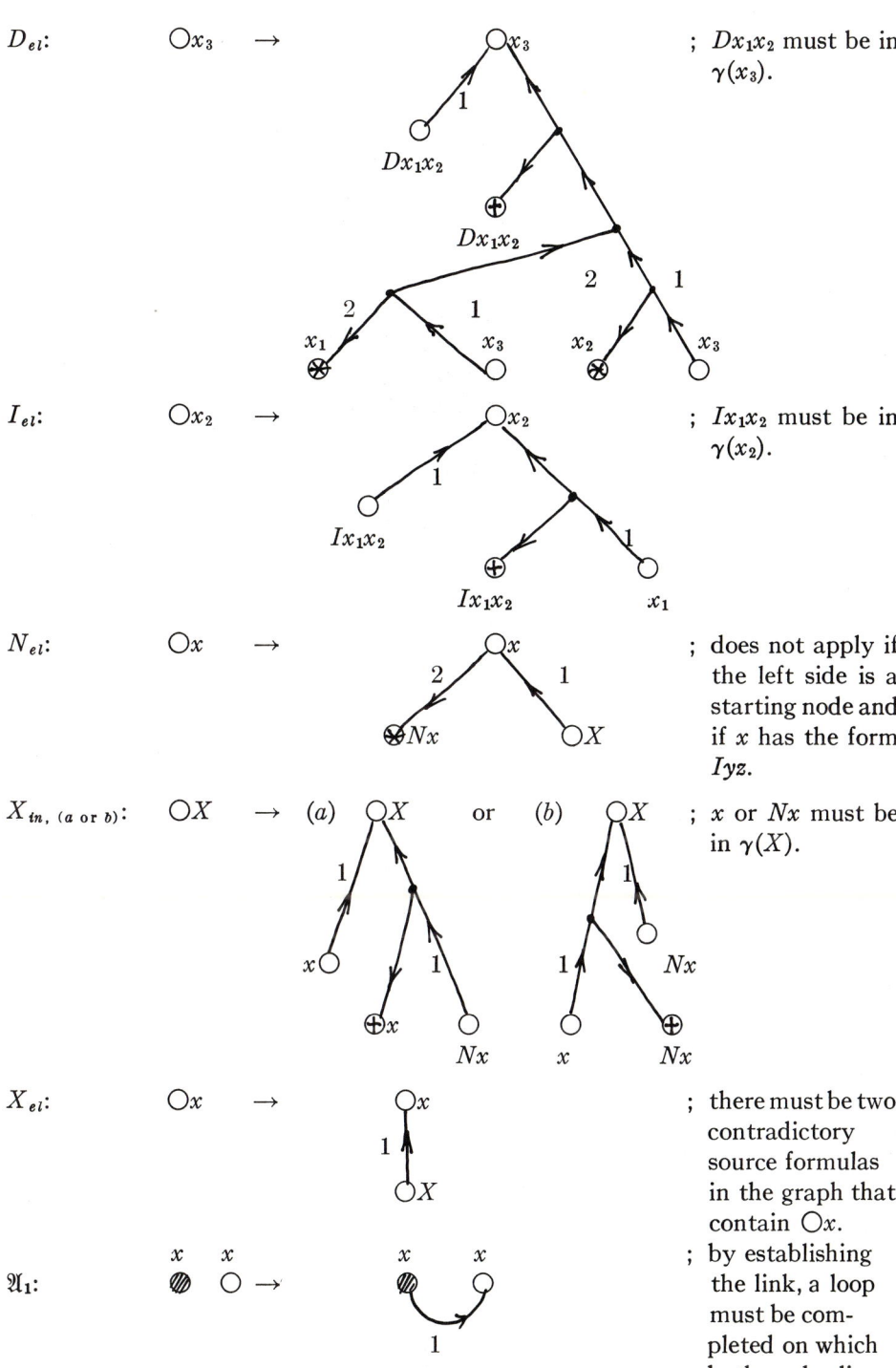

D_{el}: $\bigcirc x_3$ $\rightarrow$

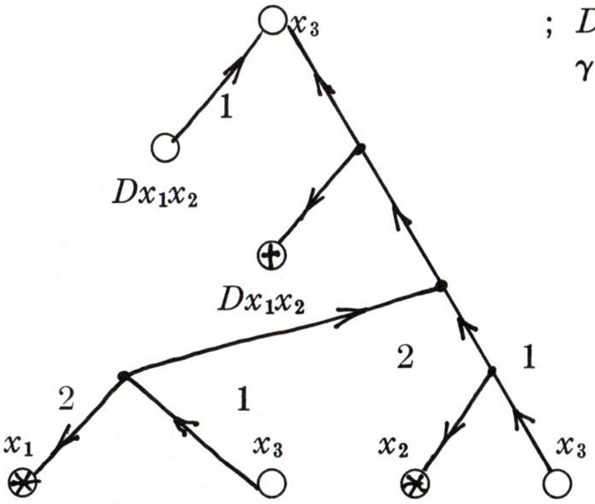

; Dx_1x_2 must be in $\gamma(x_3)$.

I_{el}: $\bigcirc x_2$ $\rightarrow$

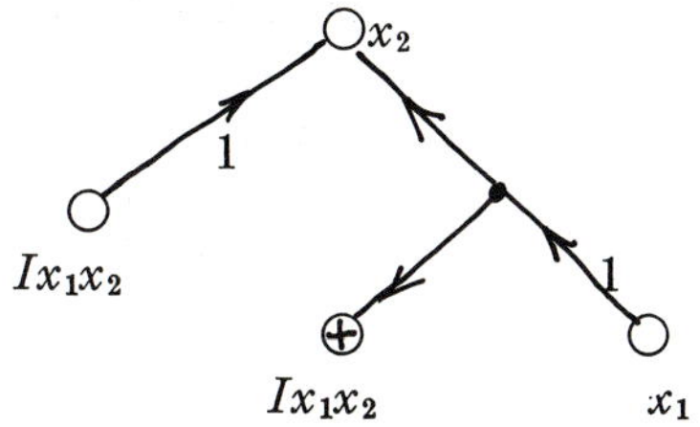

; Ix_1x_2 must be in $\gamma(x_2)$.

N_{el}: $\bigcirc x$ $\rightarrow$

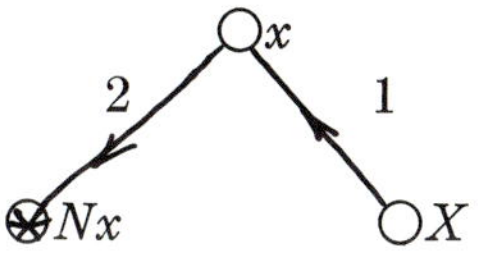

; does not apply if the left side is a starting node and if x has the form Iyz.

$X_{in,\ (a\ or\ b)}$: $\bigcirc X$ $\rightarrow$ (*a*) or (*b*)

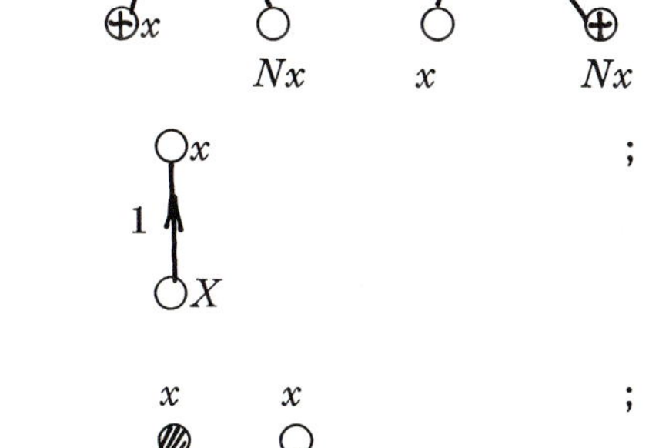

; x or Nx must be in $\gamma(X)$.

X_{el}: $\bigcirc x$ $\rightarrow$

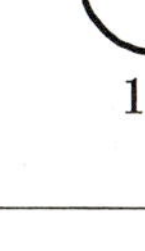

; there must be two contradictory source formulas in the graph that contain $\bigcirc x$.

$\mathfrak{A}_1$:

$\rightarrow$

; by establishing the link, a loop must be completed on which both nodes lie.

in the form of transition schemata between graphs. By convention, the left side of a transition schema is a subgraph (not necessarily proper) of an arbitrary graph G. Furthermore, a destination node appearing in the left side of a schema is an *unprocessed* destination node. The transition schema specifies a *possible replacement* in G of the subgraph that appears in the left side by the graph that appears in the right side. Note that in each transition schema an unprocessed destination node becomes a processed node. Together with this change, new unprocessed destination nodes appear, except in the case of the $\mathfrak{A}_1$ rule. The restrictions imposed on the relevant moves of $\Pi_r(PC, 1)$ are reflected here as contextual constraints on the rules of generation. They are given in the form of comments at the right of transition schemas. The condition associated with $\mathfrak{A}_1$ is essential for the construction of well-formed graphs. We denote by $\gamma(x)$ the set of source formulas or subformulas of source formulas that appear in the graph that contains the node $\bigcirc x$.

A comparison between Table 6.1 and Table 4.1, where the rules of inference of $N(SU)$ are shown, will clarify the basis of the new formulation. Source nodes correspond to information formulas and destination nodes correspond to goal formulas. While Table 4.1 specifies inferential couplings between sequents, the rules in Table 6.1 show the specific formula changes that correspond to these inferential couplings.

(c) Set of Starting Nodes of $\Gamma(SU)$

A starting node of $\Gamma(SU)$ is an unprocessed destination node, $\bigcirc x$, where x is a theorem to prove.

(d) Proofs in $\Gamma(SU)$

A *proof* of a theorem x in $\Gamma(SU)$ is a graph generated in $\Gamma(SU)$ as follows.

(1) The starting node is $\bigcirc x$.

(2) A rule of generation is applied to the starting node, producing new nodes, some of which are unprocessed destination nodes.

(3) A rule of generation is applied to each new unprocessed destination node, and the process is repeated until no unprocessed destination nodes remain.

The resulting graph is a proof graph of x in $\Gamma(SU)$; its *weight* is the sum of branch weights in the graph. A simplest proof of x corresponds to a proof graph of minimal weight. It can now be seen more clearly that our notion of a minimal proof reflects the idea that the proof is constructed by the least possible number of formulas in the simplest possible connecting pattern.

As an example of a proof in the present system, we show in (6.1) the minimal proof of Peirce's law in $\Gamma(SU)$. Proofs of this theorem in sequential form and in tree form are given in (3.2) and (4.4) respectively.

(6.1)

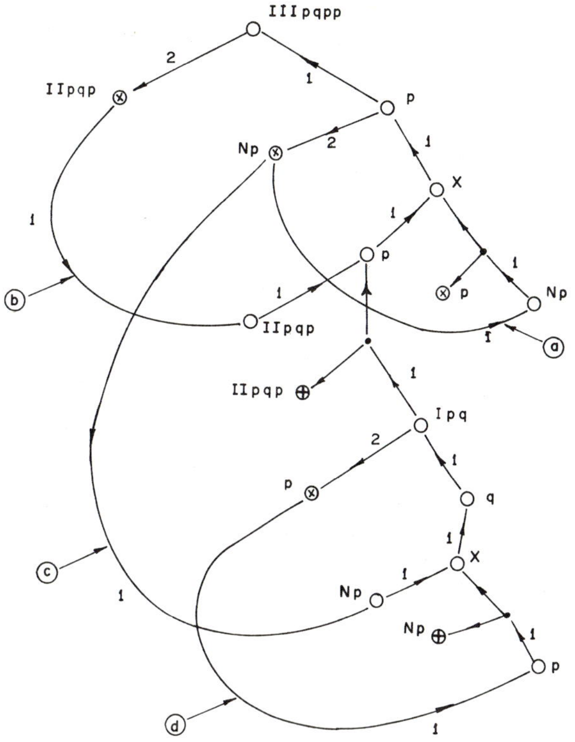

It is easy to see the one-to-one correspondence which exists between a proof graph in $\Gamma(SU)$ and a proof tree in $N(SU)$. A proof graph in the process of growth corresponds to a growing proof tree, i.e., to a proof tree that has some non-conclusive sequents at its terminals. A sequent in $N(SU)$ corresponds to a subgraph of a proof graph that consists of an unprocessed destination node and of the set of all source nodes that can be reached from this destination node via a directed path in the graph. The unprocessed destination node corresponds to the goal formula of the sequent, and the set of source nodes corresponds to the set of its information formulas. The reverse application of a rule of inference from $N(SU)$ on a sequent corresponds to the application of a rule of generation from $\Gamma(SU)$ on the unprocessed destination node that corresponds to the goal of the sequent. A comparison of the tree proof and the graph proof of Pierce's law (given in (4.4) and (6.1) respectively) will help to clarify these correspondences.

An examination of graph proofs suggests that a reasonable unit of construction is a *loop*. Each loop contains a single source node and one or more destination nodes. In the course of graph construction, a sequence of destination nodes is built on a directed path (against the arrows). A loop is formed by establishing a direct link (according to the generation rule $\mathfrak{A}_1$) between a source node that is reachable from the destination nodes along the directed path and the last destination node which was built on the directed path. Since there is a single $\mathfrak{A}_1$ link in each loop, one can designate a loop by pointing to its $\mathfrak{A}_1$ link. As an example, in the graph proof of (6.1) there are four loops. We designate them by labelling their $\mathfrak{A}_1$ as a, b, c, d. The order implied by the loop labels indicates a reasonable sequence of loop construction actions. Note that a loop in a $\Gamma(SU)$ proof corresponds to a *tree chain* in the corresponding tree proof in $N(SU)$. This chain terminates with an axiom link which corresponds to the characteristic link of the loop.

By regarding proof construction as graph construction in $\Gamma(SU)$, it is natural to view our task as that of finding a configuration with the least number of "lightest" loops, which constitutes a proof graph. We can regard the formation of a loop in a proof graph as a significant goal for a new type of "molecular argument." If such a "molecular argument" does not conclude the proof graph (i.e., if more unprocessed destination nodes remain), then it is an important additional goal of the argument to prepare some nodes that will be required for the closure of subsequent loops. Thus, early "molecular arguments" have sometimes to foresee the requirements of subsequent arguments and to prepare appropriate material for them. Examine, for example, the proof graph of Pierce's law in (6.1). The "argument" ⓐ prepares ⊛Np for the "argument" ⓔ, and the "argument" ⓒ prepares ⊛p for the "argument" ⓓ. The study of conditions for loop formation in $\Gamma(SU)$ suggests new global methods for proof construction that would improve the procedures $\Pi_r(PC, 1)$ in ways that are different from those based on the system $\mathfrak{N}(SU)$. We will not pursue here, however, a line of development that is strictly based on $\Gamma(SU)$. Rather, we will take as a basis for the new class of procedures a graph generating system $\Gamma'(SU)$ that is closely related to $\Gamma(SU)$ but is more suitable for our purposes. Our main reason for introducing the system $\Gamma(SU)$ has been to establish a clear correspondence between our previous notions of proofs and the notions of proof graphs—as an intermediate step for the specification of $\Gamma'(SU)$.

B. THE GRAPH GENERATING SYSTEM $\Gamma'(SU)$

It is important to note that our basic approach to proof construction in the systems $N(SU)$, $\mathfrak{N}(SU)$, and $\Gamma(SU)$ has been to work backwards from goal formulas to information formulas. There are many advantages to such an approach—especially if the number of information formulas is large; most heuristic theorem provers have used such a strategy ([3], [15], [18]). However, a strategy of combined development from both information formulas and goal formulas promises a more efficient proof construction procedure in a SU formulation of the propositional calculus (where the set of information formulas is, in most cases, quite managable). The deduction rules of SU treat information and goal formulas in an essentially symmetric fashion. However, rules of introduction can be associated more naturally to a backward development from goal formulas and rules of elimination can be associated more naturally to a forward development from information formulas. The non-symmetric mode of applying rules of deduction that is implied by these preferred associations is captured by the graph generating system $\Gamma'(SU)$. In $\Gamma'(SU)$, a proof grows both from goal formulas and information formulas. One of the important problems in such a two-sided approach, is to find an appropriate method for co-ordinating the forward and backward developments during the proof construction process. Such a method is embodied in the proof construction procedure $\Pi_r(PC, 3)$ that is developed in the present stage, and it enables the procedure to use the system $\Gamma'(SU)$ as its linguistic basis.

The vocabulary of $\Gamma'(SU)$ is the same as that of $\Gamma(SU)$, except for one addition: a new source node, denoted by ⊙, is introduced, and it is called an *intermediate source node*. Thus, a node ⊘ stands now for any of the nodes, ⊛, ⊕, ⊙.

The systems $\Gamma(SU)$ and $\Gamma'(SU)$ differ mainly in some of their rules of generation. The rules of generation of $\Gamma'(SU)$ are shown in Table 6.2. The last rule in this table corresponds to a combination of X_{el}, X_{in}, and $\mathfrak{A}_1$ of $\Gamma(SU)$; the links ⊘^{y}—○^{y} and ⊙^{X}—○^{X} should satisfy the condition specified in $\mathfrak{A}_1'$.

(a) Additional Conditions on Generation in $\Gamma'(SU)$

A ⊛ or a ⊕ node can be the source of more-than-one links (i.e., application of $\mathfrak{A}$ rules); a ⊙ node can only be the source of a single link. Similarly, any rule of generation can be applied more than once on the same source node if the node is of ⊛ or ⊕ type; only a single rule of generation can be applied on a source node of ⊙ type. As an example, a triple application of C'_{el} on a node ⊛Cx_1x_2 is permissible and it results in a graph which is represented as follows:

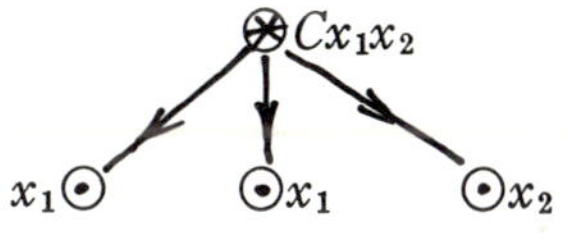

The notion of a *starting node* is identical in $\Gamma(SU)$ and $\Gamma'(SU)$. We introduce here the additional notion of a *candidate destination node*: this is an unprocessed destination node in a $\Gamma'(SU)$ graph, from which there exists a directed path to the starting node.

A *proof* of a theorem x in $\Gamma'(SU)$ is a graph generated in $\Gamma'(SU)$ as follows.

(i) The starting node is ○x.
(ii) A rule of generation is applied on ○x, producing new nodes, some of which (at least one) are candidate destination nodes.
(iii) A rule of generation is applied on a candidate destination node, or on a source node; this process is repeated until no candidate destination nodes remain.

The graph so constructed is a proof graph of x in $\Gamma'(SU)$. The weight of this

TABLE 6.2

RULES OF GENERATION OF Γ' (SU)

The rules C'_{in}, D'_{in}, I'_{in} and N'_{el} are identical with the rules C_{in}, D_{in}, I_{in}, N_{in} and N_{el} respectively of $\Gamma(SU)$ (see Table 6.1).

C'_{el} (a or b): ◍Cx_1x_2 → (a)

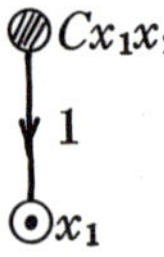

or (b)

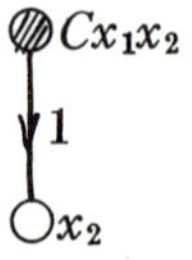

D'_{el} : ◍Dx_1x_2 →

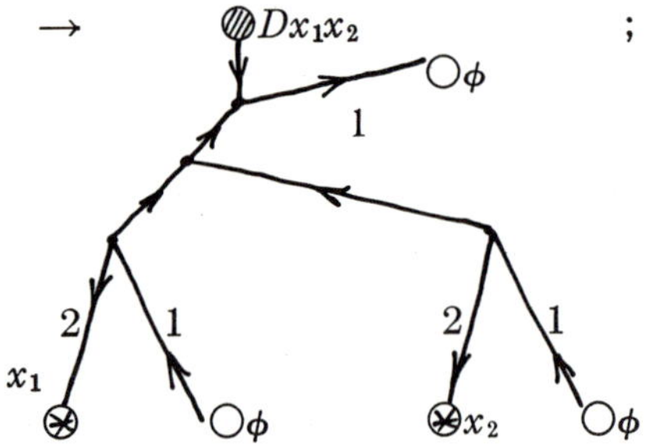

; ϕ stands for any formula

I'_{el} : ◍$Ix\ x_2$ →

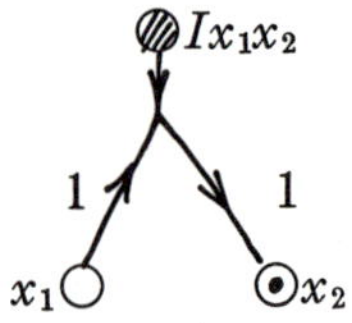

X'_{in} : ◍Nx →

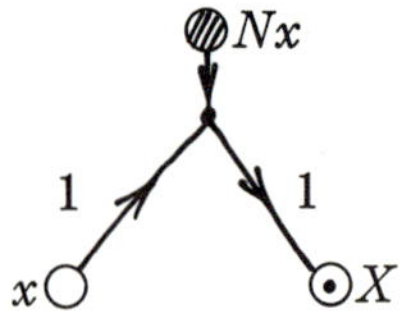

$\mathfrak{A}'_1$: ◍x ○x →

; by establishing this link, a new loop must be completed which goes through a ⊛ or a ⊗ node.

$\mathfrak{A}'_{1,\phi}$: ◍ϕ ○x →

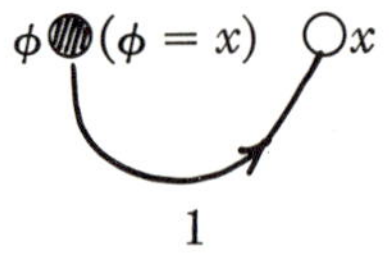

; ϕ is a formula variable which is part of a D'_{el} growth; the formula x is substituted for all occurrences of ϕ in the D'_{el} growth. The condition on the link is as in $\mathfrak{A}^1_1$.

$\mathfrak{A}'_{1,1}$: ◍y ◍Ny ○x →

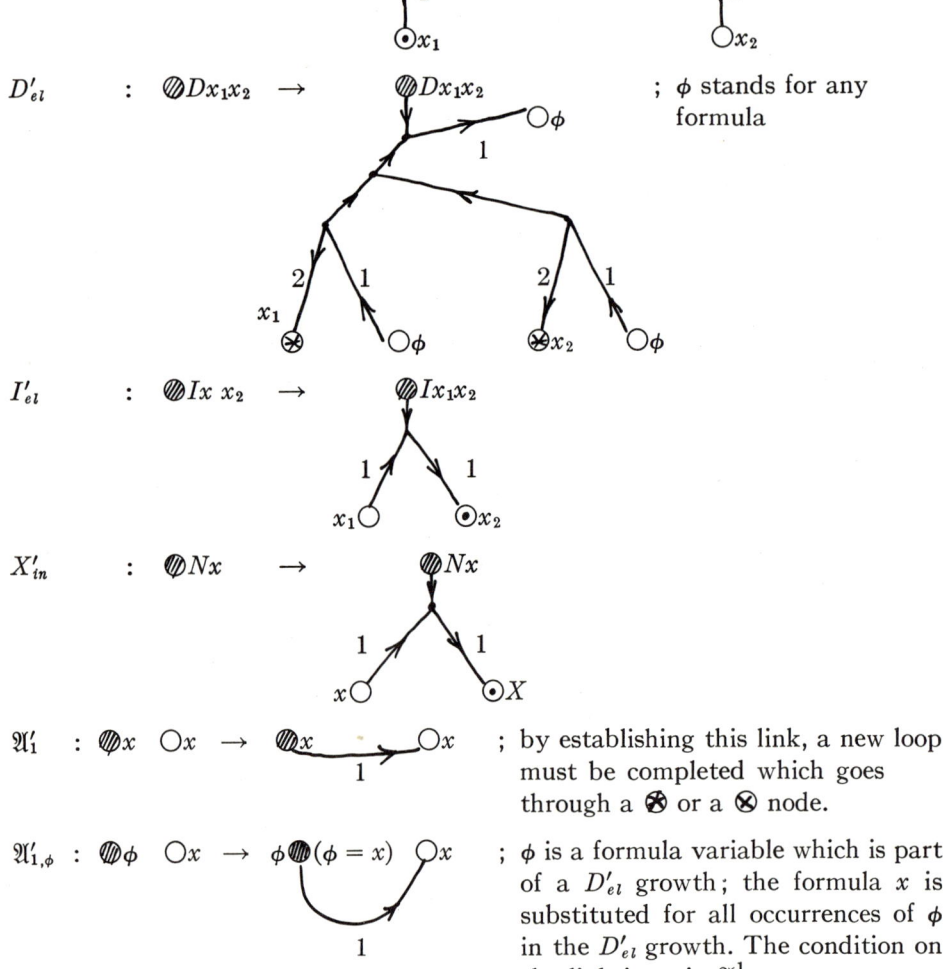

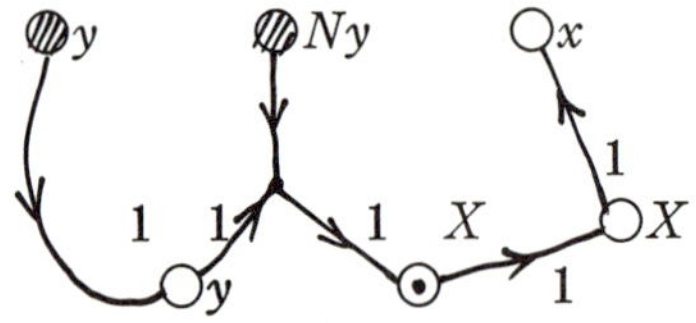

proof is the sum of branch weights in the graph. As an example of a proof in the present system, we show in (6.2) the minimal proof of Pierce's law in $\Gamma'(SU)$.

(6.2)

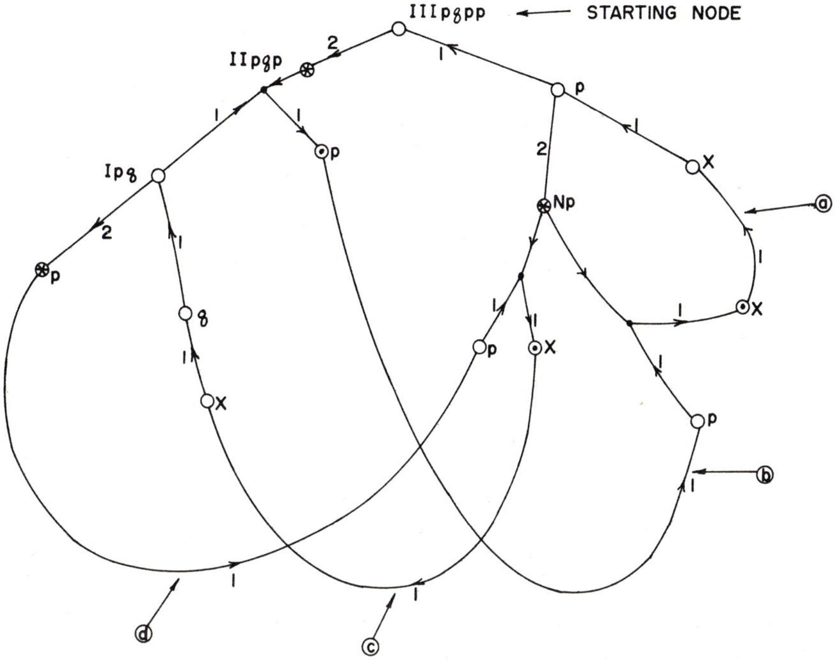

A comparison of this proof graph and the proof of Pierce's law in $\Gamma(SU)$, which is given in (6.1), shows that these proofs are essentially in one-one correspondence. The loop structures are the same, and the relative location of all the distinct formulas is conserved. The differences are due to the location of the links. In general, corresponding loops in the two proofs are formed by establishing links in different places on them. As in the case of $\Gamma(SU)$, a link in a $\Gamma'(SU)$ proof characterizes a loop. In the $\Gamma'(SU)$ proof of (6.2), there are four loops that we have labelled with the same labels that identify the corresponding loops in the $\Gamma(SU)$ proof.

In the rules D'_{el}, I'_{el}, and X'_{in} of $\Gamma'(SU)$, we have eliminated the occurrence of $\oplus$ nodes that appear in the corresponding rules of $\Gamma(SU)$. This is based on our experience which strongly suggests that these nodes are not needed for the task of constructing minimal proofs. While it can be easily demonstrated that these nodes are not needed for completeness, we do not have a proof as yet that they are not needed for proof minimality. Aside from this discrepancy in the correspondence of $\oplus$ nodes, there exists, in general, a one-one correspondence between proof graphs in $\Gamma(SU)$ and $\Gamma'(SU)$. The correspondence is as follows: consider proof graphs in $\Gamma(SU)$ and $\Gamma'(SU)$ with all their links collapsed. In these *collapsed proof graphs*, each link and its two incident nodes is replaced by one of the incident nodes (the "survival" of the nodes is ruled by the ordering ⊛, ⊕, ○, ⊙). For each collapsed proof graph in $\Gamma(SU)$, there exists an isomorphic collapsed proof graph in $\Gamma'(SU)$ and *vice versa*. The isomorphism takes each node of one graph to a node in the other that is labelled by the same formula (the node type is preserved if it is a ⊛ or ⊕), and it preserves all ordering relations between the formulas in the proof graphs.

Because of the correspondence just described, a proof graph in $\Gamma'(SU)$ can be converted unambiguously to a proof graph in $\Gamma(SU)$ and (because of the direct correspondence between proofs in $\Gamma(SU)$ and $N(SU)$) to a proof in $N(SU)$, and from it to a proof in SU. The conversions can be carried out by relatively simple procedures of the list-processing type.

C. THE NATURAL DECISION SYSTEM $\mathfrak{M}(SU)$

While the system $\Gamma'(SU)$ provides the elementary proof construction language for the procedure $\Pi_r(PC, 3)$, the overall logic of the procedure is based on a system of natural decision, that we shall call $\mathfrak{M}(SU)$. This system has certain similarities with the systems of natural decision that we have discussed previously, but it also differs from them in important respects.

The formulas in $\mathfrak{M}(SU)$ are sequents of the form $(\eta \Rightarrow x)$, as in $N(SU)$. (We will not designate suppositions in a special way here.) The axiom schemata of $\mathfrak{M}(SU)$ are as follows: *for validation* $\mathfrak{A}_1$ and $\mathfrak{A}_{1,1}$ as in $\mathfrak{N}(SU)$; *for refutation* $\mathfrak{A}_0$ as in $\mathfrak{N}(SU)$, and also $\mathfrak{A}_{0,X}$: a sequent $(\eta \Rightarrow X)$ is not valid, if its logically irredundant (l.i.) information formulas are all atomic, and if the axiom schema $\mathfrak{A}_{1,1}$ does not apply.

In the system $\mathfrak{M}(SU)$, a decision about the validity of a propositional formula x is based on a reduction process that takes an initial sequent $(\Rightarrow x)$ to a logically equivalent set of sequents, the validity of which can be directly tested via the axiom schemata of the system. This is similar to the overall approach used in the system $\mathfrak{N}(SU)$ (see section 5.A). However, the mode of reduction is different in the two systems.

In $\mathfrak{N}(SU)$, the reduction process consists of a nested sequence of transitions between sequents, where each transition corresponds to the application of a "prefabricated molecular argument" of *fixed structure*, i.e., a macromove, on a sequent. The transitions can be regarded as leaps between boundary sequents; these are sequents with certain characteristic features that define the terminal points of macromoves. The macromoves of $\mathfrak{N}(SU)$ are functionally homogeneous. This means that each macromove is designed to contribute to the same fixed extent in the overall task of reduction: A macromove reduces one main connective in the set of l.i. formulas of its boundary sequents, while preserving logical equivalence in the transition.

In the $\mathfrak{M}(SU)$, the reduction process is realized in a more dynamic and flexible manner. The transitions between boundary sequents in the decision trees of

$\mathfrak{M}(SU)$ do not have an *a priori* fixed structure. They are composed from combinations of rule applications in $\Gamma'(SU)$ that are selected and assembled (in a manner which is sensitive to the specific situation at hand) *during the growth* of the decision tree. This dynamic process of assembly is guided by the global logical requirements that are imposed by the desired reduction process. In the present system, our approach to the satisfaction of the global requirements is not through the imposition of functional homogeneity on the transitions. Each transition should satisfy the requirement of logical equivalence between its boundary sequents, but the reduction in the length of l.i. formulas is not fixed and limited to a single connective but it is open-ended, and it is determined locally during the process of assembling the transition.

It is difficult to view the transitions between boundary sequents in the decision trees of $\mathfrak{M}(SU)$ as playing the role of "molecular arguments," as in the case of $\mathfrak{N}(SU)$. If we were to identify "molecular arguments" in $\mathfrak{M}(SU)$, it would be more natural to consider them as consisting of combinations of rule applications from $\Gamma'(SU)$ whose goal is to form a loop in a proof graph. These "molecular arguments" are variable in size and they are sensitive to the properties of the evolving proof. They are realized by sequences of moves of $\Pi_r(PC, 3)$, where each move is made of one or more rule applications (we shall discuss these moves shortly). We regard a transition between boundary sequents in a decision tree of $\mathfrak{M}(SU)$ as reflecting a *logical plan* which is realized sequentially by the "molecular arguments" of $\mathfrak{M}(SU)$.

Rather than using "prefabricated" macromoves to effect transitions between sequents, the reduction process in $\mathfrak{M}(SU)$ synthesizes its transitions "on line" in a manner designed to satisfy simultaneously both logical completeness and proof minimality. Because of its dynamic nature, we cannot describe the structure of a logical plan in $\mathfrak{M}(SU)$ in the manner that we have described a transition schema—macromove in $\mathfrak{N}(SU)$. Such a plan can be described only in terms of the procedure that realizes it. We shall do this next by discussing the procedure $\Pi_r(PC, 3)$ that carries out the reduction process of $\mathfrak{M}(SU)$.

D. THE PROCEDURE $\Pi_r(PC, 3)$

(a) General Description

Two information structures are associated with a proof procedure $\Pi_r(PC, 3)$.

(i) A $\Gamma'(SU)$ graph that represents the current stage of construction of a proof; it is called the *construction map*.

(ii) A search tree of the type that we have used in previous reduction procedures for the overall administration of decisions in the proof construction process.

The construction map provides an instantaneous description of the *global state* of construction. Given a theorem x to prove, the initial global state is represented by a construction map with the single node $\bigcirc x$. The construction map that corresponds to the last global state portrays the graph proof of x in $\Gamma'(SU)$.

In the present procedure, the construction of a proof proceeds simultaneously, both as a graph in the construction map and as a decision tree in the search tree. There is vital interaction between these two representations in the course of construction. Roughly, the graph representation gives an overall view of the interactions (actual and potential) between the formulas, and this provides a basis for decisions. The tree representation helps in the organization of the sequences of decisions and in the search for alternative proofs.

States in the search tree correspond to sequents of $\mathfrak{M}(SU)$. The moves of $\Pi_r(PC, 3)$ are composed of elementary moves, each of which is an application of a rule of generation from $\Gamma'(SU)$. Thus, they have a certain structural similarity with the macromoves of $\mathfrak{N}(SU)$. However, the moves of the present system have no fixed structure at the fine level. They are made of sequences of steps that are taken simultaneously from both sources and destinations, in the form of exploratory thrusts whose immediate objective is to form loops. They can be considered as variable macromoves that are applied in a co-ordinated manner at different parts of the construction map. Actually, they can be best characterized as *logical manœuvres*. The moves-manœuvres of $\Pi_r(PC, 3)$, together with the rules that order their applications, define the decision processes that correspond to the transitions between boundary sequents in the decision trees of $\mathfrak{M}(SU)$. The effect of each move is to add a piece of a graph to the construction map.

A procedure $\Pi_r(PC, 3)$ proceeds roughly as follows. Selected features of the current global state are used as a basis for the application of a move-manœuvre that results in a new global state, and the process is repeated until a global state is attained such that the current construction map contains an acceptable proof—or until a refutation is obtained. Alternative moves are considered sequentially on the basis of a given order of preference, and the general orientation of the search is depth-first. When an acceptable proof is found in the construction map, a cleanup process is initiated that eliminates auxiliary structures in the map and yields a final global state whose corresponding construction map is the output of the procedure, i.e., the proof. The overall strategy is to attempt first, with the least possible effort, the construction of a proof which is most likely to be minimal. This is followed by a process of successive improvements, which will either yield a better proof or will strengthen the confidence about the minimality of the current best proof.

(b) Source Extensions

One of the important notions that enter in the specification of moves for $\Pi_r(PC, 3)$ is that of a *source extension*. This is a growth from a source node that develops according to the rules of generation of $\Gamma'(SU)$ in a manner devised to satisfy the desired global properties of logical plans in $\mathfrak{M}(SU)$. We can think of a source extension as a controlled deployment from a source node, intended to unravel the essential formula structure inherent in that node in a way that would facilitate the process of finding links for bridging this "forward" growth from sources with a controlled "backward" growth from destinations.

A source extension is built of *elementary source extensions.* The elementary source extensions are given as transition schemata in Table 6.3.

TABLE 6.3

ELEMENTARY SOURCE EXTENSIONS

Elementary extensions from nodes ◍ Dx_1x_2 and ◍ Ix_1x_2 are applications of the rules of generation D^1_{el} and I^1_{el} respectively.

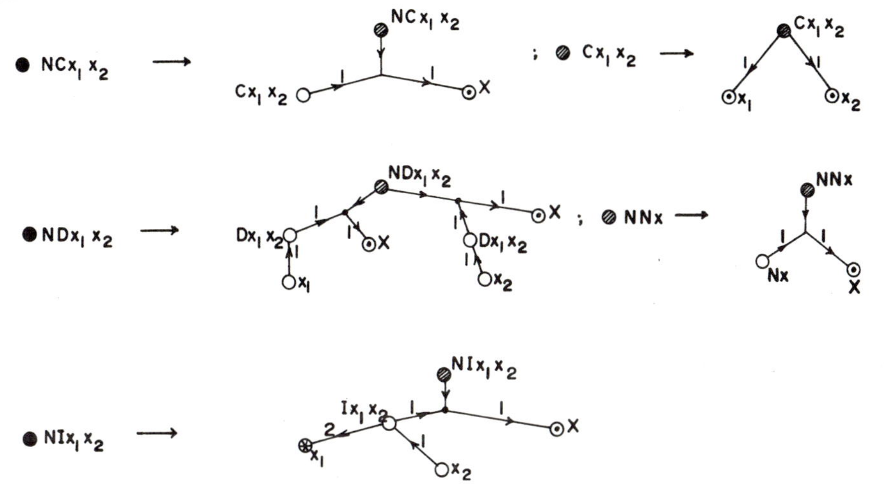

A source extension is built of elementary source extensions as follows.

(i) Generate the elementary extension from the node in question (if such an extension exists).

(ii) If any of the new nodes that are created by an elementary extension have themselves extensions, carry out the above process (i) on them, and proceed with the extensions of newly generated source nodes as long as possible.

Clearly, a source extension of a node ◍x has the form of a (finite) tree that is rooted at the node ◍x. Each subtree of a source extension tree is itself a source extension of the node at which it is rooted.

As we shall see shortly, the information formulas of a state correspond to certain source formulas in the evolving proof graph. The source extension process realizes a systematic fragmentation of source formulas into their subformulas, so that the length of the l.i. information formulas of successive states is monotonically decreased. Note that this process terminates when a source formula is reached which is either atomic (negated or not) or it is X. Thus, the source extension process effects a reduction of information formulas up to the point where the axioms of the system can decide directly on the validity of a state.

A source extension has some similarity with a macromove of $\mathfrak{N}(SU)$ whose key formula is an information formula. An important structural distinction, however, is that source extensions are not fixed, and they can develop as long as the formulas involved can allow them, performing this way as much fragmentation as possible on information formulas. Source extensions are constituents of moves in the procedure $\Pi_r(PC, 3)$. They develop from sources, usually in conjunction with some other development that is based on a destination.

(c) States

The initial state is given by $(\Rightarrow x)$, where $\bigcirc x$ is the starting node in the construction map (x is the candidate theorem). In a construction map that represents a global state other than the initial, each candidate destination node determines a state in the search tree. More specifically, a candidate destination node corresponds to the *goal formula* of a state.

The set of *information formulas* of a state that is determined by a candidate destination node $\bigcirc y$ corresponds to a set of source nodes in the construction map that are called *candidate sources* of $\bigcirc y$. These nodes form a subset of the set of *potential linking mates* of $\bigcirc y$. A potential linking mate of $\bigcirc y$ is a source node that is reachable from $\bigcirc y$ via a directed path in the construction map, and it is such that if linked to $\bigcirc y$ by a new branch, it will yield a new loop that has an incident node of $\otimes$ or $\oplus$ type.* The potential linking mates of $\bigcirc y$ can be obtained from the construction map via the following simple procedure. Starting from $\bigcirc y$, trace on the construction map all the paths that can be taken by following the arrows, and mark the paths traversed until a $\otimes$ or a $\oplus$ node is reached on each path. These nodes, as well as the source nodes that follow them along the continuation of the paths, are the potential linking mates of $\bigcirc y$. These sequences of source nodes should be uninterrupted. A path can continue until the sequence of source nodes is terminated (either no branch leaves a node or the next node is not a source node) or until the path encounters a previously traversed part of the graph.

In order to specify the set of candidate sources, we need to introduce first the notion of the *line of ancestry* of a node in a construction map. Each application of a rule of generation on a destination node—with the exception of the linking rules $\mathfrak{A}'_1$ and $\mathfrak{A}'_{1,1}$—and also each development of a source extension establishes an *ancestry relation* between the "old" formula from which generation or extension takes place and the "new" descending formulas. Since the ancestry relation is transitive, it is possible to determine in a straightforward way, on the construction map, the set of nodes that are ancestors of a given node up to the starting node of the map; this set constitutes the line of ancestry of the given node. A slight complication arises in the case of a descendant from a node ◍Dx_1x_2. An elementary

*The set of potential linking mates may be empty; if this is the case, then the set of information formulas of the state will be empty also.

extension from ◍Dx_1x_2 amounts to an application of the rule of generation D'_{el} (see Table 6.2). In this rule, two lines of ancestry merge temporarily and then they separate again. The line from ◍Dx_1x_2 descends to ⊛x_1 and ⊛x_2, and the line from the top node ⊙ϕ is transmitted to the two bottom ○ϕ nodes. The bottom formula variables (the ϕ's) are specified only after a connection is established (via the rule $\mathfrak{A}_{1,\phi}$) between the top ⊙ϕ and a destination node ○y. Thereafter, the bottom ϕ's are identified with y and they adopt its line of ancestry.

Now, given a candidate destination node ○y, the set of its candidate sources is obtained by excluding from the set of its potential linking mates all the source nodes that are on its line of ancestry.

Note that the complete determination of a state at any stage of construction is obtainable (via relatively simple processes) from the graph that has been grown in the map over the entire previous construction activity. In the present procedure, a state appears clearly as a partial description of the total problem situation; such a description includes just the essential features that are required for the selection of a next move.

(d) Moves

We shall distinguish between opening moves that apply mainly at the initial state and between two types of moves (linking moves and development moves) that apply at states that have information formulas.

(α) *Opening moves.* Opening moves apply at states that do not have information formulas, i.e., at states with a form ($\Rightarrow x$). Thus, opening moves always apply at the initial state (and this is the origin of their name). Given a state ($\Rightarrow x$), the relevant opening moves at this state are as follows.

(1) (i) If $x = Cx_1x_2$, apply one of the two alternative forms of the rule of generation C'_{in}; (ii) If $x = Dx_1x_2$, apply N'_{el}; (iii) If $x = Ix_1x_2$, apply I'_{in}; (iv) If $x = Nx_1$, apply N'_{in}.
(2) Continue the growth started in (1) by generating the extensions of the newly created source nodes (when possible).

It should be evident that an opening move has the nature of a manœuvre or of a *flexible development* whose extent and depth depends on the structure of the goal formula x in the state. If $x = Cx_1x_2$, there are two alternative opening moves. In the remaining cases, there is a single move.

As an example of an opening move in $\Pi_r(PC, 3)$, consider again Pierce's law for which we have the starting node ○$IIIpqpp$. The opening move-manœuvre from this node is as follows:

(6.3)

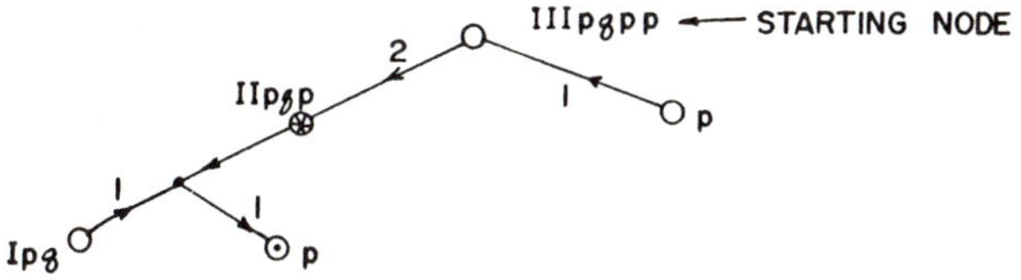

The rule I'_{in} applied at the starting node produces two new nodes. This is then followed by the elementary extension of the source node ⊛$IIpqp$, from which no further extension is possible.

Two types of moves apply at "states with information": linking moves and development moves. Linking moves attempt to close a loop in a proof graph, and development moves attempt to bring the situation to a point where linking moves are applicable. Relevant moves are tried from a state S in a *specified order*. Linking moves are tried first with the intention of rapidly establishing a loop. If a loop is possible, then before establishing it, care is taken to generate information formulas that may be required by other states that enter in the logical plan of which the loop closure process is a part. Such information may be required for conserving the logical equivalence of the state transition that corresponds to the plan. (This point will be discussed further below, and it will be illustrated in the example of this section.)

(β) *Moves from states that have information formulas.* For each state $S = (\eta \Rightarrow x)$, where x corresponds to a candidate destination node, and η to the non-empty set of candidate sources of ○x, we have a *complete* set of relevant construction moves such that if a move from this set is applied at S, then a certain progress towards the solution of S will always be made. If, for a given unresolved S, the set of relevant moves is empty, then S has no solution. If a linking move is not applicable at S, then development moves are attempted with the objective of either fragmenting the goal formula x or of bringing it into an "interaction" with information formulas whose form is disjunction or negation. Again, the moves are designed so that subsequent states will have available a complete set of relevant moves and so that decisions about the validity of these states (and about directions of subsequent development) can be made.

We shall present next the relevant construction moves for "states with information," together with an indication of the specific order in which these moves should be attempted. Only the main decisions for ordering moves will be specified. These decisions reflect our present stage of thinking about a heuristic strategy that was found (through experimentation) appropriate for the efficient construction of minimal proofs.

ORDERED LIST OF RELEVANT MOVES FOR AN INTERMEDIATE STATE

$$S = (\eta \Rightarrow x)$$

(I) *Linking moves.* (A) If there exists among the candidate sources a node ●x (i.e., the information formula is identical with the goal formula), then a direct linking move is possible. There are two alternative cases.

Case 1. If $x \neq X$, then establish an $\mathfrak{A}'_1$ link between ○x and ●x, with the following provisions (*a*) If the candidate source node ◍x to which the link is being established is part of the elementary extension of a node ◍Iyx, then generate a linking structure of the following type.

(6.4)

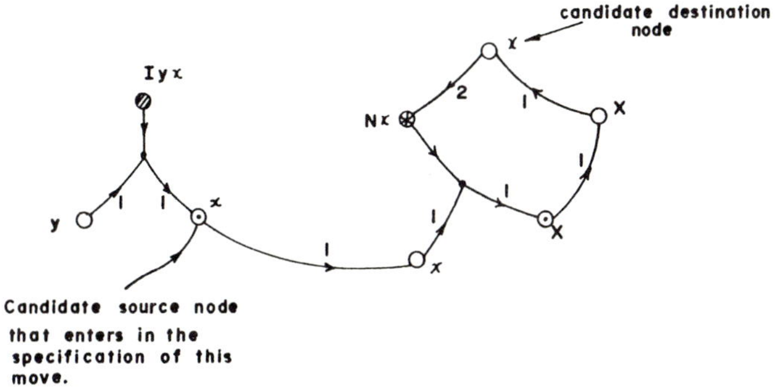

(In this diagram, we are assuming that x is a non-negated formula. If x is a negated formula [say, $x = Nv$], then substitute x by Nv and Nx by v in the linking structure.) Following the generation of the linking structure, develop the extension of the newly created source node ⊛Nx (or of ⊛v if $x = Nv$), if the node has an extension. (The linking structure is introduced because Nx is a logically required information formula for the state that has y as a goal formula. This can be seen by considering the (part of the) decision tree that corresponds to the construction in (6.4).

(6.5)

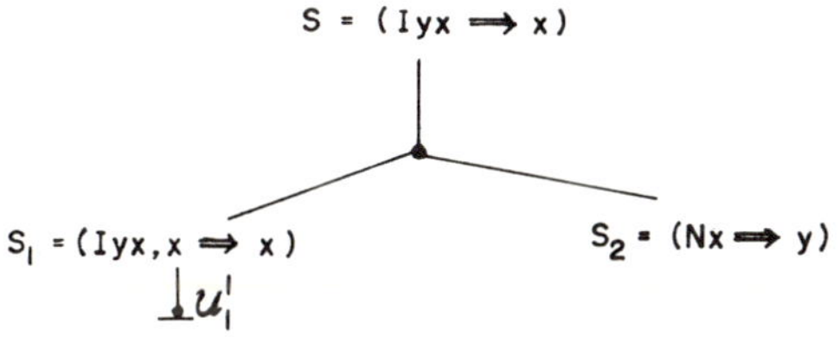

The state transition shown in (6.5) should conserve logical equivalence. This requires that Nx be an information formula in the state S_2. After the direct linking move is applied at S_1, the node ○y becomes a candidate destination node, and it determines the state S_2. It is important, therefore, for S_2 to have available, from this point on, source formulas that may be required for a decision on its validity. Note that if the node ⊛Nx or its extension are not involved in further construction, then the linking structure can be collapsed and replaced by a simple branch between ⊙x and ○x during the subsequent clean-up phase (this would result in a weight saving of 6).)

(*b*) If the candidate source node ◍x to which the link is to be established is of type ⊙, and if it is already linked to another destination node, then generate a new copy of the extension that leads to it from the nearest $\circledast$ or $\oplus$ node above it; use then the node $\oslash x$ in the new extension for establishing the link. (This rule is introduced in order to satisfy the conditions of generation in $\Gamma'(SU)$ according to which only a single rule is applicable on a $\odot$ node, while more-than-one rules are applicable from nodes of $\circledast$ and $\oplus$ type [see Table 6.2 and associated conditions].)

(*c*) If there exist more than one candidate sources for which a direct linking move can be established, then a preference ranking should be based on the minimization of move weight (e.g., a single-branch link is to be considered before links that require linking structures, or links that require copies of extensions). In general, any relevant direct linking move has preference over any other relevant move.

Case 2. If $x = X$, then establish an $\mathfrak{A}'_1$ link between $\bigcirc X$ and $\oslash X$, with the following provisions.

(*a*) If there exist more than one candidate sources for which the link can be established, then consider these sources in the following order of preference.

$$X(NNx),\ X(NIx_1x_2),\ X(NDx_1x_2),\ X(NCx_1x_2), \tag{6.6}$$

where $X(N\alpha)$ stands for a node $\odot X$ that appears in the elementary extension of a node of type $\oslash N\alpha$. In the present relatively simple procedure, we assume no preference ranking between alternative sources that are of the same type $X(N\alpha)$, except for cases where $N\alpha = NDx_1x_2$. A source node $\oslash NDx_1x_2$ has two nodes $\odot X$ in its elementary extension; if one of these has already entered in a linking move, then the other one is a preferred source among the sources of type $X(NDx_1x_2)$. In more elaborate procedures, it is possible to introduce further order among sources of the same type on the basis of chosen features of α. (The general rationale for our ordering is to cause an early "unraveling" of information formulas and a postponement of appearances of goal formulas—especially of double goal formulas that result from applying a generation rule at a $\bigcirc Cx_1x_2$ candidate destination node.)

(*b*) Here also, if the candidate source node to which the link is to be established (note that in the present case such a node is always of type $\odot$) is already linked to another destination node, then generate a new copy of the extension that leads to it from the nearest $\circledast$ or $\oplus$ node above it; use then the node $\odot X$ in the new extension for establishing the link.

(B) If there exist among the candidate sources, two contradictory formulas y, Ny, then an *indirect linking move* is possible. There are two alternative cases.

Case 1. If $x \neq X$, then on the basis of $\mathfrak{A}'_{1,1}$, establish a link as follows.

(6.7)

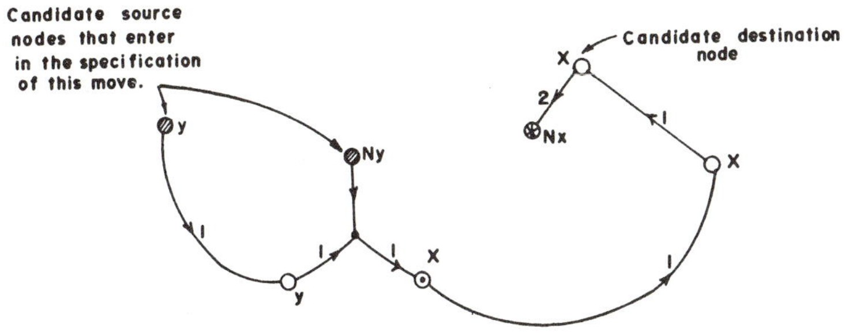

In this diagram we assume that x is a non-negated formula. If x is a negated formula (say, $x = Nv$), then substitute Nx by v in the linking graph. Following the generation of the indirect link, develop the extension of the newly created source node $\otimes Nx$ (or of $\otimes v$, if $x = Nv$), if the node has an extension. (The source node $\otimes Nx$ is introduced because Nx is a logically required information formula for subsequent states. The argument here is similar to the one given previously for the case of linking structures. If the node $\otimes Nx$ or its extension is not involved in further construction, then the branch leading to $\otimes Nx$ (and the extension of this node) can be eliminated from the proof graph at the clean-up phase).

Case 2. If $x = X$, then establish a link as follows.

(6.8)

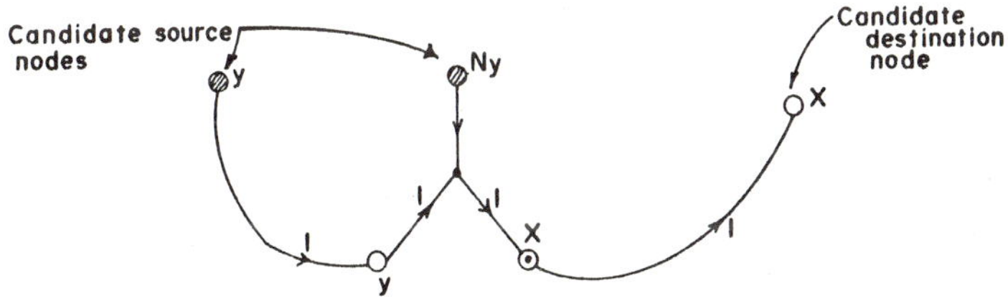

Again, in both cases of (B), if the sources $\oslash y$, $\oslash Ny$ are of type $\odot$, test for the possible need to generate copies. If there are alternative indirect linking moves, their preference ranking is arbitrary. In general, any relevant indirect linking move has preference over any other relevant development move.

(II) *Development moves.* These moves are classified by the form of the goal formula. In each case, they are given in the order in which they are to be considered.

(A) $x = Cx_1x_2$.

(1) If there exists among the candidate sources a node $\odot X$ which is not of type $X(NCyz)$, then apply the rule of generation N'_{el}; then develop the extension of the newly created source node.

(2) Apply one of the two alternative forms of the rule of generation C'_{in} (there is no *a priori* preference for one of the two forms). Generate a

source node ⊛NCx_1x_2, which is to be used as an auxiliary test node and associate it with ○Cx_1x_2. This situation can be represented in the construction map* as follows:

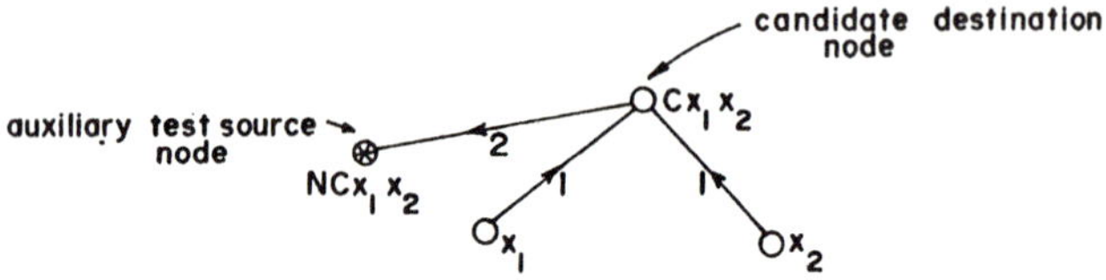

If it is found at a subsequent stage of proof development from this point, that the test node ⊛NCx_1x_2 *can* participate in a linking move, then a certain move is introduced as a relevant alternative from the present state, with the same preference rank as the last move used at the state. This conditional move is as follows: "apply the rule of generation N'_{el} and develop the extension of the newly created source node." (Thus, the test node ⊛NCx_1x_2 is not actually used in the potential linking move, but it serves to introduce an alternative move of high rank at the appropriate state, on the basis of a cue derived from "experimentation" in the current line of proof, which suggests that the new alternative move is promising. This approach increases the flexibility of the process of search in the space of proofs.)

(3) If there exists among the candidate sources a node ⊙ϕ which is part of the elementary extension of a node ◍Dyz, then establish an $\mathfrak{A}'_{1,\phi}$ link between the nodes ○x and ⊙ϕ (and then perform the appropriate substitutions in the extension of ◍Dyz). No preference for alternatives here.

(B) $x = Dx_1x_2$.

(1) If there exists among the candidate sources a node ⊙ϕ which is part of the elementary extension of a node ◍Dyz, then establish an $\mathfrak{A}'_{1,\phi}$ link between ○x and ⊙ϕ (and then perform the appropriate substitutions in the extension of ◍Dyz).

(2) Apply the rule of generation N'_{el}; then develop the extension of the newly created source node.

(C) $x = Ix_1x_2$.

(1) Apply the rule I'_{in}; then develop the extension of the newly created source node. Generate a test source node ⊛NIx_1x_2, and use it as in the previous case (II)(A)(2) of the present list for introducing the following conditional move: "apply the rule of generation N'_{el} and develop the

*In the examples of this paper, we will not show explicitly this operation on the maps since there will not be occasion to utilize its effect.

extension of the newly created source node." This move is to have the same preference ranking with the last move used at the state.

(2) If there exists among the candidate sources a node $\odot X$, then develop as follows,

(6.9) 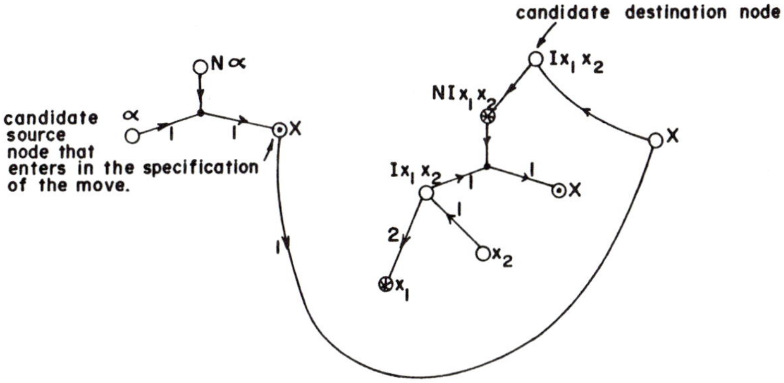

and continue the extension of the newly created source node $\otimes x_1$. If there exist more than one nodes $\odot X$ among the candidate sources, then the coupling to a $X(N\alpha)$ with an α of least "complexity" is to be preferred. The number of binary connectives in α can be used here as a measure of complexity. (The rationale for this move is an attempt to explore the possibility of building a simpler proof that is based on a *reductio* argument where the structure of the formula Ix_1x_2 is not involved.)

(D) $x = Nx_1$. Apply the rule N'_{in}; then develop the extension of the newly created source node.

(E) $x = p$ (p is a propositional variable).

(1) If there exists among the candidate sources a node $\odot\phi$ which is part of the elementary extension of a node ⊘Dyz, then establish an $\mathfrak{A}'_{1,\phi}$ link between $\bigcirc x$ and $\odot\phi$ (and then perform the appropriate substitutions in the extension of ⊘Dyz).

(2) If there exists among the candidate sources a node $\odot X$, then apply the rule of generation N'_{el}.

(3) If no linking move and no development move apply, then the state whose goal is $x = p$ has no solution. (This last rule is an application of the refutation axiom $\mathfrak{A}_0$ of $\mathfrak{M}(SU)$. Note that if we have arrived at the present point in the process of move application, then all logically irredundant formulas are atomic and the validation axioms $\mathfrak{A}_1$, $\mathfrak{A}_{1,1}$ do not apply [since previous attempts to apply linking moves must have failed].)

(F) $x = X$.

(1) If there exists among the candidate sources a node $\odot\phi$ which is part of the elementary extension of a node ◍Dyz, then establish an $\mathfrak{A}'_{1,\phi}$ link between $\bigcirc X$ and $\odot\phi$ (and then perform the appropriate substitutions in the extension of ◍Dyz).

(2) If no linking move and no development move apply, then the state whose goal is $x = X$ has no solution. (This last rule is an application of the refutation axiom $\mathfrak{A}_0$ of $\mathfrak{M}(SU)$.)

(e) Attention Control for the Construction of a Proof

The sequencing of moves in the course of proof construction is managed by the following rules:

(i) If, after the application of a move, no candidate destination node is left on the construction map, then the construction is completed and a proof graph is contained in the map. Proceed to the clean-up phase.

(ii) If, after the application of an opening move or a development move, there are candidate destination nodes in the construction map (there may be one or two), then if there is only one, take it to determine the next state, and if there are two, choose arbitrarily among them. Focus attention on the next state and proceed to the selection of a move and to its application on that state.

(iii) If, after the application of a linking move, one or more candidate destination nodes appear in the construction map, determine the next state on the basis of the candidate destination node that is closest to the source node to which the previous link was established. Focus attention on this next state and proceed to move selection and application.

The effect of these rules is to provide a "line of attention" that determines the sequence of states and the construction moves that the procedure follows in the process of constructing a proof. If no move is possible from a state, then no proof exists, the candidate theorem is not valid, and the procedure stops.

(f) The Clean-Up Process

In the course of developing a proof, the construction map plays the role of a combination sketch-pad and map of operations, where several auxiliary structures are tentatively developed in order to help the decisions that are taken in the construction process. When a solution is attained on the construction map, i.e., when we have a graph without candidate destination nodes, then a proof in $\Gamma'(SU)$ is easily obtainable from this solution-graph by eliminating the auxiliary structures and by simplifying parts of the graph. This is achieved via a clean-up process whose rules are as follows (the rules are applied in the order indicated).

(i) All branches that are not incident on loops of the graph are eliminated, with the exception of "supposition branches" produced below a node Ix_1x_2, i.e.,

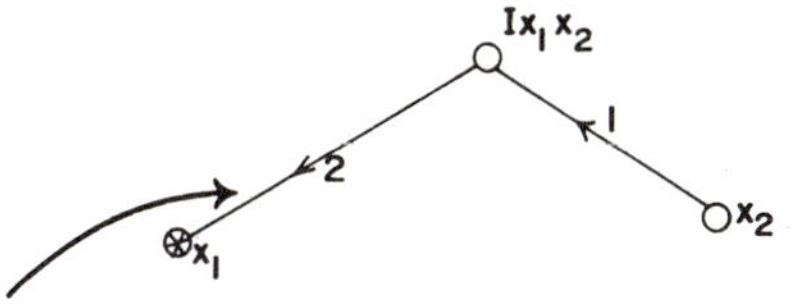

this branch is not eliminated even if it is not incident on a loop.

(ii) Consider each linking structure that was built for a direct linking move. If the source node of this structure, or its extension, is not involved in the construction (they are not linked to other parts of the graph), then the linking structure is eliminated and it is replaced by a link consisting of a single branch (this case occurs in example 1 of Appendix II).

(iii) Consider each development below a $\bigcirc Dx_1x_2$ node which is based on N'_{el} and a source extension, and perform the following condensations if applicable.

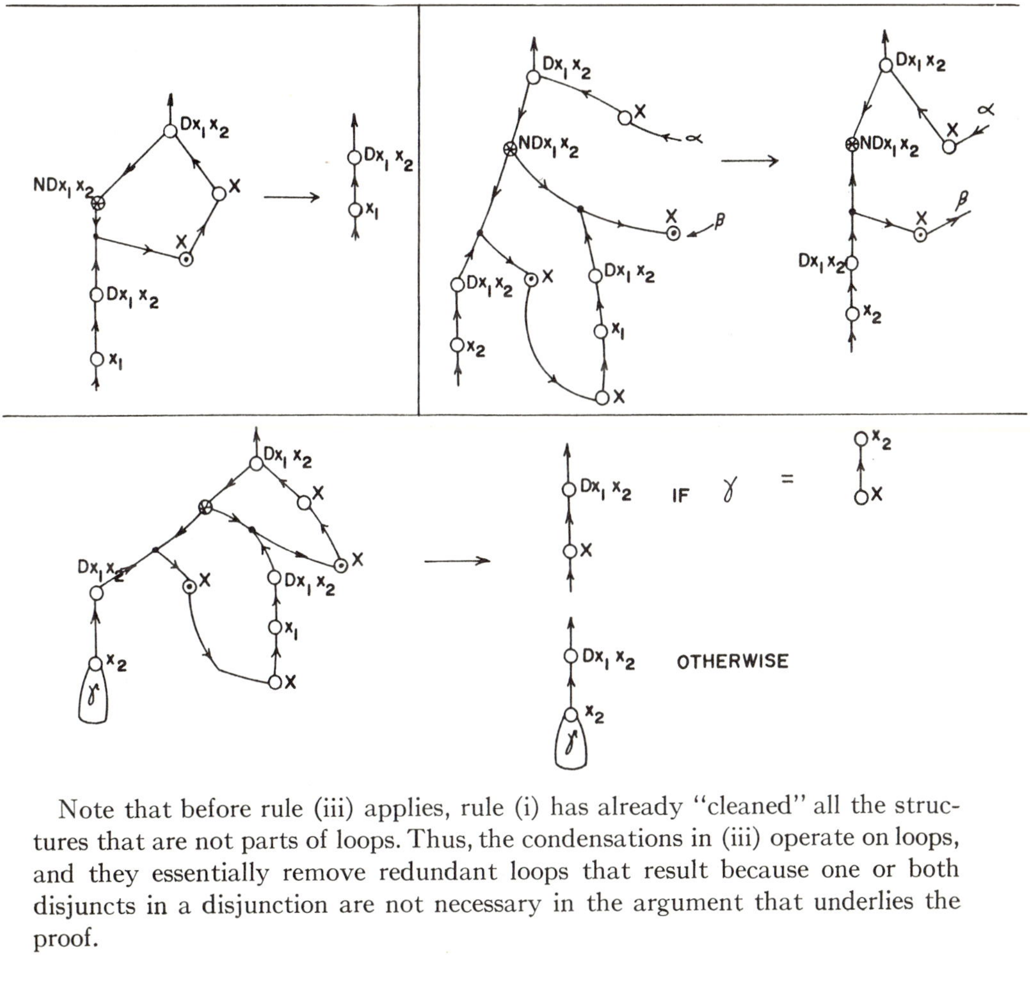

Note that before rule (iii) applies, rule (i) has already "cleaned" all the structures that are not parts of loops. Thus, the condensations in (iii) operate on loops, and they essentially remove redundant loops that result because one or both disjuncts in a disjunction are not necessary in the argument that underlies the proof.

(g) An Example of Proof Construction—Proof Representations

The properties of the procedure $\Pi_r(PC, 3)$ that we have discussed so far will become clearer if we illustrate them by means of an example. Let us consider then the construction of a proof to Pierce's law in the present system.

The construction map, as it appears after the construction process is completed, is shown in (6.10). There are four stages of construction, and they correspond to the four subgraphs that are shown in the map (for purposes of this illustration, the subgraphs are shown separated). Each subgraph is marked with a number that indicates the order of generation of the subgraph; the number points specifically to the destination node from which the generation of the subgraph is initiated. Except for one branch (shown enclosed in dashed lines) which is removed during the clean-up process, the map in (6.10) represents the minimal proof of Pierce's law in $\Gamma'(SU)$ (see (6.2)).

(6.10)

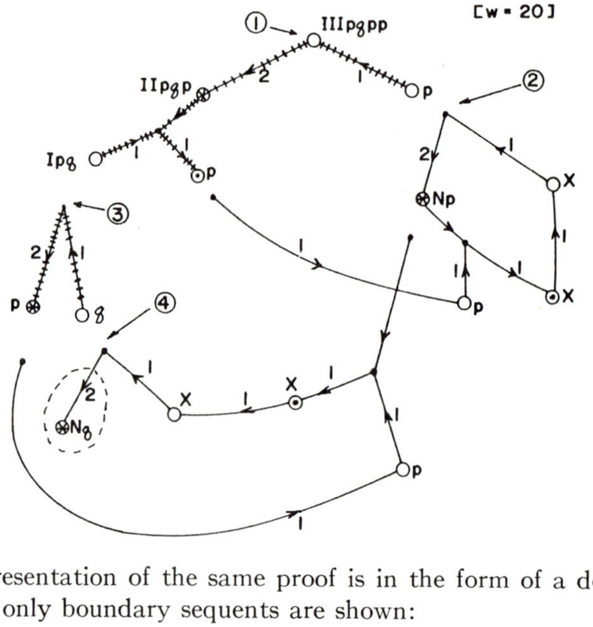

Another representation of the same proof is in the form of a decision tree in $\mathfrak{M}(SU)$, where only boundary sequents are shown:

(6.11)

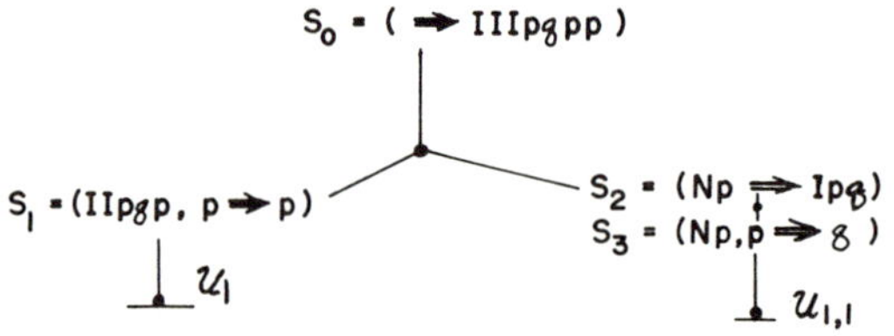

It is easy to see that the state transitions in this decision tree have the desired logical properties. Consider, for example, the transition from S_0 to S_1, S_2: the transition conserves logical equivalence and it reduces the length of logically irredundant formulas. It is interesting to compare this proof, which is formulated in $\mathfrak{M}(SU)$, and the proof obtained in the decision system $\mathfrak{N}^*(SU)$ (see (5.21)). The $\mathfrak{N}^*(SU)$ proof, which relies on transitions that have fixed structure, has a weight of 40; the $\mathfrak{M}(SU)$ proof, which is based on flexible transitions that are developed "on line," has a weight of 20—the minimal weight!

A state transition in a $\mathfrak{M}(SU)$ decision tree is built sequentially (usually this is an interrupted sequence) via a process that we have characterized as the execution of a logical plan. Let us examine this process in the case of the transition from S_0 to S_1, S_2 in our example. (See the construction map in (6.10).) Following the opening move (which builds the subgraph ① by I'_{in} and source extension), there is a single candidate destination node $\bigcirc p$, with candidate source nodes $\circledast IIpqp$, $\odot p$; these nodes completely specify the state S_1. The opening move produces also a non-candidate destination node $\bigcirc Ipq$. This destination node determines a *conditional descendant state*, S_2, whose information formulas are incompletely specified. The state is conditional until $\bigcirc Ipq$ becomes a candidate node. Unless the source $\odot p$ is involved in the construction, the destination $\bigcirc Ipq$ is not a candidate.

In general, an opening or a development move M in our present system may elicit conditional descendant states that originate from destination nodes in source extensions. Not all of these conditional states will enter necessarily in the proof. However, they are all considered as potential participants in the logical plan whose execution is initiated by the application of the move M. Those conditional states that enter in the proof will be specified sequentially in the course of the execution of the plan.

Let us return now to our example. After the opening move, attention focuses on S_1. The only possible move now is a direct linking move; the move constructs the subgraph ② by $\mathfrak{A}'_1$ and the requirement of a linking structure (since $\odot p$ is part of the elementary source extension of an implication). Now the node $\bigcirc Ipq$ becomes a candidate destination node and S_2 is no more a conditional state. The potential linking mates of $\bigcirc Ipq$ are $\circledast Np$ and $\circledast IIpqp$; however, since $\circledast IIpqp$ is in the line of ancestry of $\bigcirc Ipq$, only $\circledast Np$ is a candidate source in S_2. The state S_2 is completely specified at this point, and hence the transition from S_0 to its descendants S_1, S_2 is also completed. The process that we have just described, i.e., the execution of the logical plan that started with the opening move, has involved activity from S_0 to S_1, then activity below S_1 (that completed a loop in the construction map), and then a return to S_2.

Let us follow the construction process after the initial logical plan. A development move I'_{in} is the only possible move at S_2. It builds the subgraph ③ and it yields a single next state, S_3, with candidate destination $\bigcirc q$. Now, the candidate sources are $\{p, Np\}$ (again, $\circledast IIpqp$ is excluded because it is in the line of ancestry of q). The only possible move at S_3 is an indirect linking move. Its application produces the subgraph ④, and no other candidate destinations are left. This terminates the construction. In the clean-up phase, the branch that leads to $\circledast Nq$ is eliminated (according to the rule (i) of the clean-up process) and the remaining graph in the construction map is the minimal proof of Peirce's law in $\Gamma'(SU)$. Note that at no point during the development of this minimal proof have we had any choice of moves. There was a unique line of construction.

Consider now in a more direct way (i.e., not via an interpretation which is based on $\mathfrak{M}(SU)$) the nature of the construction process in our example. We have, essentially, two "molecular arguments." The first argument closes a main loop with a link $p\odot$—$\bigcirc p$, and it establishes an auxiliary loop while building a linking structure for the production of the source node $\circledast Np$ (which will be needed for the second argument). The construction of the first argument is carried out by the first two moves. Each move resembles a manoeuvre which involves several elementary moves (i.e., applications of rules of generation from $\Gamma'(SU)$) that develop both from destination and source nodes. The second "molecular argument" closes two loops simultaneously by using two moves, the second of which establishes an indirect link that completes the construction.

The direct language of construction includes the notion of a molecular argument that is executed by a sequence of moves-manoeuvres and that are themselves carried out by applications of rules of generation. The notion of a logical plan, which is derived from the system $\mathfrak{M}(SU)$, enters here only in an indirect way. This notion is used to guide the detailed specification of the elements of the construction process, so that the process will satisfy global properties of logical completeness. Thus, we have two languages of construction that correspond to the two information structures that are associated with our procedure. The correspondence between representations in the two languages is known, but is not straightforward. Each representation, however, has a different range of usefulness. In combination, these representations provide a good basis for evolving a procedure, such as $\Pi_r(PC, 3)$ whose objective is to satisfy in an efficient way both logical completeness and proof minimality.

The tree representation of (6.11) shows clearly the logic of the proof in our example. However, in constructing the proof, this tree was formed in a certain sequence of steps, with the "line of attention" starting at S and going through the states in a specific order. This is shown schematically in the following diagram, where the dashed line represents the "line of attention."

(6.12)

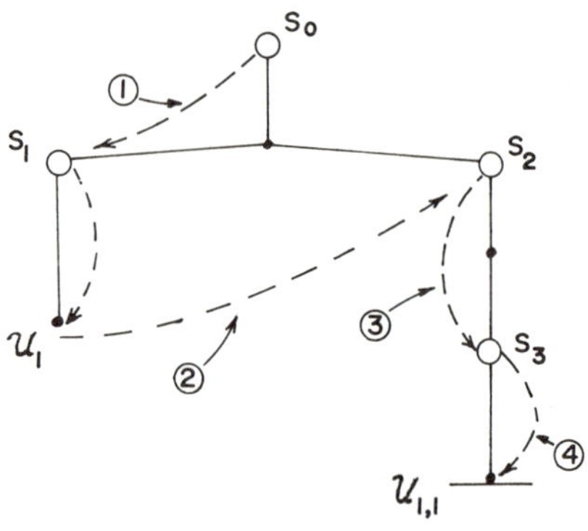

Clearly, we can represent the proof by a single chain which is determined by the path of the "line of attention" over the proof tree. In the case of the present example, this will result in the following representation of the proof:

$S_0 = (\Rightarrow IIIpqpp)$

(1) $\downarrow$ Opening move (I'_{in} + source extension), [5]

$S_1 = (IIpqp, p \Rightarrow p)$

(6.13) (2) $\downarrow$ Direct linking move ($\mathfrak{A}'_1$ + linking structure), [7]

$S_2 = (Np \Rightarrow Ipq)$

(3) $\downarrow$ Development move (I'_{in}), [3]

$S_3 = (p, Np \Rightarrow q)$

(4) $\downarrow$ Indirect linking move ($\mathfrak{A}'_{1,1}$). [5]

A number in brackets associated with a move denotes the move weight—alternatively, the weight of the subgraph constructed by the move.

Since the chain representation of the proof (or rather the decisions made in proof construction) is a natural representation in our present system, we shall adopt it as a basis for representing search trees in $\Pi_r(PC, 3)$. Thus, branching in the search tree occurs only when alternative moves are possible from a state. This means that search trees will have only disjunctive branching. Note that in our present example the search tree and the proof chain (6.13) are identical since there is a unique line of proof construction.

Examination of the proof graph in (6.10) shows that it is made of a supporting structure (shown by $\overset{\perp\perp\perp\perp}{\gamma}$ lines) and of a set of linking structures. The supporting structure was built in stages 1 and 3, while the two linking structures were built in stages 2 and 4. It is interesting to note that the supporting structure corresponds to the conventional tree representation of the formula *IIIpqpp*. Such correspondences between subformulas of the theorem to prove and substructures of the proof appear often in graph proofs constructed by $\Pi_r(PC, 3)$. The examples given in Appendix II are intended to illustrate further properties of the present system.

(h) Control of the Search Over the Space of Alternative Proofs

As indicated during the discussion in our previous example, search trees are made of proof chains, and each proof chain corresponds to the scan of a decision-proof tree in $\mathfrak{M}(SU)$ by a "line of attention." The branching in a search tree is due to alternative relevant moves from states. Since all the moves of $\Pi_r(PC, 3)$ are designed to satisfy logical equivalence of state transitions in decision trees, then all the alternative chains in a search tree are logically equivalent. This means that it suffices for one chain to be stopped by a refutation axiom in order to stop the process of proof search and to output a "not valid" result for the candidate "theorem." Also, after one proof chain is completed, then we know that if alternative moves exist and if they are pursued, they will yield alternative proofs.

Our general approach to the control of search is to attempt first the growth of a single proof chain which is most likely to correspond to a minimal proof. In the development of such a chain, only the most preferred relevant move is applied at each state. If a refutation is offered, then the search stops. If a proof is obtained, then its weight is designated as the current best weight w^*. If no alternative relevant moves exist from the states of the current search tree, then the process stops and it outputs the proof that has the current best weight. In the example of Peirce's law, there are no alternative relevant moves, and the process stops after generating a single proof— which is the minimal proof. This is also the case in the first example of Appendix II. However, in the two last examples of the Appendix, alternative proofs do exist, and the search procedure continues beyond the formation of the first proof.

If alternative relevant moves exist, then they are tried according to their preference ranking, first from states that are close to the top of proof chains, and then, progressively, from states that are further down the chains. After the completion of a proof chain, a new direction of search is chosen by returning to the top of the search tree and by scanning it systematically down to the first state where an untried alternative move exists that has the *same* preference ranking with the move that was last applied at this state (the state can be the top state itself). A new development is now initiated at this state; it starts with the alternative move just found and it proceeds via the application of most preferred moves to completion of a new proof chain. When untried alternative moves with a preference ranking identical to the last used no longer exist in the search tree, a similar process is initiated that seeks alternative moves of the next lower preference rank. The approach to the order of search which we have just outlined is the same with the approach used in the processes of "successive improvements" in our previous two procedures.

Each new development goes in depth in an attempt to obtain a complete proof. The development stops when a weight is attained which exceeds w^*. If a proof is found with a lesser weight than the current best weight, then the weight of this proof establishes a new standard of proof quality for the subsequent stages of search. The process of search for alternative proofs stops either when a given limit of effort is exceeded or when there are no alternative lines of search that yield better proofs than the current best proof. The search over alternative proofs which we have just outlined is illustrated in the example 3 of Appendix II.

It is possible to introduce improvements in the process of search over alternative proofs that would increase the rate of convergence towards the minimal proof and that would therefore decrease the amount of total search. One approach that deserves further study is the one that we have used for the conditional development moves from $\bigcirc Cx_1x_2$ and $\bigcirc Ix_1x_2$ (see (II)(A)(2) and (II)(C)(1) respectively in the ordered list of relevant moves for states with information). In this approach, information that is obtained in the course of developing a proof chain is used to control the identity and order of preference of alternative moves that have to be considered in the subsequent search. This way, the search will be more sensitive to the properties of the specific problem at hand. An important question here is to find methods for gathering information about the proof space while constructing a specific proof. In the case of our conditional development moves, this involves the generation of a test source node and the carrying out of "linking experiments" with this node.

Another important idea towards improving efficiency of search is to introduce processes for the recognition of certain syntactic symmetries in proof graphs, so that the growth of alternative proofs that are identical—except for a re-labelling of nodes—could be avoided. A situation where such a recognition of syntactic symmetries would be clearly useful is illustrated in the example 3 of Appendix II.

E. PROPERTIES OF THE PROCEDURE $\Pi_r(PC, 3)$

THEOREM. *The procedure* $\Pi_r(PC, 3)$ *is a decision procedure for the propositional calculus.*

Proof Sketch. From a detailed (and relatively lengthy) case analysis, it can be verified that the effect of a sequence of move applications in $\Pi_r(PC, 3)$, starting from any initial state $(\Rightarrow x)$, is to realize a transition from the initial state to a logically equivalent set of states each of which can be tested directly for validity. The process of transition terminates after a finite number of move applications, since each move monotonically reduces the length of the logically irredundant formulas in a state (both goal formulas and information formulas). When all the logically irredundant formulas are either atomic or X, then the procedure can decide directly about the validity of a state via the moves that realize the axiom schemata of the decision system $\mathfrak{M}(SU)$.

Thus, given any theorem in the propositional calculus, the procedure $\Pi_r(PC, 3)$ will generate its proof. The proof will appear initially as a graph in $\Gamma'(SU)$. This graph can be converted into the form of a proof in SU (as discussed in section 6.B).

CONJECTURE. *The procedure* $\Pi_r(PC, 3)$ *produces the minimal proof for any theorem in the propositional calculus.*

This conjecture is strongly supported by considerable experimental evidence (many hand simulations), and by some arguments that are not completely conclusive as yet. Furthermore, our work to date suggests that the graph representation of proofs provides an appropriate basis for proofs of minimality.

The procedure $\Pi_r(PC, 3)$ appears to be considerably *efficient*. This has been illustrated in this paper by a few examples. However, as shown in example 3 of Appendix II, there is room for improvement in the efficiency of the procedure. We expect that improved versions of $\Pi_r(PC, 3)$ will include refinements in the rules for move selection and in the rules for sequencing the search over alternative proofs. It would also be interesting to explore other types of approaches to the construction of minimal proof graphs that might capture in a better way the ease with which people find ways of closing such graphs.

7. SUMMARY AND DISCUSSION OF THE EVOLUTION OF PROOF CONSTRUCTION SYSTEMS

Let us summarize the main ideas that enter into the evolution of the systems for the solution of our proof construction problem. The problem is to find the simplest proof of a theorem in the propositional calculus. This problem was first posed, in its initial raw form, within a system of suppositions.

In order to enable the application of a known problem solving mechanism, i.e., a reduction procedure, to its solution, the problem is reformulated in a system of natural inference $N(SU)$. This marks the beginning of the first stage of development, or the first level of evolution. In the transition $SU \rightarrow N(SU)$, a specific decision is made about the mode of applying deduction rules of SU in the development of proofs in $N(SU)$. This decision induces the (commonly used) problem solving approach of working backwards from goal formulas to suppositions and deduced formulas. It should be emphasized that it is an important choice point for the designer to specify the way in which rules of a "rule system" are to be used in the formulation of rules of inference (or moves) of a system of reduction. During the evolution discussed in this paper, such a choice point occurs at the beginning of the first and third stages.

The representation of the problem in $N(SU)$ permits the formulation of reduction procedures $\Pi_r(PC, 1)$ where a systematic search in width takes place over the solution space. Relevant moves are restricted by a small set of "reasonable" heuristic rules which is initially suggested from experience. The usefulness of such restrictive rules for complex problems of search is well known. The open problem at present is to understand their "life cycle", i.e., the processes of their birth (must their origin be exclusively human?), their maturation and their ultimate survival (what inferential processes should be used for evaluating their strength and for ultimately proving or refuting their value?). In our case, the heuristic rules for move restriction were born from paper and pencil experimentation, they were shown in the second stage of evolution not to spoil logical consistency, and they have given strong empirical evidence to date that they do not affect adversely proof minimality. They now have reached the status of strong heuristic rules.

A wide class of refinements that can be found in several heuristic search procedures that are characteristic of the present state of the art* apply to the procedures $\Pi_r(PC, 1)$: stronger recognizers for terminations, avoiding obvious circular developments, recognizing and transferring solutions of "similar" problems, and improving the attention control mechanisms. It is an important

*A survey and classification of such procedures is given by Newell in [19].

advantage of our reduction framework that these features can be naturally formulated and extended within the framework, and their relationships can be clarified. A promising possibility, not sufficiently explored as yet, is to study in our reduction framework the optimal balance of static recognition activities *versus* dynamic growth-search activities that take place in heuristic procedures of type $\Pi_r(PC, 1)$.

A more radical modification in $\Pi_r(PC, 1)$ is guided by a *strategy of successive improvements*. This strategy attempts to satisfy the requirement of getting the best possible result for any given investment of problem solving effort. According to this strategy, a single, reasonably good proof should be obtained first, and subsequent effort should be directed to possible improvements. It is important for the realization of such a strategy to have a good basis for developing a search in depth. Thus, a preference ranking over the relevant moves and a recognizer of dead ends are desirable. The problem of finding a simple ordering of the moves, when one exists in principle, is not trivial. The difficulty is compounded when even reasonable approximations to such an ordering do not exist. It is useful to recognize that such non-obliging situations are quite common. The important question is: under what conditions (at what stage of the solution, in what context) orderings of moves or other affinities and relationships between moves can be found and formulated? This is a pattern discovery and classification problem whose information basis is the space of solutions (in our case, the space of proofs). A reasonable approach to this problem must start with some *a priori* scheme for articulating the structure of a proof. The idea of looking for "molecular arguments" appears in this context. This idea, together with the need for a recognizer of dead-ends, provide the triggers for the transition to the second stage of evolution, which is based on the natural decision system $\mathfrak{N}(SU)$.

Two basic concepts are necessary for the formulation of $\mathfrak{N}(SU)$: the concept of logically irredundant formulas and the axiom of refutation. The former helps to isolate information which is "really" needed for proof construction, and the latter provides a recognizer of dead-ends. These two concepts come from logic. They subsume knowledge of the formal properties of the system in which we are trying to construct proofs. It is difficult to see how such knowledge can develop exclusively from the experience in the task of proof construction. It seems that it is necessary to have a separate system* that is specifically oriented to the finding and relating of the formal properties in the problem area under consideration. Without such a knowledge system, a strong evolutionary jump in the power of a problem solving system is not feasible. In our case, as a result of knowledge introduced in the second stage, a new type of procedure is suggested. Also, this knowledge is used to evaluate the heuristic rules on move restriction that were introduced in the previous stage. At present, the knowledge creating and theory forming activities are exclusively in the human domain. It is conceivable, however, that parts of such activities are mechanizable. It would be

*It is suggestive to regard the relationship between such a system and the proof-construction system as analogous to the relationship between "research" and "production."

extremely challenging and rewarding for machine problem solving to attempt a mechanization of such theory formation processes (see [1]).

The two new logical concepts that enter in $\mathfrak{N}(SU)$ provide the basis for a decision schema, whereby an initial state is reduced to a set of states whose validity can be directly tested. The rules of decision of $\mathfrak{N}(SU)$ are devised to carry out the overall reduction by taking one (equal) reduction "step" each. The synthesis of the decision rules of $\mathfrak{N}(SU)$ is based on this functional requirement. The decision rules that are synthesized correspond to the *macromoves* for reduction. The formulation of a specific mode for realizing the overall reduction, i.e., the choice of a functional specification for the macromoves, is an important decision point in the design. Each variation on these functional specifications will result in a new class of procedures. We can view this point differently: since a macromove has the status of a "molecular argument" that imposes a certain high-level structure on proofs, then each new way of structuring proofs that is implied from a different notion of a "molecular argument" will result in a new class of procedures.

The representation of the proof construction problem in the decision system $\mathfrak{N}(SU)$, and also in the related system $\mathfrak{N}^*(SU)$, allows the formulation of reduction procedures $\Pi_r(PC, 2a)$, where a proof is obtained with certainty (if it exists) after taking a few large leaps, with the help of the macromoves, in the search space. Thus, these decision systems provide a *planning space*. One proof, or a few alternative proofs, are obtainable rapidly in this space, but they are usually far from minimal. The strategy of successive improvements that is realized by the procedure $\Pi_r(PC, 2)$ consists of two phases: first find a proof in the planning space (find a feasible solution), and then try to simplify it (use the feasible solution as a focal point for finding an optimal solution). The idea of looking first for a feasible solution and then trying to improve it in order to arrive at the desired optimal solution is very valuable and quite old in the art of problem solving. It does not guarantee, however, that the optimal solution will be found. In our case, the approach taken in the second stage of development yields a powerful procedure of successive improvements. We have experimental evidence that this procedure requires much less effort for obtaining a minimal proof than the procedures of the first stage. It is not clear, however, whether the procedure can always find a minimal proof, or under what conditions this would be possible. Our situation here has similarities with a problem of optimization in automatic programming, where one wishes to obtain optimal object code at the output of a system that consists of a macrocompiler followed by a post-optimization programme.

The third stage of development is triggered by a search for a more uniform and direct approach to the construction of a minimal proof, and also by the emergence of a new *graph representation of proofs*. The idea of the graph representation is dominant in this stage. A proof is represented as a closed graph of a special type. Such a graph clearly shows the pattern of interactions between the formulas that enter in the proof. Proofs are regarded as words in graph languages, and these languages are described in terms of the generating systems $\Gamma(SU)$ and $\Gamma'(SU)$. The problem of finding ways for articulating the structure of a proof (e.g., of finding a useful "molecular argument") can now be approached by studying proof-graph generating systems. The graph representations immediately suggest a unit of problem solving activity, namely, the set of move applications that lead to the closure of a loop in the proof graph under construction. The proof construction problem is much easier in its graph representation; furthermore, the notion of loop closure seems to be a natural subgoal for the overall task of construction. This is an informal observation. It would be interesting to study this point more methodically by psychological experimentation.

The procedure $\Pi_r(PC, 3)$ of the third stage is based on the new idea of graph representations, on the logical and problem solving notions that already existed in the previous stages of development, and on several new problem solving ideas. The overall schema of reducing the initial state to a set of states whose validity can be directly tested is the same as in the second stage. However, this schema is now realized in a more dynamic and flexible manner. The steps of $\Pi_r(PC, 3)$ are logical manoeuvres that consist of elementary reasoning moves. These elementary moves develop both backwards from goal formulas and forwards from supposed or deduced formulas. Several of the step-manoeuvres of the system combine to close a loop in the proof graph. The sequence of loop closures is co-ordinated by a logical plan. The plan imposes on the sequence the global goals of the logic of reduction. The hierarchy defined by elementary move-manoeuvre-loop closure-plan gives considerable structure to the problem solving process. The large features of this process can be expressed within our framework for reduction procedures. This is true specifically for the reduction schema and for the logical plans. However, processes of move selection and attention control are more naturally expressed in terms of the language of proof graphs. In $\Pi_r(PC, 3)$, it is natural to work with two information structures simultaneously: the construction map, where the proof graph is manufactured, and the search tree, where the overall administration of the decisions takes place. This type of scheme, which combines a sketch-pad-experimental bench type of structure with a decision tree structure, is required by many complex problem solving systems. It appears that while our general reduction framework is useful at the global level of description in such systems, the decisions at lower levels are strongly ruled by the specific structures of the solution spaces under consideration, and specific structures cannot be fitted naturally within a single general framework. It is crucial for the efficiency of the problem solving system to have "appropriate" representations for structures in the solution space. Examples of such "appropriate" representations are the graph representations of our third stage.

The relationship between approaches to proof construction in our three stages of development is shown schematically in (7.1). Think of S as a given starting situation and G as a desired goal situation in a certain space. A minimal path is to be found from S to G.

(i) In an elementary procedure of the first stage, a systematic exploration proceeds away from S in concentric circles, looking into all possible directions since no basis for a preferred direction exists. When the moving circular front reaches G, the problem is solved.

(ii) Procedures of the second stage have a sense of direction which is acquired by the added logical knowledge that appears at this stage. A macromove guarantees one step of progress in the direction of the goal, but such a step may involve unnecessary movement in other directions. At the "planning phase" of the procedure, the path *SxaxbcdeG* is obtained; at the simplification phase, the loop *xax* is removed and the minimal path obtained.

(iii) In the procedure of the third phase, the execution of each logical plan takes us an uneven distance in the direction of the goal. The size and nature of the step depends on local conditions. The three steps shown in the figure have different sizes and they directly yield the solution. It should be emphasized again that these three illustrations are a very simplified description of the situation.

(7.1)

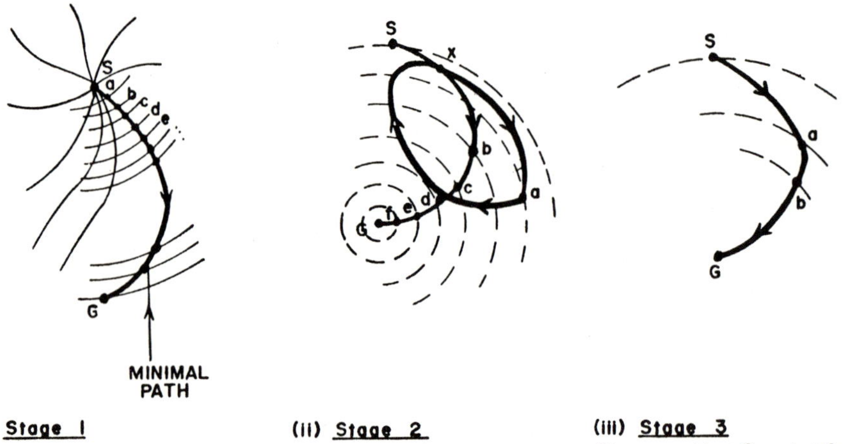

(i) Stage 1 — Moves: Sa, ab, bc, etc.

(ii) Stage 2 — Macromoves: Sa, ab, bc, etc.

(iii) Stage 3 — Plan execution: Sa, ab, bG.

8. CONCLUSIONS

The conceptual framework that we have introduced in section 2 has proved useful in the formulation of several proof construction procedures and in the clarification of key relationships between these procedures. The concept of problem solving efficiency suggested by this framework permits the evaluation of relative efficiencies for procedures that have different search spaces (i.e., they are working with different representations of the problem). By examination of the variety of relevant moves that is available to each of our three proof construction procedures, it is immediately clear that the likelihood of unnecessary generation of states in a search tree is strongly decreased by going from $\Pi_r(PC, 1)$, to $\Pi_r(PC, 2)$, and $\Pi_r(PC, 3)$. Thus, efficiency increases as we go up in the evolutionary process. Results of experimental work with out procedures confirm this point. The example that we have used throughout the paper illustrates a strong gain in problem solving efficiency as we move from $\Pi_r(PC, 1)$ to $\Pi_r(PC, 3)$.

The procedure $\Pi_r(PC, 3)$ was found (from extensive hand simulations) to be a powerful theorem prover. It also has certain structural features that may prove to be of general significance for future problem solving procedures. It would be of considerable interest to write experimental programmes for computer realizations of this procedure.

Further theoretical work with the graph representation of proofs promises to add insights into the nature of logical arguments, and it may also yield formal results on minimality properties of proofs. Extensions of the present work into (fragments of) the predicate calculus are well worth exploring. The work of Kanger [20] would be relevant here.

One of our main objectives in this work was to probe the nature of the problem of representation in problem solving procedures. Each major transition in the evolution of the proof construction systems is characterized by a reformulation of the problem, i.e., in each transition a change of representation takes place. A preliminary analysis of these representational changes indicates that they are of different types.

The effect of the first change which takes place in the transition $SU \rightarrow N(SU)$ is to establish a match between the problem and a given problem solving procedure. Here we have an encoding of the problem into the given language of the procedure. This is essentially a *language translation* task of the type performed by a compiler.

The second change which takes place in the transition from $N(SU)$ to $\mathfrak{N}(SU)$ reflects the utilization of acquired knowledge about the formal properties of the solution space towards increasing the power of the procedure. The overall form of the procedure remains essentially the same, but states and moves are redefined on the basis of the new knowledge. A more rational, planned, search for a solution can now be developed. Here we have an *intelligent adaptive process* which utilizes new information towards a redesign of parts of the procedure in order to increase its power.

In the transition to the third stage, a new language for representing the solution space is created. This has strong implications for the overall form of the procedure. Roughly, the effect of this transition is to create a "custom-made" procedure that matches best the given problem. Here, the situation is almost opposite to that of the first transition. The task is one of *finding an "appropriate" space for the treatment of the problem*, and also of designing a procedure that goes with it.

The transitions to the first and third stages involve processes of pairing a problem with a procedure. These are difficult processes that require creativity (see [1]). It appears that the transition to the third stage is substantially more difficult than the transition to the first stage. The mechanization of processes such as these is still far in the future. On the other hand, it appears that the mechanization of the process of transition to the second stage is in sight (at least parts of the process). The details of the evolution that are presented in this paper are intended to provide clues for the possible mechanization of the processes of transition.

The ability to move through an evolution of the type described here seems to be a necessary property of an ideal general purpose problem solving system. Putting it differently, an important requirement for such a system is the ability to respond immediately, but perhaps poorly, to problems from a new class (with certain restrictions on the type of the class), and to reach a point—after a certain period of activity with problems from this class—where it can perform as if it were especially designed for these problems. At present, we are far from having an automatic system of this type. Progress towards attaining such a system depends on the degree of success in transferring some of the evolutionary design processes discussed in this paper from man to machine.

ACKNOWLEDGMENTS

I am indebted to J. Guard and T. H. Mott for stimulating my interest in the logical systems that I have used in this work. The first stage of development of the proof construction procedure was carried out in collaboration with G. Cooke and D. A. Walters. I would also like to acknowledge the fruitful discussions that I have had with my students at the Carnegie Institute of Technology in a graduate course on complex information processing where this work was presented.

REFERENCES

[1] Amarel, S., "On the Mechanization of Creative Processes," *IEEE Spectrum* (April, 1966).

[1.1] ——— "Problem Solving Procedures for Efficient Syntactic Analysis," paper presented at the ACM 20th National Conference (Aug., 1965).

[2] ——— "An Approach to Problem Solving by Computer," Part II of *Final Report AFCRL-62-367*, contract no. AF19 (604)-8422 (May, 1962).

[3] Newell, A., and Simon, H. A., "The Logic Theory Machine: A Complex Information Processing System," *IRE Transactions on Information Theory*, IT-2(3) (Sept., 1956).

[4] Wang, H., "Toward Mechanical Mathematics," *IBM Journal of Research and Development*, 4(1) (Jan., 1960).

[5] Davis, M., "Eliminating the Irrelevant from Mechanical Proofs," *Proc. Symposium Applied Mathematics*, XV, (Providence, R.I.: American Mathematical Society, 1963).

[6] Wos, L., Robinson, G. A., and Carson, D. F., "Efficiency and Completeness of the Set of Support Strategy in Theorem Proving," *Journal of the ACM*, 12(4) (Oct., 1965).

[7] Beth, E. W., *On Machines Which Prove Theorems*, Simon Stevin, Wis-En Natuurkundig Tijdschrift, 32e Jaargang, Aflevering 2.

[8] Fitch, F. B., *Symbolic Logic—An Introduction* (New York: Ronald Press Co., 1952).

[9] Nidditch, P. H., *Introductory Formal Logic of Mathematics* (London: University Tutorial Press, 1957).

[10] Gentzen, G., "Untersuchungen Uber Das Logishe Schliessen," *Math. Zeit.*, 39 (1934).

[11] Jaskowski, S., "On the Rules of Suppositions in Formal Logic," *Studia Logica*, 1 (1934).

[12] Mott, T. H., Guard, J. R., Bennett, J. H., and Easton, W. B., "Introduction to Semi-Automated Mathematics," *Final Report AFCRL 63-188*, contract no. AF19 (628)-468 (April, 1963).
[13] Cooke, G., "Machine Theorem Proving," *Sc. Rep. No. 1, AFCRL 925*, contract no. AF19 (604)-8422 (Sept., 1961).
[14] Walters, D. A., "A Machine Procedure for Efficient Generation of Simplest Proofs in the Propositional Calculus," *Part I of Final Report AFCRL-62-367*, contract no. AF19 (604)–8422 (May, 1962).
[15] Gelernter, H. L., and Rochester, N., "Intelligent Behavior in Problem-Solving Machines," *IBM Journal of Research and Development*, 2 (4) (Oct., 1958).
[16] Slagle, J., "A Multipurpose Theorem-Proving Heuristic Program That Learns," *Proc. of IFIP Congress '65*, 2 (Washington, D.C.: Spartan Books, 1966).
[17] Rosenbloom, P., *The Elements of Mathematical Logic*, Dover, 1950.
[18] Slagle, J., "A Heuristic Program That Solves Symbolic Integration Problems in Freshman Calculus," *Journal of the ACM*, 10, 1963.
[19] Newell, A., and Ernst, G., "The Search for Generality," *Proc. of IFIP Congress '65*, 1 (Washington, D.C.: Spartan Books, 1965).
[20] Kanger, S., "A Simplified Proof Method for Elementary Logic," *Computer Programming and Formal Systems*, Braffort, Hirschberg, eds. (Amsterdam: North Holland Publishing Co., 1963).

APPENDIX I

DEFINITIONS OF RULES OF DECISION IN $\mathfrak{N}(SU)$ IN TERMS OF RULES OF INFERENCE IN $N(SU)$

We are giving below the detailed structural definitions of the transition schemata for the fourteen rules of decision of the system $\mathfrak{N}(SU)$ in terms of the transition schemata for the rules of inference of the system $N(SU)$. (The latter are specified in Table 4.1.) The transitions are between sequent forms. While the consequent sequent (an arrow points into it) and the antecedent sequents (an arrow points away from each of them) of a decision rule are completely specified (the parentheses enclosing such sequents are doubled), the intermediate sequents in each schema are indicated in incomplete form. In most cases we have merged a N_{in}^N or N_{el}^N inference rule with the X_{in}^N rule which is immediately associated with it (as if a decision to carry out a *reductio* argument is made simultaneously with the choice of contradictory formulas for the argument). Also, we have merged terminations that are obtained by a combination of C_{el}^N and the axiom schema $\mathfrak{A}_1$. The decision rules are named as in Table 5.1, i.e., according to the characteristic operator and the location of the key formula. Throughout the representations, the numbers in brackets denote weights. The l.i. information formulas are underlined only in completely specified sequent forms. The variables η, α, β stand for arbitrary (possibly empty) sequences of formulas, some of which may be suppositions, or l.i. information formulas, or both. If the key formula of a decision rule is an information formula, it may or may not be a supposition; we do not indicate this explicitly in our representations. The variables x, x_1, x_2, x_3 stand for any formula in SU.

All the transitions that are specified below satisfy the desired properties for decision rules, i.e., (i) a consequent sequent is valid if and only if the set of antecedent sequents is valid; (ii) the set of l.i. formulas remains the same in the transition, except for the elimination of one connective (and the fragmentation) of one l.i. formula; and (iii) the weight of the transition is minimal. The property (i) can be verified by proving the logical equivalence between a consequent sequent and its antecedent sequents in a decision rule (via the use of the deduction theorem [see (5.1)]), the property (ii) can be verified with the help of theorems 5.1 and 5.2, and (iii) can be verified by search over alternatives. (The interested reader is invited to carry out the required proofs within the systems that are discussed in this paper.)

$(C, R)_{(a) \text{ or } (b)}$, [2]

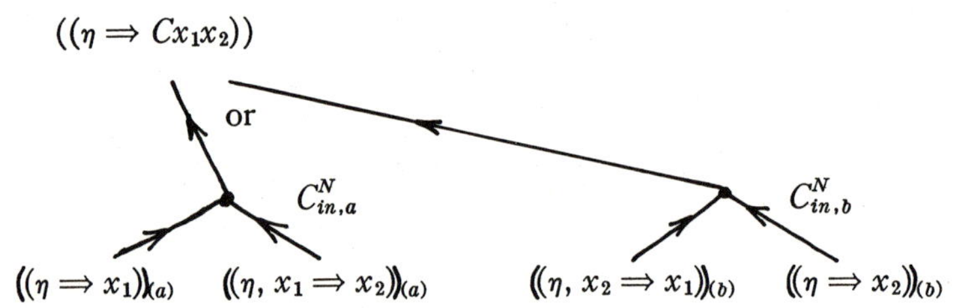

This decision rule has two structural versions; both are explicitly given here.

(C, L), [20]

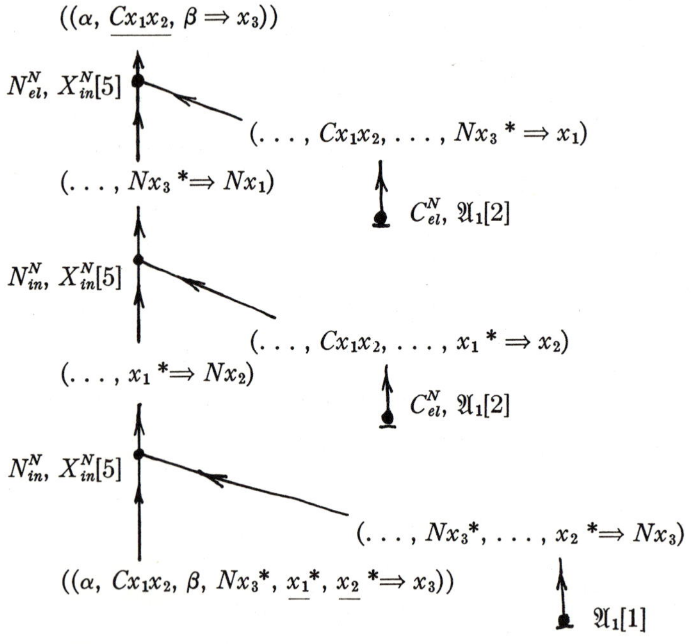

$(D, R)_{(a) \text{ or } (b)}$, [14]

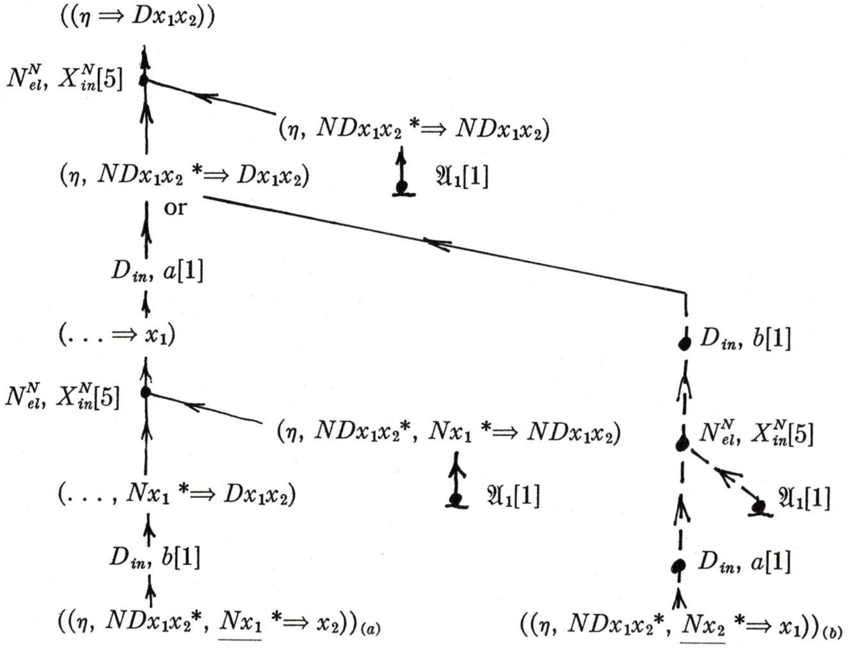

Here we have the two structural versions of this decision rule; the continuation of the (*b*) version is similar to the (*a*) version and it is sketched at right.

(D, L), [8]

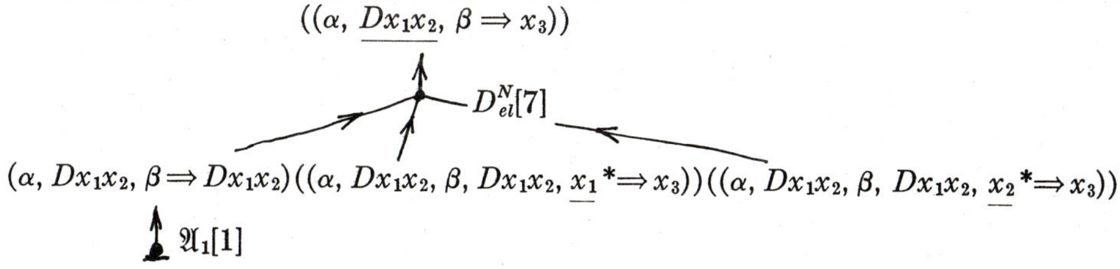

(I, R), [3]

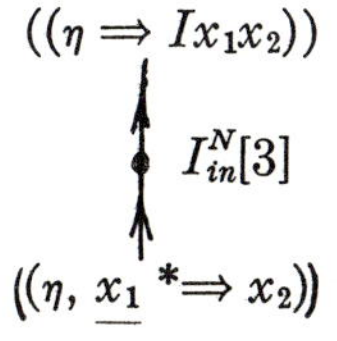

(I, L), [20]

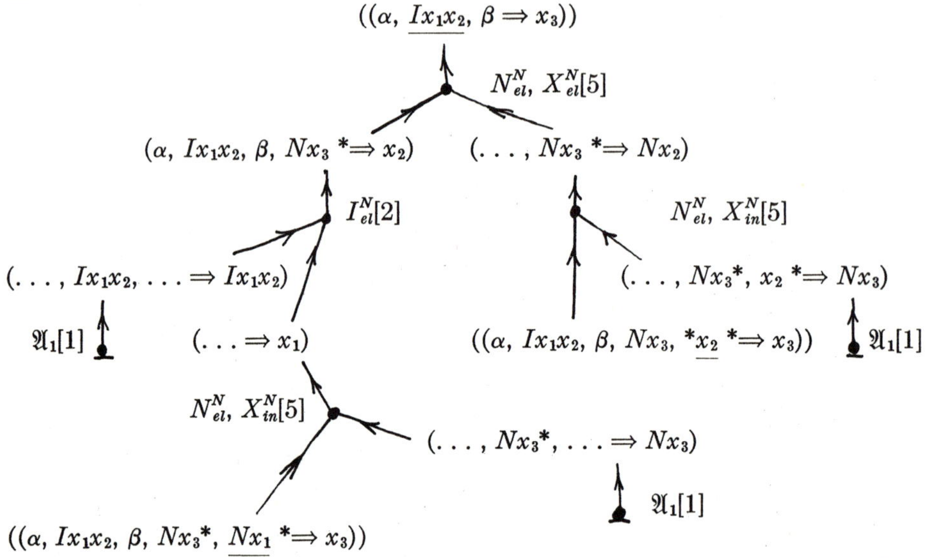

$(NC, R)_{(a) \text{ or } (b)}$, [14]

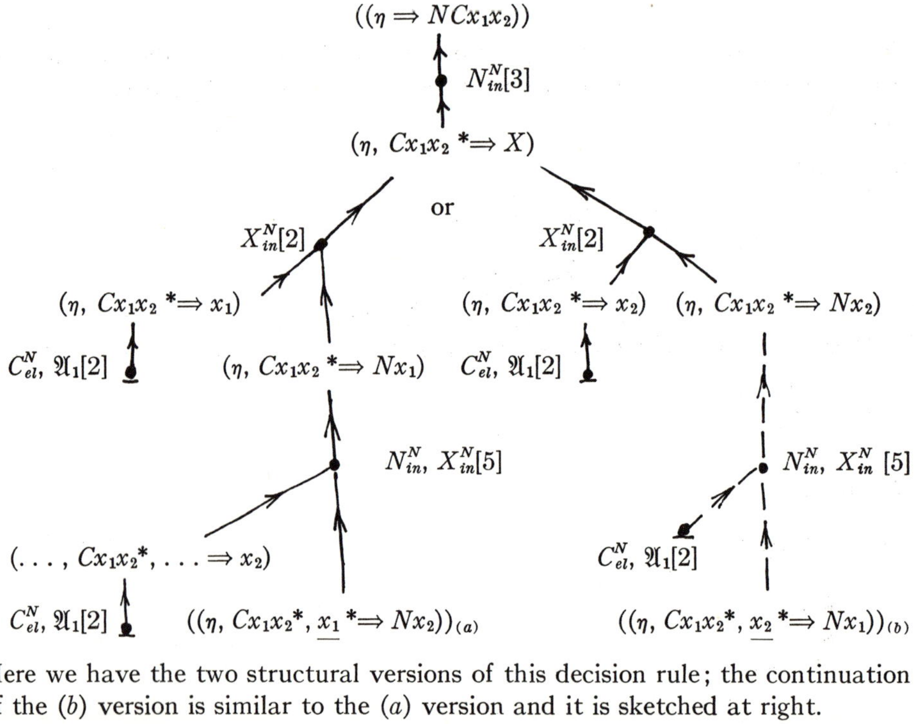

Here we have the two structural versions of this decision rule; the continuation of the (*b*) version is similar to the (*a*) version and it is sketched at right.

$(NC, L)_{(a) \text{ or } (b)}$, [20]

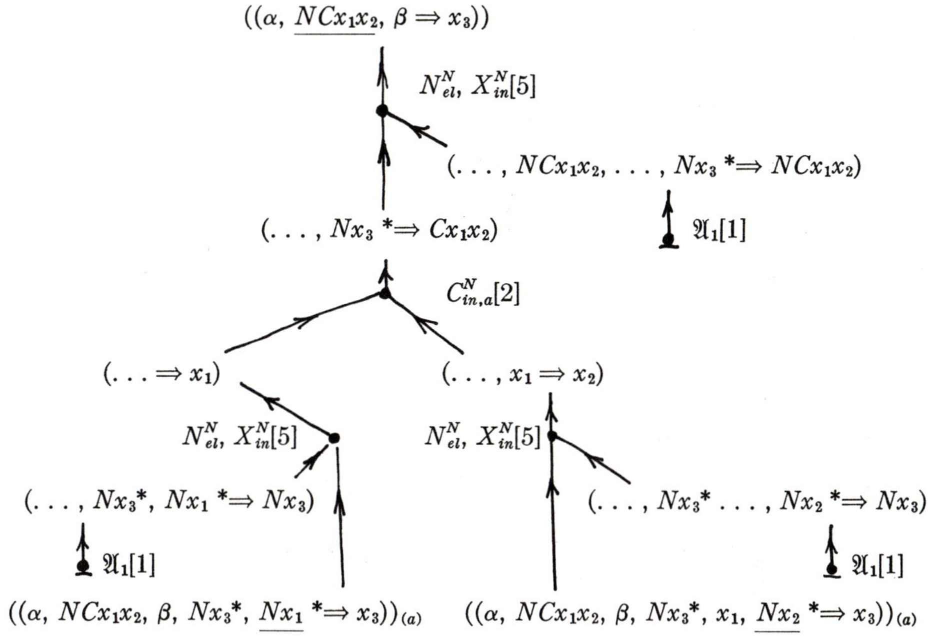

For this decision rule we have two structural versions. The (a) version is shown here; the (b) version is obtained if the $C^N_{in,a}$ rule of inference is replaced by $C^N_{in,b}$ and the schema is continued in an obvious manner; in both the (a) and (b) versions, the characteristic representations of the antecedent sequents of the rule of decision remain the same.

(ND, R), [17]

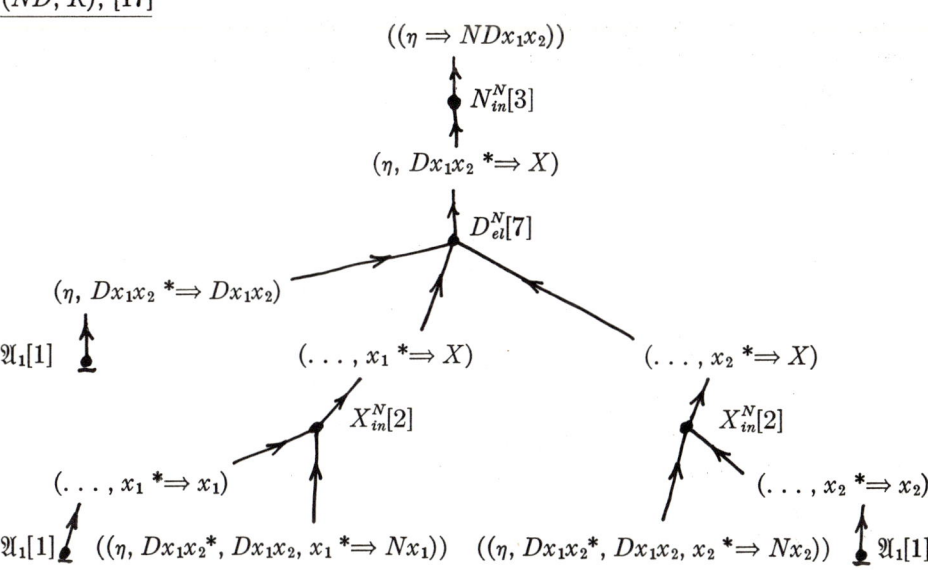

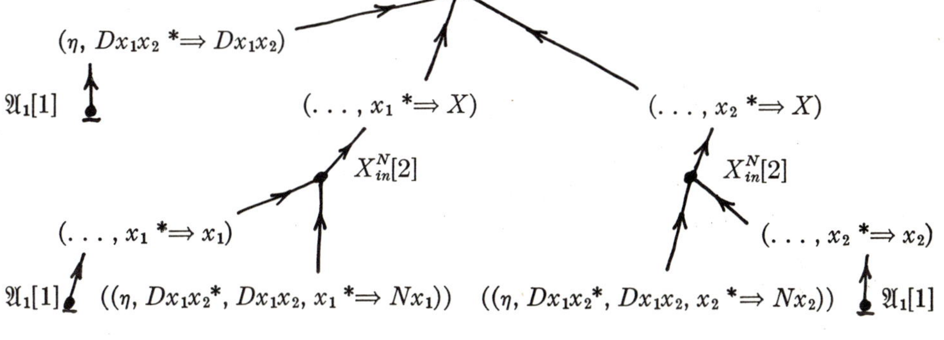

$(ND, L)_{(a) \text{ or } (b)}, [20]$

$(\alpha, \underline{NDx_1x_2}, \beta \Rightarrow x_3)$

↑ $N_{el}^N, X_{in}^N[5]$

$(\alpha, NDx_1x_2, \beta, Nx_3 * \Rightarrow NDx_1x_2)$

$(\ldots, Nx_3 * \Rightarrow Dx_1x_2)$ ↑ $\mathfrak{A}_1[1]$

↑ $D_{in,a}^N[1]$

$(\ldots \Rightarrow x_1)$

↑ $N_{el}^N, X_{in}^N[5]$

$(\ldots, Nx_1 * \Rightarrow Dx_1x_2)$ $(\ldots, NDx_1x_2 \ldots, Nx_1 * \Rightarrow NDx_1x_2)$

↑ $D_{in,b}^N[1]$ ↑ $\mathfrak{A}_1[1]$

$(\ldots \Rightarrow x_2)$

↑ $N_{el}^N, X_{in}^N[5]$

$(\ldots, Nx_3, * \ldots, Nx_2 * \Rightarrow Nx_3)$

↑ $\mathfrak{A}_1[1]$

$((\alpha, NDx_1x_2, \beta, Nx_3*, \underline{Nx_1*}, Nx_2 * \Rightarrow x_3))_{(a)}$

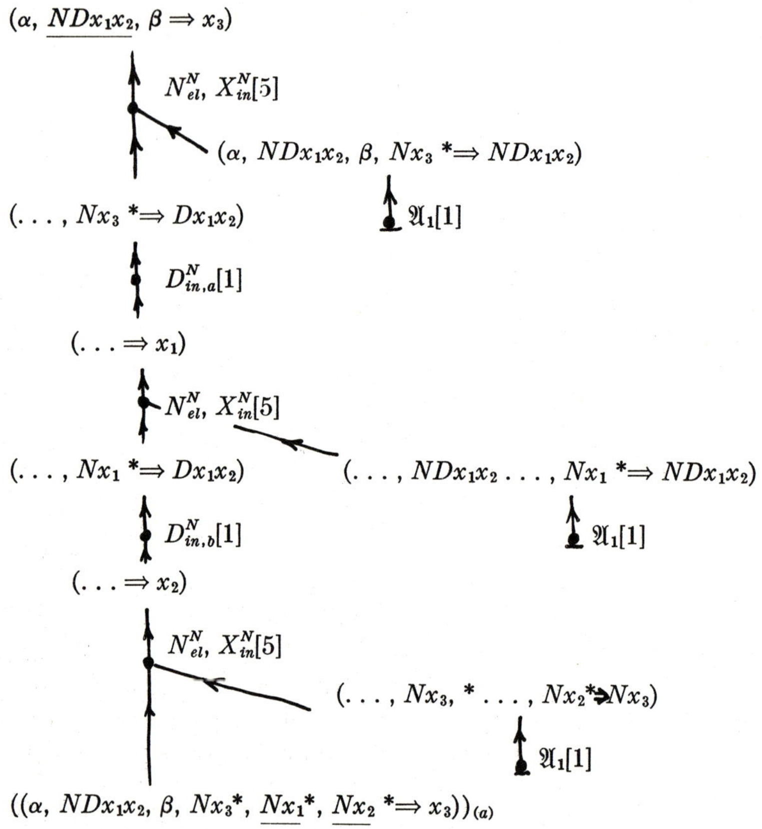

In this decision rule we have two structural versions. The (a) version is shown here. The (b) version is obtained if the $D_{in,a}^N$ and $D_{in,b}^N$ rules are interchanged in the schema.

$(NI, R), [8]$

$((\eta = NIx_1x_2))$

↓ $N_{in}^N, X_{in}^N[5]$

$(\eta, Ix_1x_2 * \Rightarrow x_2)$ $((\eta, Ix_1x_2 * \Rightarrow Nx_2))$

↑ $I_{el}^N[2]$

$(\eta, Ix_1x_2 * \Rightarrow Ix_1x_2)$ $((\eta, Ix_1x_2*, Ix_1x_2 \Rightarrow x_1))$

$\mathfrak{A}_1[1]$ ↑

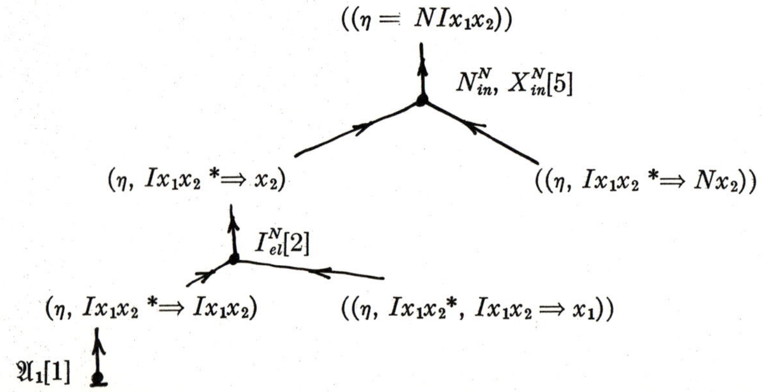

(NI, L), [15]

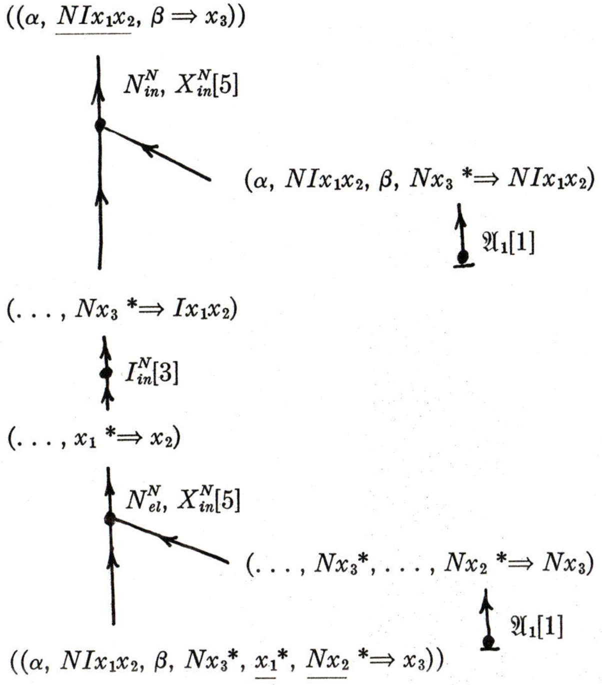

((α, NIx₁x₂, β ⇒ x₃))
N^N_in, X^N_in[5]
(α, NIx₁x₂, β, Nx₃ *⇒ NIx₁x₂)
𝔄₁[1]
(. . . , Nx₃ *⇒ Ix₁x₂)
I^N_in[3]
(. . . , x₁ *⇒ x₂)
N^N_el, X^N_in[5]
(. . . , Nx₃*, . . . , Nx₂ *⇒ Nx₃)
𝔄₁[1]
((α, NIx₁x₂, β, Nx₃*, x₁*, Nx₂ *⇒ x₃))

(NN, R), [6]

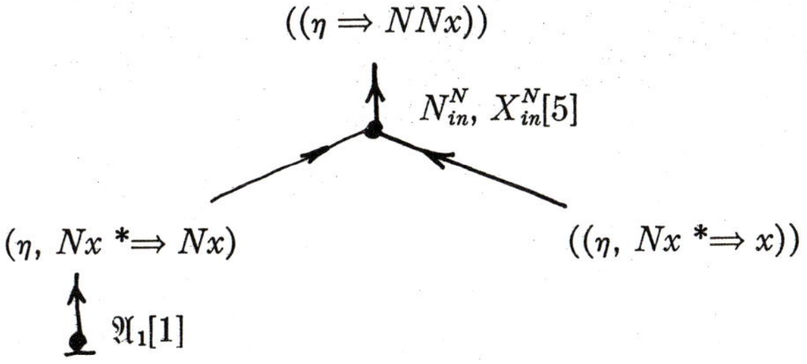

((η ⇒ NNx))
N^N_in, X^N_in[5]
(η, Nx *⇒ Nx)
((η, Nx *⇒ x))
𝔄₁[1]

(NN, L), [12]

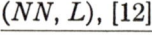

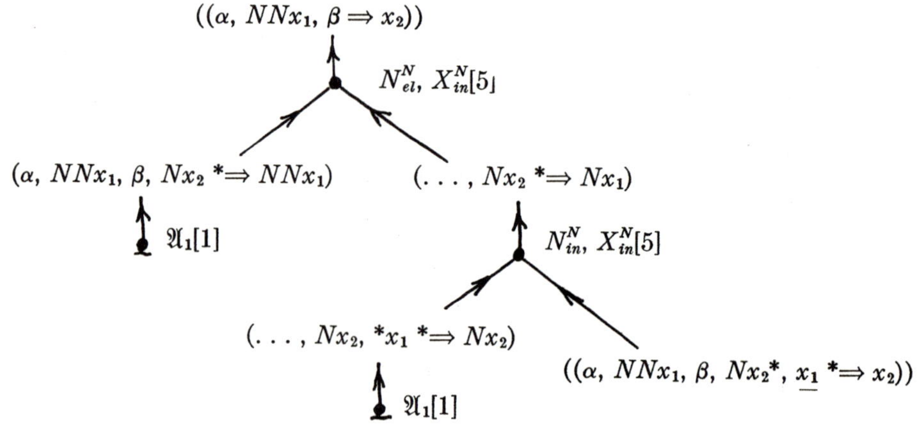

APPENDIX II

EXAMPLES OF PROOF CONSTRUCTION WITH THE PROCEDURE $\Pi_r(PC, 3)$

EXAMPLE 1. *Construction of minimal proof for the theorem IIpIqrIIpqIpr.*

The construction map is shown in (II.1). There are seven stages of construction, and they are reflected in the map by seven subgraphs of the proof graph that are built in sequence (these subgraphs are shown slightly separated). The first three stages of construction erect a supporting structure and the four subsequent stages build links betweeen parts of the supporting structure. Notice that the supporting structure (which is shown by ++++ lines) corresponds to the conventional tree representation of the theorem.

The search tree for administering the construction process is shown in (II.2). The circled number associated with each move corresponds to the subgraph that it constructs; the bracketed number denotes move weight. We distinguish different occurrences of the same formula in the construction map by marking them with different superscripts.

(II.1)

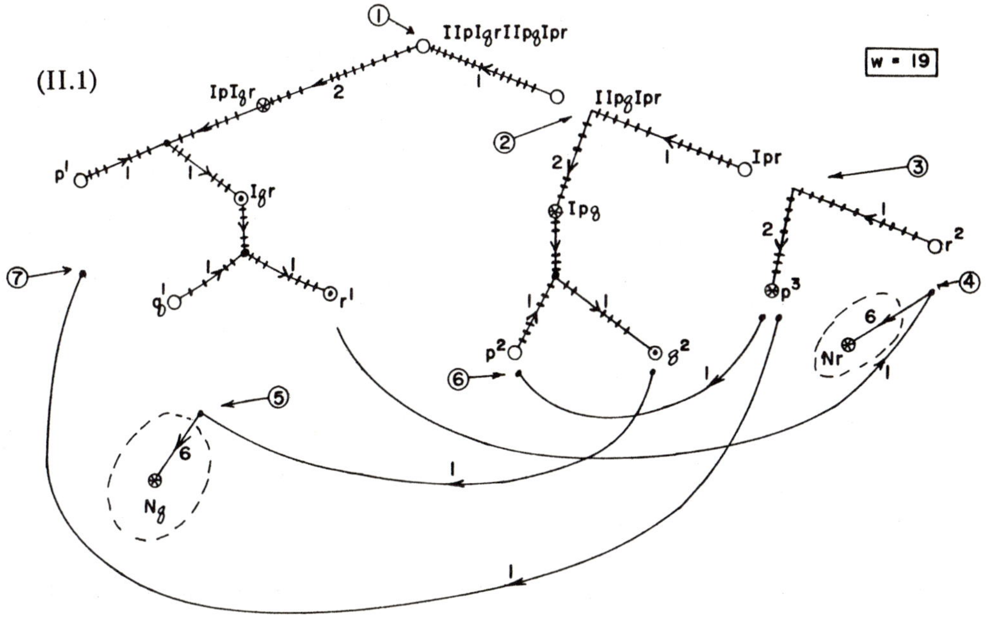

$S_0 = (\Rightarrow IIpIqrIIpqIpr)$

① Opening move (I'_{in} + source extension), [7]

$S_1 = (IpIqr,\ Iqr,\ r^1 \Rightarrow IIpqIpr)$

② Development move (I'_{in} + source extension), [5]

$S_2 = (Ipq,\ q^2,\ IpIqr,\ Iqr,\ r^1 \Rightarrow Ipr)$

③ Development move (I'_{in}), [3]

$S_3 = (p^3,\ Ipq,\ q^2,\ IpIqr,\ Iqr,\ r^1 \Rightarrow r^2)$

④ Direct linking move ($\mathfrak{A}_1^1$ + linking structure), [1]

(II.2) $S_4 = (Nr,\ p^3,\ Ipq,\ q^2 \Rightarrow q^1)$

⑤ Direct linking move ($\mathfrak{A}'_1$ + linking structure), [1]

$S_5 = (Nq,\ Nr,\ p^3,\ IpIqr,\ Iqr \Rightarrow p^2)$

⑥ Direct linking move ($\mathfrak{A}'_1$), [1]

$S_6 = (Nr,\ p^3,\ Ipq,\ q^2 \Rightarrow p^1)$

⑦ Direct linking move ($\mathfrak{A}'_1$), [1]

During the construction process, we obtain two global states that have two candidate destination nodes each. In the global state obtained after move ④ above, the candidate destinations are q^1 and p^1 (see (II.1)). We choose to consider first q^1. This is an arbitrary choice that is guided by the rule of attention control, according to which, if more than one candidate destinations appear after a linking move, that destination is chosen which is closest to the source node from which the last link has been established (here, q^1 is closer to r^1 than p^1). After move ⑤, we have a second global state with the two candidate destinations, p^2, p^1. Here we choose p^2 as a basis for the next state, since p^2 is closer to q^2 than p^1. In the moves ④, ⑤, linking structures are to be constructed. We indicate these structures in an abbreviated form on the construction map; a linking structure originating at a destination node ○x is abbreviated as:

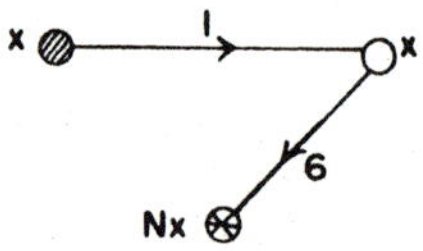

Since the nodes ⊛Nr, ⊛Nq that were grown in the linking structures of the present construction map are not needed for the proof, they are eliminated in the clean-up phase.

Note that at no point during the development of the proof did we have any choice of moves; no search was required, and we had a unique line of construction. Thus, the search tree in (II.2) is identical with the proof chain (i.e., with the unique sequence of decisions that are made in proof construction). Furthermore, we have obtained the minimal proof for the theorem *IIpIqrIIpqIpr*, $[w = 19]$, in a form that clearly shows the way in which the argument "hangs together" in terms of interactions between subformulas of the theorem.

The proof chain shown in (II.2) corresponds to the path taken by the "line of attention" of our system over a proof tree in the decision system $\mathfrak{M}(SU)$. This proof tree is shown next in (II.3); the "line of attention" is shown as a dashed line.

(II.3)

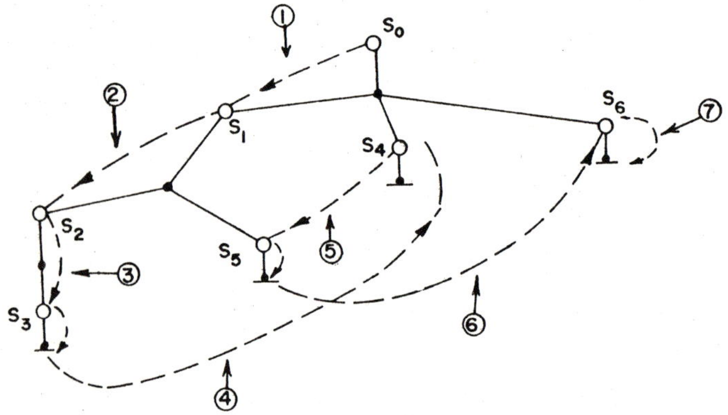

In accordance with the requirements of the natural decision system, we have the logical equivalences $S_0 \leftrightarrow S_1 \,\&\, S_4 \,\&\, S_6$ and $S_1 \leftrightarrow S_2 \,\&\, S_5$. Note that the specification of the state transition from S_0 to its descendants S_1, S_4, S_6 is done in stages. First S_1 is specified and S_4, S_6 are conditional states; then, after some activity below S_1, which results in an axiom link (i.e., a loop in the graph), S_4 is specified; then, after two more axiom links, S_6 is specified. Similarly for the transition between S_1 and its descendants S_2, S_5.

EXAMPLE 2. *Construction of minimal proof for the theorem INIpqCpNq.*

The construction map is shown in (II.4). The eight stages of construction are shown in the graph by the eight separated subgraphs.

The part of the search tree that represents the chain of decisions for constructing the proof is shown in (II.5). We shall use here the same notation as in the previous example. In the present example, the line of construction is not unique. After the opening move, two development moves are possible at S_1: one based on N'_{el} and one based on C'_{in}. We choose N'_{el} according to the rules of preference for moves. Similarly, at S_2, there are two alternative moves, one linking X^2 to X^1 and another linking it to X^3. According to the rules of preference for X links, the link to X^1 is chosen. Except for these two choice points, no other choices exist in the search tree. Our approach is first to pursue preferred moves until a proof or a refutation is obtained. In the present example, the proof that is constructed on the basis of the preferred moves is the minimal. When we pursue the other alternatives at the choice points, we find that proofs of higher weight result.

(II.4)

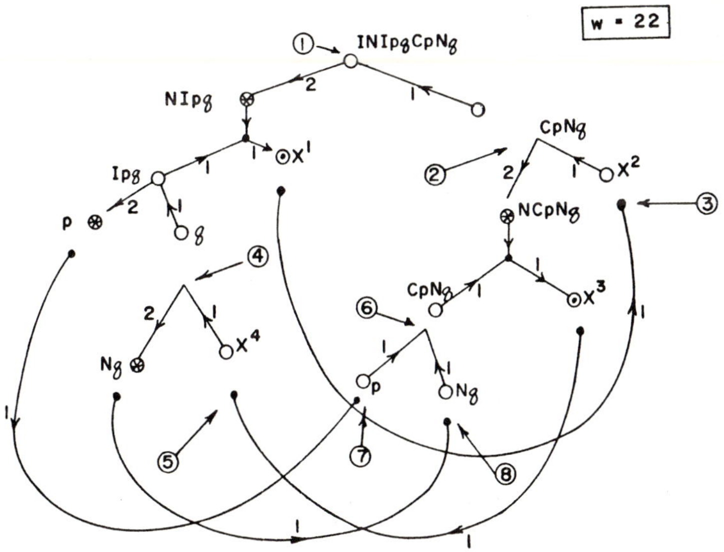

$S_0 = (\Rightarrow INIpqCpNq)$

① Opening move (I'_{in} + source extensions), [8]

$S_1 = (NIpq, X^1 \Rightarrow CpNq)$

② Development move (N'_{el} + source extension), [5]

Alternative moves

$S_2 = (NCpNq, X^3, NIpq, X^1 \Rightarrow X^2)$

③ Direct link to $X^1(\mathfrak{A}'_1)$, [1]

(II.5) $S_3 = (p, NCpNq, X^3 \Rightarrow q)$

④ Development move (N'_{el}), [3]

$S_4 = (Nq, p, NCpNq, X^3 \Rightarrow X^4)$

⑤ Direct link to X^3 ($\mathfrak{A}'_1$), [1]

$S_5 = (Nq, p, NIpq \Rightarrow CpNq)$

⑥ Development move (C'_{in}), [2]

$S_6 = (Nq, p, NIpq \Rightarrow p)$

⑦ Direct link ($\mathfrak{A}'_1$), [1]

$S_7 = (Nq, p, NIpq \Rightarrow Nq)$

⑧ Direct link ($\mathfrak{A}'_1$), [1]

The proof chain shown in (II.5) corresponds to the following tree proof in the decision system $\mathfrak{M}(SU)$ (the "line of attention" is shown as a dashed line):

(II.6)

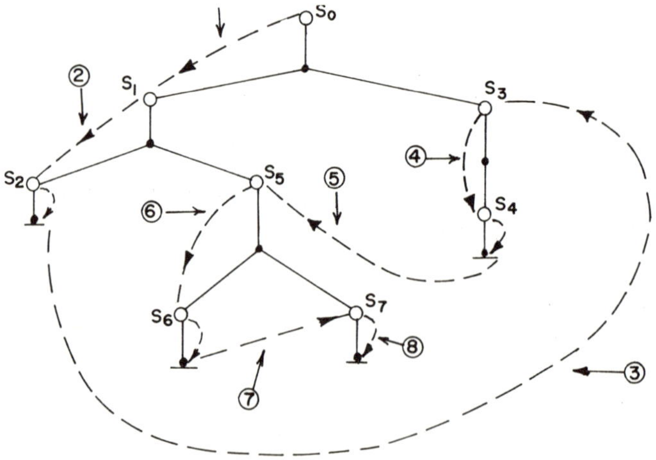

EXAMPLE 3. *Construction of minimal proof for the theorem ICIprIqrIDpqr.*

There are several alternative proofs to this theorem in our system, and the minimal proof is obtained after generating two "heavier-than-minimal" proofs.

The essential part of the search tree that is generated during the process is shown in (II.7). The construction map that corresponds to the first attempt to obtain a proof (call it proof A), is shown in (II.8). In this map, there are seven stages of construction, and they yield a proof of weight 40. The first two moves ① and ② are identical for all the alternate proofs. These are the only possible moves at states S_0 and S_1 respectively. However, at state S_2, we have three alternative moves: two direct linking moves from r^1 or r^2 to r^3, and a development move from ϕ to r^3. According to our preference ordering of moves, a direct linking move is attempted first. No preference exists between the two alternative linking moves. Suppose that a link $r^1 - r^3$ is tried first. This move constructs the subgraph ③ in the map (II.8). From the resulting state S_3, two alternative moves exist: an indirect linking move from the candidate sources Nr, r^2 (which is the preferred move), and a development move from ϕ to p. Application of the indirect linking move produces the subgraph ④ in the map. From the next state, S_4, there is a unique development that terminates with an indirect linking move at S_6; this produces the subgraph ⑦ and it closes the graph in (II.8). After clean-up, the graph in the construction map represents the proof A which has a weight of 40. This proof is then stored away.

Now, according to the search strategy outlined in section 6.D.h, the system's attention returns to the top of the proof chain that was just completed, and it slides down the chain to the first state where there exists an alternative untried move that has identical preference rank with the move which was previously applied at that state. This brings us to S_2. If we assume that the system does not recognize symmetries, then a direct linking move from r^2 to r^3 is applied at S_2. Continuing along this line, we obtain a proof of weight 40 which is similar to the proof A (call it proof A′). During this process, new states S_7, S_8, S_9, S_{10} are added to the search tree; they are the correspondents in proof A′ of the states S_3, S_4, S_5, S_6 in proof A. The proof A′ is not fully shown in (II.7).

At this stage in the construction of the search tree, there are no untried moves of identical preference ranking with those that have already been tried. The alternative of next lower rank which is closest to the top of the search tree is the development move $\mathfrak{A}'_{1,\phi}$ at S_2. This move is now taken and it leads to the proof B. The construction map for the proof B (which is the minimal proof) is shown in (II.9). The move at S_2 results in the construction of the subgraph ③$_2$ in the map of (II.9). Now, we have two candidate destinations, r^4 and r^5. Since they are equidistant from the last processed source node, one of them can be chosen arbitrarily to determine the next state. We choose r^4, and this determines the state S_{11}. At this point, there are two possible direct linking moves. Suppose that the link $r^1 - r^4$ is chosen. Further development along this line completes the proof with a weight of 24. Note that in the sequence of decisions that lead to this proof, there are alternatives at each state. The only possibility of wandering temporarily off the desired line is at S_{13} where no *a priori* preference exists for a link $r^1 - r^5$ or $r^2 - r^5$.

After the proof B is obtained, the current best weight is $w^* = 24$. Attention goes to S_{11}, and the alternative direct link $r^2 - r^4$ is tried. As soon as the development from this point goes over a weight of 4, it is stopped as it exceeds w^*. (The best proof that could be obtained by continuing in this direction has a weight of 38.) Now, attention goes either to S_3 or to S_7. But the weight of the chain from S_0 up to S_3 or S_7 is already 26, which exceeds w^*. At this point, there are no more potential alternatives in the tree that are better than the proof B. Hence, the system stops the search, and it produces at its output the proof B, which is the minimal proof of the theorem.

Note the importance of attaining early in the search a "light" proof. The search space that has to be explored in order to find a counter example that refutes the supposed minimality of the current best proof is considerably reduced as the length of the proof chain is decreased. In the present example, once the proof B is found, it is very easy to eliminate all potential challengers.

(II.7)

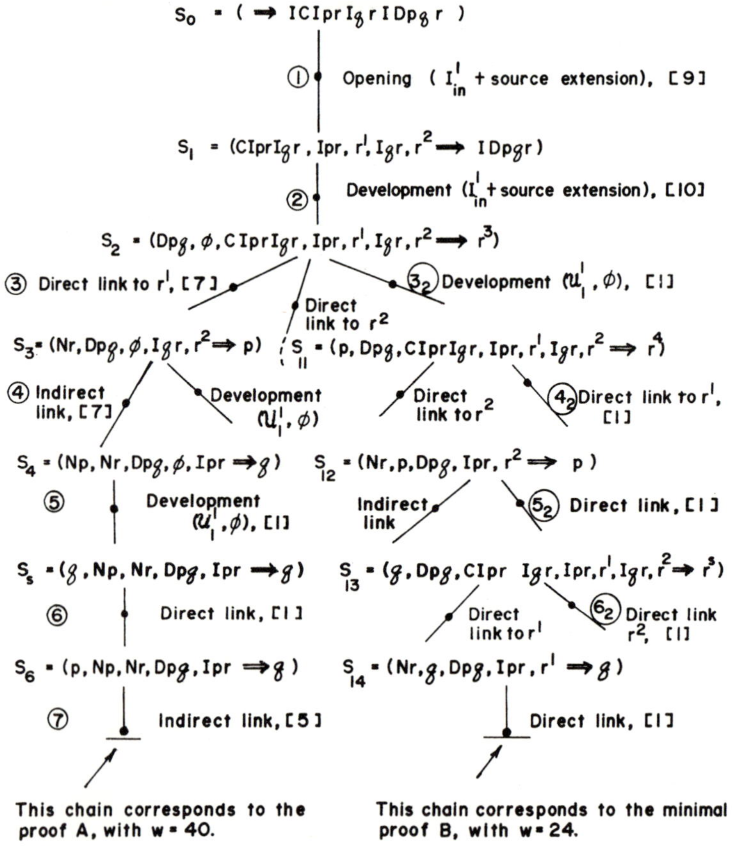

(II.8)

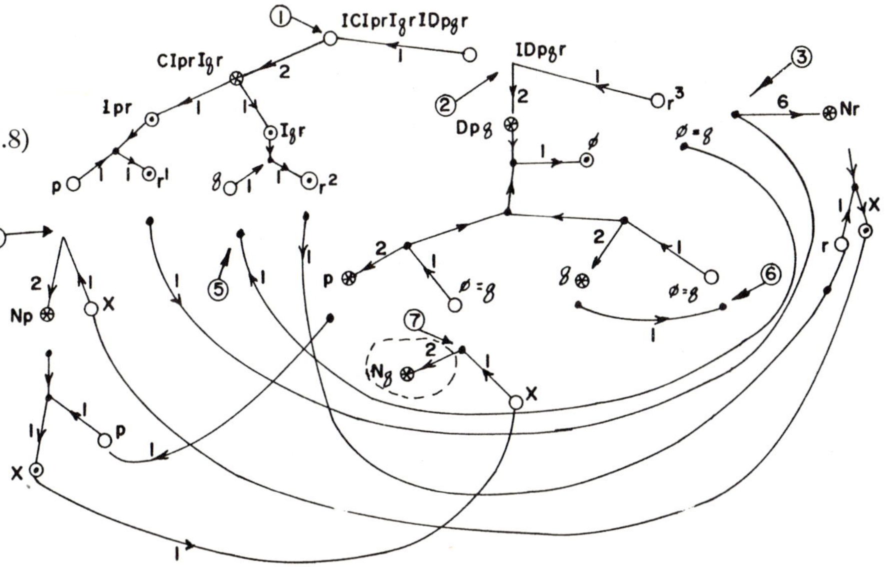

Proof A with w = 40

(II.9)

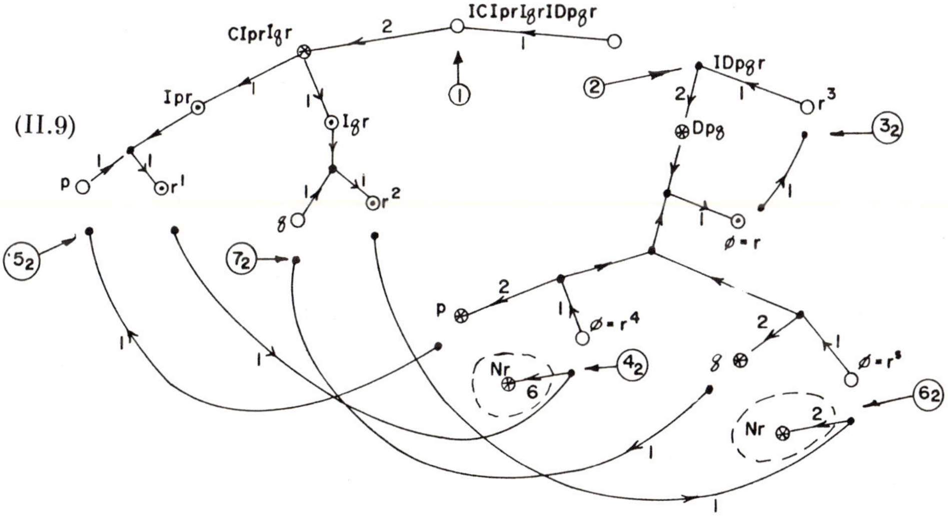

Proof B with w = 24 the minimal proof

The chain of the search tree which gives the sequence of decisions for the construction of the minimal proof B corresponds to the following tree representation of the minimal proof in the natural decision system $\mathfrak{M}(SU)$ (again, the "line of attention" is shown as a dashed line).

(II.10)

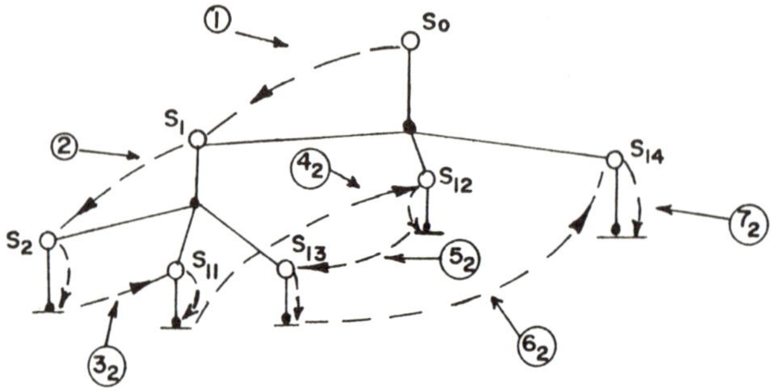

Note that the growth of the proof A′ could be completely eliminated if the system was capable of recognizing certain symmetries in the formulas, and it could then establish *a priori* that the proof A′ is isomorphic to proof A. This example illustrates the usefulness of recognizing syntactic symmetries in search processes.

The present example was brought to illustrate a situation where a minimal proof is not obtained immediately by the system $\Pi_r(PC, 3)$. A certain amount of search is necessary, and two non-minimal proofs have to be completely constructed before the minimal proof is obtained. In view of the relatively large number of alternative proofs with weight up to 50 that exist in this example, the search efficiency of the system seems exceptionally good. By adding to the present system a capability for recognizing symmetries, we can reduce the search which is currently needed for this problem by about 50 per cent.

New Directions in General Theory of Systems

Mihajlo D. Mesarovič*

1. INTRODUCTION

Systems theory is a new field and, at present, there exists a diversity of opinion regarding what constitutes the theory, the importance of the theory and the role it should play in science and engineering, the methodology which should be used to develop the field further, and so on.

Systems theory actually has two aspects: (*a*) The formal theory is concerned with the forms of the relations between the elements of the system and disregards the meaning or the interpretation of these forms in the context of a given subject matter; in particular, the formal theory, of course, uses mathematical apparatus and is also called mathematical systems theory. (*b*) The informal theory is concerned with the interpretation of the formal theory in any given class of real systems.

We define formal systems theory as the theory of formal mathematical models of real or conceptual systems. A system is then defined as a relation among the formal objects (section 2). Systems theory is concerned with the properties of these relations, and with the methods for effective description of these relations. In this paper, we shall be concerned solely with the formal systems theory and for simplicity we shall omit the adjective "formal."

Of particular relevance for this conference is the fact that systems theory is intimately related with the field of computer sciences. There are several important reasons for this. First of all, an important topic in systems theory is the study of the properties and behaviour of computer systems. The objective of the study is either to assess the computation and/or the information processing capabilities of a certain type of computers or to enable the study of large-scale systems in which the computers are elements or subsystems of the system under consideration [1].

The second important reason for the intimate relation between systems theory and the computer sciences stems from the use of computers for the problem solving (theorem proving, game playing, optimization, etc.). To fully justify

*Systems Research Center, Case Institute of Technology, Cleveland, Ohio.

this statement, one would have to go into the more precise concept of problem solving by computers. It would suffice here, however, to indicate that for effective use of a computer in the problem solving situation, it is of considerable advantage to formalize the problem to be solved as well as the problem solving process, regardless of what components or systems will be used in the process, and only then to assign the appropriate specific tasks to the computer itself. This is even more important if one wants to apply the experience gained in a given case to other similar cases. (Parenthetically, we might add that this is precisely what is presently missing in the field of artificial intelligence.) To that end, a formal systems theory of problem solving systems is needed [2].

2. GENERAL SYSTEMS THEORY

The diversity of opinions regarding general systems theory is even greater than for systems theory. These opinions can be grouped into three schools of thought. (*a*) In the first, general systems theory is considered basically as scientific philosophy and its main characteristic is belief in the existence of some fundamental laws which hold for social and biological as well as physical systems [3, 4, 5]. (*b*) The second school has its roots in engineering [6, 7] and considers general systems theory essentially as comprising a set of concepts and tools which have an especially wide domain of applicability and also the methodology of how to apply these tools to large-scale systems engineering problems. (*c*) The third school is formal, mathematical, and is actually based on the notion of systems theory defined in the introduction. General systems theory is regarded as a theory of abstract mathematical models which are general in the sense that they have very little structure and reflect the formal structural properties of a large class of real systems.

Such a theory has been proposed, and the basic concepts were outlined as early as 1959 [8]. An important role of the theory for science, as well as for engineering, was predicted on the ground that it provides the framework for a unification, on a formal basis, of many diverse concepts and, furthermore, that it provides a framework for the study of complex large-scale systems.

Presently, two different approaches are being used in the attempt to develop further such a formal general systems theory. In the first approach, one essentially proceeds by formalization from the real (or conceptual) systems [9]. Starting from the intuitive systems concepts (which concepts reflect real-life complex situations), one introduces the most basic and general formal concepts and investigates the basic properties of general systems. Subsequently, by adding more structure to the constituents of the systems, more specific theories can be developed [10]. Using formalization approach, a number of important systems concepts related to input-output description, as well as to the goal-seeking description of systems, have been developed [11, 12, 13, 14, 15]. In the second approach, one proceeds via abstraction. Starting from some specific, relatively well-known mathematical models of systems (such as found in automata theory, control theory, etc.), some more general systems are obtained by abstraction, i.e., by using more abstract mathematical concepts that describe only those properties of the system which are considered as essential [6, 7, 8, 9, 20]. Apparently, these two approaches are complementary and eventually will meet on a given level of abstraction.

In this paper, we shall not try to review the results obtained so far in general systems theory. They are all essentially related to the study of systems from the input-output viewpoint, i.e., by considering them as relations or as a union of functions. Rather, the objective of our presentation is to indicate some new directions in which general systems theory should broaden its conceptual basis if it is really going to live up to its expectations. However, for the sake of illustration, some of the results in the area will be briefly outlined.

Basic building blocks of general systems theory are formal objects which are defined as sets with appropriate structure needed for the introduction of the appropriate systems concept.

Let a family of objects be given $V = \{V_i : i \in I\}$ where I is the index set for the family V. A system S is simply a relation defined on V, i.e.,

$$S \subset \times \{V_i : i \in I\}$$

where $\times$ indicates Cartesian product.

On the basis of this definition, general systems theory becomes simply a general theory of relations. Following the formalization approach, one starts from such a general notion of a system and then proceeds to assume more structure for the objects $V_1, \ldots, V_n$ and investigates the properties induced by the relation S.

For example, an object V is called an algebraic object if there exists at least one function $p: V^k \to V$, where k is the order of p, which is closed in V.

A system defined on algebraic objects is called an algebraic system. A characteristic of an algebraic system is that the elements of the systems can be combined (as specified by the algebraic operation) to produce other elements of the same object. (A detailed account of an algebraic theory of systems is given in [11] and [2].)

An alternative approach to develop a more specific theory of systems is via general topology. An object for which there is defined a set-valued set function $f: P(V) \to P(V)$, where $P(V)$ is the family of all subsets of V, is called a topological object. f induces an extended topology and a minimal general topology for V. Instead of indicating how the elements of an object V can be combined (as in algebraic approach), topological function f indicates the neighbourhood of the elements of V or of the subsets of V. A system defined on a family of topological objects is called a topological system (see section 3).

By introducing more specific algebraic (or topological) operations, systems with more structure can be defined and the results which are applicable for that particular class of systems can be obtained. It is important to realize, however, that a number of very important systems properties (e.g., stability, consistency and completeness, reproducibility, etc.) can be investigated even on the basis of a very general set-theoretic notion of a system. As an illustration, we can cite some recent investigations on interactions in a system.

Assume that a system $S \subset X \times Y$ is given together with its state representation $S_z: X \times Z \rightarrow Y$, where X represents input, Z state, and Y output of the system. Let S be a multivariable system such that the output object Y has m components, $Y = \{ Y_i: 1 \leqslant i \leqslant m\}$, the state representation is then a union of functions,

$$S_z = \cup_{i \leqslant m} S_{zi}; \qquad S_{zi}: X \times Z \rightarrow Y_i.$$

Let the input objects be partitioned by an equivalence relation,

$$X = \cup \{X^q: q \in Q\},$$

where Q is an index set. This induces a family of subsets in Y,

$$Y = \cup \{ Y^q: q \in Q\}.$$

Subsets of Y^q do not necessarily form a partition in Y.

Define two sets of relations:

$$R[S; z, q] = \{S_z(x, z): x \in X^q\}$$

$$R[S_i; z, q] = \{S_{zi}(x, z): x \in X^q\}.$$

Given a pair (z, q), $R[S_i; z, q]$ is the set of all ith components of the output which can be obtained starting from the state z and using the inputs from X^q. On the other hand, $R[S; z, q]$ is the set of all outputs (i.e., all m components considered simultaneously) which can be obtained starting, again, from z and using X^q. In other words, an element of $R[S; z, q]$ is a set of m output components $(y_1, \ldots, y_m)$ while an element of R_i is only the ith component of the output.

We can now introduce the following definitions.

(1) A general system S is *non-α-cohesive* with respect to $z \in Z$ if for each $q \in Q$,

$$R[S; z, q] = \mathop{\times}_{i=1}^{m} R[S_i; z, q].$$

(2) A general system S is *non-β-cohesive* if for each $q \in Q$

$$R[S; q] = \mathop{\times}_{i=1}^{m} R[S_i; q],$$

where

$$R[S_i; q] = \cup \{R[S_i; z, q]; z \in Z\},$$

$$R[S; q] = \cup \{R[S; z, q]; z \in Z\}.$$

Conceptually, a system is non-cohesive if it is possible to obtain as an output the sequence of any output components which can be obtained separately. If this is not possible, there apparently exist some internal constraints in the system. The system is then considered to possess cohesion.

Conditions for the cohesion in a system can now be given in terms of the relations R and the types of the systems function S. Typical of the results are the following two theorems.

THEOREM. *Assume that for some* $(z, q) \in Z \times Q$ *there exists a function*

$$\phi\colon P \times \{z\} \to Y_r,$$

where

$$P = \bigcup \left\{ \mathop{\times}_{\substack{i=1 \\ i \neq r}}^{m} S_i(x, q) : x \in X^q \right\}$$

and furthermore,

$$\phi(p, z) = S_r(x, z)$$

where $p = [S_1(x, z), \ldots, S_{r-1}(x, z), S_{r+1}(x, z), \ldots, S_m(x, z)]$. *A sufficient condition for* S *to be* $\mathfrak{a}$*-cohesive with respect to* z *is that*

$$K[R[S_r; z, q]] \gg 1,$$

where $K(A)$ *is the cardinality of* A.

(The proof of the theorem is given in [14].)

The theorem shows that cohesion in a system depends upon the relation which exists among the output components of the system and upon the cardinality of the relation R.

A more restrictive condition is obtained for a type of generalized linear system. Let Z, Y_i for every i, X^q, and $Y_i(q)$ for every q and i, be groups. General system S is then quasi-linear if

$$S(\bar{x} + \tilde{x}, \bar{z} + \tilde{z}) = S(\bar{x}, \bar{z}) \circ S(\tilde{x}, \tilde{z})$$

where $+$ and $\circ$ are group operations.

THEOREM. *Let* S *be a quasi-linear system. If* S *is non-*$\mathfrak{a}$*-cohesive with respect to any state* $z \in Z$, *it is non-*$\mathfrak{a}$*-cohesive with respect to the entire state object* Z.

(The proof of the theorem is given in [14].)

A number of other results on cohesion as well as reproducibility are also obtained. Since these results are derived for a class of very abstract systems, they can be applied to a variety of systems which are used more often. For example, the results on cohesion are shown to imply conditions on finite controlability for differential equation systems, on connectedness of finite automata, and on computability of Turing machines [14].

3. NEW DIRECTIONS IN GENERAL SYSTEMS THEORY

Current research in general systems theory is primarily directed to the study of the properties of the relations of some rather general types. Actually, the research is still in its early stage, and the results are scanty and not quite complete. Yet, it is important to consider whether there are some alternative directions in which this theory should be developed. Some of these new directions which have special conceptual import and which seem to be of special promise will be outlined in this paper.

(a) *Implicitly Defined (Inductive) Systems*

A major problem in any non-trivial systems study is to find a method for constructive specification of systems. A standard way to arrive at such specification is to introduce a new object, state object, so that whenever an input-state pair is given, it is possible to determine the output, and, in particular, to do this by using an algorithm presented in terms of a set of equations. In other words, the input to the system is always considered to be given beforehand and the problem of constructive specification is how to determine the respective output. The important class of systems for which the inputs also have to be constructed (as, e.g., in the systems used in axiomatic logic, symbol manipulation, and the study of algorithms) has been completely left out of the systems theory. These systems can be brought into the general formalism of systems theory via the notion of an implicitly defined (inductive) system.

Assume that for a system $S: X \to Y$ (or $S: X \times Z \to Y$) the systems function S is specified as a set of functions $S = \{S'\}$ and furthermore, S is an inductive class of functions. This means that there exists a subset $A \subset S$ and a function $r: S^n \to S$ such that $S' \in S$ if and only if either $S' \in A$ or there exists n elements of S, i.e. $(S^1, \ldots, S^n)$, such that $r(S^1, \ldots, S^n) = S'$.

Elements of A are called atoms (or atomic functions) while r is called the deductive function of the set S. Of course, in general, one has a family of deductive functions $\bar{r} = \{r\}$, and the elements of S are generated by applying any $r \in \bar{r}$ on the appropriate sequence of atoms or previously generated elements.

Systems defined as an inductive class are called implicitly defined systems. A typical example of implicitly defined systems is the propositional calculus. Axioms of the calculus and all of its theorems are elements of the system objects while the inductive class of systems functions consists of all the sequences of deductive rules for the calculus (*modus ponens* and substitution).

For any given S defined as an inductive class, it is possible, of course, to define a new system S^c which is specified as a function (rather than inductive class) and which is equivalent in a given sense to S. To that end, one selects a new object U

such that there exists a one-to-one function on S onto U, $f: S \to U$. A new system S^c is then

$$S^c: X \times U \to Y \qquad (\text{or } S^c: X \times Z \times U \to Y),$$

and the behaviour of S can be inferred by studying S^c. However, in any further study of such systems, cognizance should be taken of the fact that a constructive specification of the system is originally given as an inductive class of function.

In view of the practical importance of symbol manipulating systems and the conceptual importance of formal, axiomatic, mathematical systems, a theory of implicitly defined systems in the framework of the systems theory is highly desirable.

(b) Open Systems

Another important class of systems which need investigation are so-called open systems. Consider a state representation of a system, $S_z: X \times Z \to Y$. Let the two families of subsets $P_x \subset P(X)$ and $P_z \subset P(Z)$ be given, where $P(X)$ and $P(Z)$ are power sets of X and Z respectively. Furthermore, let P_x and P_z be covers for X and Z. A system is called open if it is specified by a mapping

$$S_p: P_x \times P_z \to P(Y).$$

In other words, all one knows about the system S is that whenever $x \in X'$ and $z \in Z'$ are given (where $X' \in P_x$ and $Z' \in P_z$), the output y will be an element of a given subset $Y' \subset Y$.

There is a basic uncertainty in the system S since the output of the system can be predicted only up to the elements of a family of subsets $P_y = S_p(P_x \times P_z)$. Conceptually, this basic ignorance results either from the fact that one is not able to isolate the system from the environment or because one is able to observe the system only indirectly, e.g., over an insufficient period of time. Typically, the observation results in an open system under the following conditions. (1) The system is influenced by disturbances from the environment, which disturbances are only partially known. (2) The system is adaptive (in a given sense) so that any future behaviour depends upon the adaptation process. In order to properly understand this process, a very long past history of the systems behaviour has to be accounted for, which history might not be available. (3) The observation process intimately relates the system and the observer so that, during the observation process, both the system and the observer are changed.

It is apparent from these examples that the open systems represent an extremely important class since they appear in large number of fields ranging from engineering (disturbances) to biological and psychological systems (adaptation).

In order to develop a more complete theory of open systems, it is desirable to find a way to describe systems as closed systems on some newly derived objects, e.g., in terms of the power sets of the objects. Often, the systems with uncertainties are closed via probabilistic approach, i.e., by assigning probabilities to the elements of the objects. This effectively partitions an object into the classes of elements with equal probability. The difficulty in this approach is that the probabilistic structure added to the objects is often too restrictive and the information needed to specify it completely is not available.

A more appropriate approach to the study of open systems is provided by the extended and/or general topologies.

For the sake of illustration, consider an example. Let the object V be given such that V can be either an input or an output of a system. Given a set-valued set function p: $P(V) \to P(V)$, let p be an isotonic mapping, i.e.,

$$[P' \subset P''] \to [p(P') \subset p(P'')].$$

On the basis of p, one can introduce a neighbourhood function in the sense of Hammer's extended topology [22]. Given any subset $B \subset U$, a neighbourhood $u(B)$ is determined now by the function p in the following way:

$$u(B) = cpc(B)$$

where c is the complement function on V, i.e., for any $A \subset V$, $c(A) = V - A$. Assume, furthermore, that the object V is decomposable, i.e., there exists a family of subsets $P \subset P(V)$, such that P_x is a cover for V, and for all P_x

$$[y \in P_x] \leftrightarrow [y \in u(P_x)].$$

A system defined on V, as the input object, induces now an open system defined on P. Instead of being given an input, one is given a subset from the family P, $P_x \in P$, which indicates that any element in P_x might be expected to influence the system.

Properties of such a system can be now studied with reference to the property of the family P. If the elements of P are completely ordered, it is possible to establish a function on P into the set of real numbers, F: $P \to R$, and to use the numbers in the range of F for a description of the system. If the set P is a partition of V, a basis exists for introducing a probabalistic description by mapping P into the interval (0, 1). Apparently, in order to justify the introduction of the probabalistic structure on the systems objects, it is necessary to have rather detailed information about the system, which information might not be available. Therefore, a more realistic study of open systems should be based on the less structured objects.

(c) *Goal-Seeking Approach*

There is a number of reasons why general systems theory has to be concerned with decision making systems, problem solving systems, and the like. First of all, since general systems theory is based on abstraction of the behaviour of real systems, it has to be concerned with the characterization and properties of real-life goal-seeking systems. However, there also exist some formal, mathematical reasons for which one might want to describe a system via the goal-seeking approach. To appreciate that, let us first introduce the notion of a goal-seeking system on a formal basis.

Let a system $S\colon U \to Y$ be given. Introduce a new object M and represent S as a relative product of two functions $P\colon U \times M \to Y$ and $D\colon Y \times U \to M$, i.e.,

$$[(u, y) \in S] \leftrightarrow [(u, m, y) \in P] \wedge [(y, u, m) \in D].$$

Function P is usually referred to as the process while D represents a problem solver or a decision maker.

Consider the subsystem D. Again, decompose D in two subsystems $D = \{L, N\}$

$$L\colon Y \times U \to U^*, \qquad N\colon U^* \times Y \to M.$$

Under interpretation, L is the learning function while N is the selection strategy. Notice that, in general, the co-domain of L is the power set of U^*, but for the sake of simplicity it would suffice here to consider U^* as being the co-domain. Of course, this assumption affects all subsequent mappings.

Finally, take N and define the following two mappings.

$$P^*\colon U^* \times M \to Y^*, \qquad G_u\colon Y_u{}^* \times M \to Q$$

where $Y_u{}^*$ is the range of the restriction of P^* on a single element of $u \in U$, i.e.,

$$p^* \mid \{u\}\colon \times M \to Y_u{}^*.$$

P^* is to be interpreted as a model of P while G_u is the preferance (or performance) function. Given P^* and G_u the mapping D is now specified by the selection mapping

$$H\colon \{G_u\} \to M.$$

In summary then, subsystem D is specified indirectly by a family of mappings $\{L, P^*, G_u, H\}$ and, for any pair (u, y), the output of D is obtained by applying L on (u, y), $u^* = L(y, u)$, and then by applying H on u^* via P^* and G_{u^*}, i.e., the output of D is equal to the output of H where the argument for H is generated by P^* and G_{u^*}

$$H(G_{u^*}) = D(u)$$

where $G_{u^*}\colon Y^*_{u^*} \times M \to Q$ and $Y^*_{u^*} = P^*(\{u^*\} \times M)$.

We can now define a more general notion of a goal-seeking system. A general system S will be called a goal-seeking system if it can be decomposed into a family of subsystems $S = \{S_i\}$ such that at least one of its subsystems is specified as a goal-seeking subsystem $\{$i.e., in terms of $(L, P^*, G_u, H\})$.

It should be noticed that the notion of a system is not changed when the goal-seeking approach is used. Namely, the system is still a relation on $U \times Y$, i.e., $S \subset U \times Y$ and the goal-seeking subsystem D is similarly $D \subset (U \times M \times Y)$. The function D, however, is specified indirectly, i.e., whenever the arguments of D are given, (u, y), the value of D is obtained by applying L and H, which is usually referred to as solving a goal-seeking problem.

It can be shown that a goal-seeking specification can be significantly simpler than input-output description. This is apparently the case if the system is a goal-seeking system and the attempt is made to describe it on the input-output basis. Detailed discussion of this, however, will lead us outside of the scope of this paper.

REFERENCES

[1] Mesarovič, Mihajlo D., "Systems Approach to Theory of Computing Systems," reprinted from *IEEE Transactions on Military Electronics*, MIL-8, no. 2 (April, 1964).

[2] ——— "Toward a Formal Theory of Problem Solving," *Comp. of Augumentation of Human Reasoning* (Washington, D.C.: Spartan Press, 1964).

[3] Bertalanfy, L., "An Outline of General Systems Theory," *British Journal of the Philosophy of Science*, 1, no. 2 (1950).

[4] Boulding, K., "General Systems Theory-Skeleton of Science," *General Systems Yearbook*, 1 (Ann Arbor: University of Michigan, 1956).

[5] Gerard, R. W., "Entitation, Animorgs, and Other Systems," *Views on General Systems Theory* (New York: John Wiley & Sons, 1964).

[6] Chestnut, H., "The Automatic Control Business as a Controlled Process," *Neue Technik* (Zurich, 1963).

[7] Hall, A. D., "A Methodology for Systems Engineering" (New York: Van Nostrand, 1962).

[8] Mesarovič, Mihajlo D., and Eckman, Donald P., "On Some Basic Concepts of the General Systems Theory," *SRC Report 1-A-61-1*. Also appeared in *Proceedings of the 3rd International Congress on Çybernetics* (Namur, Sept. 11–15, 1961).

[9] Mesarovič, Mihajlo D., "Toward the Development of a General Systems Theory," *Neue Technik* (Zurich, 1963).

[10] ——— "Foundations for a General Systems Theory," *Views on General Systems Theory* (New York: John Wiley & Sons, 1964).

[11] Windeknecht, Thomas, "Concerning An Algebraic Theory of Systems," see this volume.

[12] Windeknecht, Thomas, and Mesarovič, Mihajlo D., "On General Dynamical Systems," to appear in "Proceedings of the Symposium on Differential Equations and Dynamical Systems" (Mayaguez, Puerto Rico, Dec. 27–30, 1965).

[13] Banerji, Ranan, "The State Space in Systems Theory," *Joint Symposium on Logic Computability and Automata Theory* (Oriskany, New York, Aug., 1965).

[14] Birta, Louis, "A Formal Approach to Concepts of Interaction," *Systems Research Center Report 81-A-65-31* (Cleveland, Ohio: Case Institute of Technology, Oct., 1965).

[15] Macko, Donald, "Formal Theory of Multi-Level Systems" (to appear).

[16] Zadeh, Lofti, "From Circuit Theory to Systems Theory," *IRE Proceedings* (May, 1962).

[17] ——— "The Concept of State in System Theory," in *Views on General Systems Theory* (New York: John Wiley & Sons, 1964).

[18] Zubov, V. I., "The Methods of A. M. Lyapunov and Their Applications" (Izdat. Leningrad. Univ. 1957). (In Russian.)

[19] Kalman, R. E., "Algebraic Theory of Linear Systems," *Proceedings of the Allerton Conference* (Illinois: Urbana, Oct. 20–22, 1965).

[20] Arbib, Michael, "A Common Framework for Automata Theory and Control Theory," *SIAM Journal on Control*, ser. A, 3, no. 3 (Philadelphia, 1965).

[21] Hammer, P. C., "Extended Topology: Set-Valued Set-Functions," *Nieuw Archief voor Wiskunde*, 10 (1962), 55–77.

Concerning an Algebraic Theory of Systems

Thomas G. Windeknecht*

INTRODUCTION

At the present time many considerations suggest the need for a theory of systems possessing considerably greater generality and conceptual power than existing theories. The present trend in engineering is to undertake the analysis and design of man-machine systems of greater and greater complexity. As larger and larger systems are modeled and studied by traditional methods, there is an increasing danger that the engineer will in the end be overwhelmed *conceptually* by the existence of profound mathematical complexity in the system model—complexity that is "inessential" for the formulation and, perhaps, for the solution of the engineer's problem concerning the system. Thus, there exists a need for the development (on the theoretical level) of greater understanding of the "essential" structure and behaviour of systems. Judging from the size and complexity of some systems of current concern to engineers, the need for greater theoretical understanding is rapidly becoming urgent.

If it is agreed that present theories do not suffice for the study of very large systems, we must inquire more deeply into available theoretical avenues that have not been fully explored. An often-heard conjecture in the systems field is relevant in this regard: "The essential structure of systems has an *algebraic* character. Furthermore, when the abstract structure of systems is so formalized, an essential understanding of the behaviour of systems can be gotten from algebraic considerations."

This paper represents an attempt to establish in some measure the validity of the above conjecture. In particular, it is our purpose here to present a rigorous basis for an algebraic theory of systems. Only the most rudimentary concepts of the proposed theory can be developed in a work of this length. Hence, it is necessary here to address ourselves only to the problem of demonstrating that a unified theory of systems is possible on an abstract level. Detailed considerations of the proposed theory must necessarily be presented elsewhere.

Our approach to the subject of systems theory, the reader will find, is com-

*Systems Research Center, Case Institute of Technology, Cleveland, Ohio.

pletely consistent with the approach of Mesarovič [1–4]. In particular, the approach is formal, definite, and mathematical and the most basic positions, e.g., a system is a relation on formal objects, the attributes of a system are logical assertions about the system relation, cause and effect objects are essential for the description of "functional systems," etc., are due to Mesarovič.

σ-NORMAL SYSTEMS

A principal object to be studied in the proposed algebraic theory of systems is the "σ-normal system," an algebraic relation defined in the following manner:

Definition 1 (Definition of a σ-normal system)

(i) A *cause object* is any non-empty semi-group.*

(ii) An *effect object* is any non-empty semi-group having left-cancellation.

(iii) A *system function on* $I \times \Omega$, where I is a cause object and Ω is an effect object, is any function $f: I \rightarrow \Omega$ (into) which has the property: for every ordered pair (x, y) of elements from I, there exists an element $\alpha \in \Omega$ such that

(1) $$f(xy) = f(x)\alpha.$$

Note. The set of system functions on $I \times \Omega$ is in general non-empty. For example, every homomorphism $h: I \rightarrow \Omega$ (into) is a system function on $I \times \Omega$. That is, since h is a homomorphism, for all $x, y \in I$,

(2) $$h(xy) = h(x)h(y).$$

Also, if $f: I \rightarrow \Omega$ and $g: I \rightarrow \Omega$ are functions such that for all $x, y \in I$,

(3) $$f(xy) = f(x)g(y),$$

then, as is easily proved, f is a system function on $I \times \Omega$.

(iv) A non-void set $\mathfrak{F}$ of a system functions on $I \times \Omega$ is said to be *complete* if for every $f \in \mathfrak{F}$ and every $x \in I$ there exists a $g \in \mathfrak{F}$ such that for all $y \in I$,

(4) $$f(xy) = f(x)g(y).$$

(v) A *σ-normal system* is the union of any complete set of system functions. That is, let $\bar{\mathfrak{F}}$ be the set of all system functions on $I \times \Omega$, where I is a cause object and Ω is an effect object. Then, if $\mathfrak{F}$ is any complete subset of $\bar{\mathfrak{F}}$,

(5) $$\Sigma = \cup_{f \in \mathfrak{F}} f$$

is a σ-normal system.

*A set A is called a *multiplicative set* if there is associated with every ordered pair (x, y) of elements of A a unique element $xy \in A$. xy is called the *product* of x and y $(x, y \in A)$. A multiplicative set A is a *semi-group* if $(xy)z = x(yz)$ for all $x, y, z \in A$. A semi-group A is said to have *left-cancellation* if for all x, y, z in A, the condition $xy = xz$ implies $y = z$. For a basic treatment of the theory of semi-groups, see Ljapin [5].

It is obvious that every σ-normal system Σ is a relation on $I \times \Omega$ where I is a cause object and Ω is an effect object, i.e.,

$$\Sigma \subseteq I \times \Omega. \tag{6}$$

TRANSITION FUNCTIONS

The concept of "transition" from one system function to another is of basic interest and importance. This concept can be conveniently developed as follows.

Let $\mathfrak{F}$ be any complete set of system functions on $I \times \Omega$. Define a relation $J \subseteq \mathfrak{F} \times I \times \mathfrak{F}$ such that for all $f, g \in \mathfrak{F}$ and all $x \in I$, $(f, x, g) \in J$ if and only if for all $y \in I$, $f(xy) = f(x)g(y)$.

THEOREM 1. *J is a function mapping $(\mathfrak{F} \times I)$ into $\mathfrak{F}$ and J has the property*

$$J(f, xy) = J(J(f, x), y) \tag{7}$$

for all $f \in \mathfrak{F}$ and all $x, y \in I$.

Proof. Since $\mathfrak{F}$ is a complete set of system functions, for every $f \in \mathfrak{F}$ and every $x \in I$, there exists a $g \in \mathfrak{F}$ such that $(f, x, g) \in J$. Suppose $(f, x, g) \in J$ and, as well, $(f, x, h) \in J$. That is, assume for all $y \in I$ that

$$f(xy) = f(x)g(y), \qquad f(xy) = f(x)h(y). \tag{8}$$

Clearly,

$$f(x)g(y) = f(x)h(y) \tag{9}$$

for all $y \in I$. However, since Ω has left-cancellation, for all $y \in I$,

$$g(y) = h(y), \tag{10}$$

i.e., $g = h$. This proves J is a function with domain $(\mathfrak{F} \times I)$ and co-domain $\mathfrak{F}$. To show that J has the given property, consider $J(J(f, x), y)$ for arbitrary $f \in \mathfrak{F}$ and $x, y \in I$. Clearly, $J(f, x)$ is defined and therefore $J(J(f, x), y)$ is defined and both are elements of $\mathfrak{F}$. Suppose that

$$J(f, x) = g \tag{11}$$

and

$$J(J(f, x), y) = h, \tag{12}$$

i.e.,

$$J(g, y) = h. \tag{13}$$

From (11), we have

$$f(xu) = f(x)g(u) \tag{14}$$

for all $u \in I$, and by (13) we have

$$g(yv) = g(y)h(v) \tag{15}$$

for all $v \in I$. Therefore, for arbitrary $z \in I$,

$$
\begin{aligned}
(16) \qquad f((xy)z) &= f(x(yz)) && (I \text{ is associative}),\\
&= f(x)g(yz) && \text{by (14)},\\
&= f(x)(g(y)h(z)) && \text{by (15)},\\
&= (f(x)g(y))h(z) && (\Omega \text{ is associative}),\\
&= f(xy)h(z) && \text{by (14)}.
\end{aligned}
$$

That is,

$$
\begin{aligned}
(17) \qquad J(f, xy) &= h && \text{by (16)},\\
&= J(J(f, x), y) && \text{by (12)}.
\end{aligned}
$$

Hence the theorem is proved.

Definition 2. Given any σ-normal system Σ, the relation J as defined above is called the *transition function* of Σ. Clearly, associated with every σ-normal system Σ is a transition function J. The identity

$$(18) \qquad J(f, xy) = J(J(f, x), y),$$

which holds for every transition function J, is the *semi-group property** of J.

COROLLARY. *Let $\mathfrak{F}$ be any complete set of system functions on $I \times \Omega$, let Σ be the σ-normal system corresponding to $\mathfrak{F}$, and let J be the transition function of Σ. Then, for all $f \in \mathfrak{F}$ and all $x, y \in I$,*

$$(19) \qquad f(xy) = f(x)J(f, x)(y).$$

SUCCESSOR TRANSFORMATIONS

The nature of transition from one system function to another can be fruitfully viewed from a slightly more abstract point-of-view.

Let $\mathfrak{F}$ be a complete set of system functions on $I \times \Omega$ whose union is the σ-normal system Σ, and let J be the transition function of Σ. For every element $x \in I$, define a function L_x: $\mathfrak{F} \rightarrow \mathfrak{F}$ such that for all $f \in \mathfrak{F}$,

$$(20) \qquad L_x(f) = J(f, x).$$

Clearly, each function L_x is a transformation on $\mathfrak{F}$. Let $\mathfrak{L}$ denote the set of all transformations on $\mathfrak{F}$ thus defined. Each element $L_x \in \mathfrak{L}$ will be called a *successor transformation* on $\mathfrak{F}$.

THEOREM 2. *$\mathfrak{L}$ is an anti-homomorphic image of I.*

Proof. We are to show that (i) $\mathfrak{L}$ is a semi-group and (ii) there exists a function Φ on I onto $\mathfrak{L}$ such that for all $x, y \in I$,

$$(21) \qquad \Phi(xy) = \Phi(y)\Phi(x).$$

*A similar property mentioned in connection with the concept of system "states" and "state-transition" is frequently referred to in the literature. For examples, see Bellman [6], Zubov [7], Kalman [8], Zadeh [9], Ginsburg [10], and Banerji [11].

It is well known [5] that any set of transformations (on an arbitrary set) which is closed under product of transformations (composition) is a semi-group. It suffices, therefore, to show that the function $L_y L_x$ belongs to $\mathfrak{L}$ for all choices of L_y and L_x in $\mathfrak{L}$, where for all $f \in \mathfrak{F}$,

$$(L_y L_x)(f) = L_y(L_x(f)). \tag{22}$$

Concerning this, we have for all $f \in \mathfrak{F}$,

$$\begin{aligned} (L_y L_x)(f) &= L_y(L_x(f)) && \text{by (22),} \\ &= L_y(J(f, x)) && \text{by (20),} \\ &= J(J(f, x), y) && \text{by (20),} \\ &= J(f, xy) && \text{by Theorem 1,} \\ &= L_{xy}(f) && \text{by (20),} \end{aligned} \tag{23}$$

i.e., $L_{xy} = L_y L_x$. Clearly, $L_{xy} \in \mathfrak{L}$ since $xy \in I$. Hence, in every case, $L_y L_x \in \mathfrak{L}$ and it is proved that $\mathfrak{L}$ is a semi-group. To complete the proof, define a function $\Phi: I \rightarrow \mathfrak{L}$ such that for all $x \in I$,

$$\Phi(x) = L_x. \tag{24}$$

Clearly, Φ is onto $\mathfrak{L}$. Also, we have for all $x, y \in I$,

$$\begin{aligned} \Phi(xy) &= L_{xy} && \text{by (24),} \\ &= L_y L_x && \text{by (23),} \\ &= \Phi(y)\Phi(x) && \text{by (24).} \end{aligned} \tag{25}$$

Hence, Φ is an anti-homomorphism. Finally, since Φ is onto $\mathfrak{L}$, $\mathfrak{L}$ is an anti-homomorphic image of I.

COROLLARY. *Let $\mathfrak{F}$ be any complete set of system functions on $I \times \Omega$ and let Σ be the σ-normal system corresponding to $\mathfrak{F}$. Then, for all $f \in \mathfrak{F}$ and all $x, y \in I$,*

$$f(xy) = f(x)L_x(f)(y) \tag{26}$$

where L_x is the successor transformation on $\mathfrak{F}$ corresponding to x.

TERMINAL SYSTEMS

Nearly all of the σ-normal systems of interest in applications are defined indirectly through some method or other of constructive specification [4]. (Various methods for such constructive specification will be our major concern in the remainder of this article.) Essentially, all of these methods for the constructive specification of σ-normal systems are such as to permit two distinct choices to be made for the associated effect object. One of these choices (on the surface) may appear to lack an appropriate semi-group operation and, in fact, may be simply a set. Either of the two choices may be the more "natural" of the two for the investigator to make, depending upon his objectives in the formal study of the system. It is well, therefore, to demonstrate that with the σ-normal system there is the possibility of studying both parts of this dichotomy of systems. The following lemma and definition make clear our concern for those systems whose outputs (i.e. "effects") are of "point-wise" interest, and show their place in the proposed theory.

LEMMA. *Every non-empty set is an effect object.*

Proof. Let Ω be an arbitrary non-empty set. For all $\alpha, \beta \in \Omega$ define the product of α and β to be

$$\alpha\beta = \beta. \tag{27}$$

Then, as is easily shown, Ω is closed and associative under this notion of product and the product (27) permits left-cancellation, i.e., Ω is an effect object. (We leave the details to the reader.)

An effect object Ω shall be called *trivial* if and only if for all $\alpha, \beta \in \Omega$, $\alpha\beta = \beta$.

Definition 3. A σ-normal system Σ is said to be *terminal* if and only if the effect object Ω associated with Σ is trivial. If a σ-normal system is not terminal, it is a *non-terminal system.*

COROLLARY. *Let $\mathfrak{F}$ be any complete set of system functions on $I \times \Omega$ and let Σ be the σ-normal system corresponding to $\mathfrak{F}$. If Σ is terminal, then for all $f \in \mathfrak{F}$ and all x, $y \in I$,*

$$f(xy) = J(f, x)(y), \qquad f(xy) = L_x(f)(y), \tag{28}$$

where J is the transition function of Σ and L_x is the successor transformation on $\mathfrak{F}$ corresponding to x.

We turn now to the important task of demonstrating the relevance of the study of σ-normal systems to various areas relating to systems theory. In particular, we consider areas in which the investigation of systems through the use of mathematical models is firmly established, and which have within them some quite general concepts of system.

ABSTRACT MACHINES

The abstract machine is a generalization of the sequential machine introduced some time ago by Ginsburg [10]. The concept of the abstract machine is closely related to the concept of a σ-normal system. In developing the relationship, our intent is to show the relevance of the proposed theory of systems to the theory of sequential machines [12–14].

Definition 4. An *abstract machine* M is a 5-tuple $\langle S, I, \Omega, \delta, \omega \rangle$ satisfying the following axioms:

(i) S is a non-empty set.
(ii) I and Ω are non-empty semi-groups; Ω has left-cancellation.
(iii) δ is a mapping of $S \times I$ into S such that

$$\delta(s, xy) = \delta(\delta(s, x), y), \tag{29}$$

for all $s \in S$ and all $x, y \in I$.

(iv) ω is a mapping of $S \times I$ into Ω such that

$$\omega(s, xy) = \omega(s, x)\omega(\delta(s, x), y), \tag{30}$$

for all $s \in S$ and all $x, y \in I$.

The elements of S, I, and Ω are called states, inputs, and outputs, respectively.

Let M be an arbitrary abstract machine. Define a relation $\Sigma \subseteq I \times \Omega$ such that

$$\Sigma = \bigcup_{s \in S} \bigcup_{x \in I} (x, \omega(s, x)). \tag{31}$$

THEOREM 3. *Σ is a σ-normal system.*

Proof. For every $s \in S$, define a function $f_s: I \to \Omega$ such that for all $x \in I$,

$$f_s(x) = \omega(s, x). \tag{32}$$

Let $\mathfrak{F}$ be the set of all functions f_s thus defined. We shall prove that $\mathfrak{F}$ is a complete set of system functions on $I \times \Omega$. Note, first, by definition 4, I is a cause object and Ω is an effect object. Choose any $s \in S$ and any $x \in I$, and consider $f_s(xy)$ for arbitrary $y \in I$. By part (iv) of definition 4, we have

$$\omega(s, xy) = \omega(s, x)\omega(\delta(s, x), y). \tag{33}$$

Therefore, using (33) we have for any $y \in I$,

$$f_s(xy) = f_s(x)f_{\delta(s,x)}(y). \tag{34}$$

That is, for every $f_s \in \mathfrak{F}$ and every $x \in I$, there exists a function $f_{s'} \in \mathfrak{F}$ such that $f_s(xy) = f_s(x)f_{s'}(y)$ for all $y \in I$. In particular, $f_{s'} = f_{\delta(s,x)}$. Therefore, $\mathfrak{F}$ is complete. Now, by our definition, Σ is given by

$$\Sigma = \bigcup_{s \in S} \bigcup_{x \in I} (x, \omega(s, x)). \tag{35}$$

Clearly,

$$f_s = \bigcup_{x \in I} (x, \omega(s, x)). \tag{36}$$

Therefore,

$$\Sigma = \bigcup_{s \in S} f_s = \bigcup_{f_s \in \mathfrak{F}} f_s. \tag{37}$$

The latter is by construction. Therefore, since Σ is the union of a complete set of system functions on $I \times \Omega$, Σ is a σ-normal system.

Theorem 3 establishes a close connection between mathematical machine theory and the algebraic theory of systems being discussed. In particular, theorem

3 makes it apparent that every abstract machine defines a σ-normal system. The converse is also true, and the proof is left as an exercise. It should be made clear that we do not wish to equate the proposed algebraic theory of systems with mathematical machine theory. Our interest here roams far afield from mathematical machine theory, e.g., to systems described by differential equations. Our intention is to demonstrate that a unified theory of systems is possible on an abstract level. Certain general results of methematical machine theory may be applicable to areas not generally thought to be connected with it, and conversely.

It will be evident to the reader that a fundamental distinction between mathematical machine theory and the proposed theory of systems lies in our emphasis on the relation Σ is a fundamental object to be studied. The reason for this should be recognized. It appears to us that the objects of study in systems theory (and, in fact, in mathematical machine theory) are the properties such systems can and do have in applications. In general, the properties of systems are formal statements about the relation Σ and its transition function J. An undue emphasis on arbitrary 5-tuples, etc., tends to obscure the basic formal issues.

For example, at the start in the theory of sequential machines and abstract machines, the 5-tuple introduces a concept of identity of machines which is *not* the intuitive one (or ones) of importance in applications. This is evidenced by the fact that presentations of length in this field normally begin with the introduction of another, more intuitive concept of equivalent machines.* For example, in sequential machine theory, the results on minimal machines are based upon this intuitive notion of equivalence. With the σ-normal system, this problem is avoided as can be demonstrated as follows.

Definition 5. (Ginsburg) Let $M = \langle S, I, \Omega, \delta, \omega\rangle$ and $M' = \langle S', I, \Omega', \delta', \omega'\rangle$ be abstract machines. A state $s \in S$ is said to be equivalent to a state $s' \in S'$, i.e., $s \equiv s'$, if and only if for all $x \in I$,

$$\omega(s, x) = \omega'(s', x). \tag{38}$$

The abstract machines M and M' are said to be *equivalent*, i.e., $M \cong M'$, if and only if for every state $s \in S$ there exists a state $s' \in S'$ and for every state $s' \in S'$ there exists a state $s \in S$ such that $s \equiv s'$. (It can be proved that the relations $\equiv$ and $\cong$ are equivalence relations.)

Let Σ and Σ' be the σ-normal systems corresponding to M and M', respectively. That is, define

$$\Sigma = \bigcup_{s\in S} \bigcup_{x\in I} (x, \omega(s, x)), \tag{39}$$

$$\Sigma' = \bigcup_{s'\in S'} \bigcup_{x\in I} (x, \omega'(s', x)). \tag{40}$$

THEOREM 4. *If* $M \cong M'$ *then* $\Sigma = \Sigma'$.

Proof. As in the proof of theorem 3, let $\mathfrak{F}$ be the complete set of system

*Indeed, a number of theorems are required in order to make up for the original definition.

functions on $I \times \Omega$ corresponding to M, i.e., for every $s \in S$, define a function $f_s \in \mathfrak{F}$ (f_s: $I \to \Omega$) such that for all $x \in I$,

$$f_s(x) = \omega(s, x). \tag{41}$$

Similarly, let $\mathfrak{F}'$ be the complete set of system functions on $I \times \Omega'$ corresponding to M', i.e., for every $s' \in S'$, define a function $g_{s'} \in \mathfrak{F}'$ ($g_{s'}$: $I \to \Omega'$) such that for all $x \in I$,

$$g_{s'}(x) = \omega'(s', x). \tag{42}$$

Choose any $s \in S$. Since $M \cong M'$, there exists a state $s' \in S'$ such that $s \equiv s'$, i.e., such that for all $x \in I$,

$$\omega(s, x) = \omega'(s', x). \tag{43}$$

Combining (41), (42), and (43), this yields for all $x \in I$,

$$f_s(x) = g_{s'}(x), \tag{44}$$

i.e., $f_s = g_{s'}$. That is, for every $f_s \in \mathfrak{F}$, there exists a function $g_{s'} \in \mathfrak{F}'$ such that $f_s = g_{s'}$. However, as was mentioned in the proof of theorem 3,

$$\Sigma = \bigcup_{f_s \in \mathfrak{F}} f_s, \tag{45}$$

and

$$\Sigma' = \bigcup_{g_{s'} \in \mathfrak{F}'} g_{s'}. \tag{46}$$

Hence, we have proved $\Sigma \subseteq \Sigma'$. Conversely, it can easily be seen that for every $g_{s'} \in \mathfrak{F}'$, there exists an $f_s \in \mathfrak{F}$ such that $g_{s'} = f_s$. Hence, $\Sigma' \subseteq \Sigma$. Therefore, $\Sigma = \Sigma'$.

To summarize: the concept of identity on the class of σ-normal systems with a common cause object I corresponds precisely to the notion of equivalence of abstract machines.

GENERALIZED AUTOMATA

The study of σ-normal systems can be conveniently related to automata theory [13] in the following manner.

Definition 6. A *generalized automaton* is an abstract machine $\langle S, I, \Omega, \delta, \omega \rangle$ with the following additional properties.

(i) Ω is trivial.

(ii) $S = \Omega$.

(iii) $\delta = \omega$.

The consistency of these latter conditions (i)–(iii) with the former conditions (namely, $\delta(s, xy) = \delta(\delta(s, x), y)$ and $\omega(s, xy) = \omega(s, x)\omega(\delta(s, x), y))$ can be readily established. The following corollary to theorem 3 is immediate.

COROLLARY. *If $\langle S, I, \Omega, \delta, \omega\rangle$ is any generalized automaton, the relation*

(47) $$\Sigma = \bigcup_{s \in S} \bigcup_{x \in I} (x, \omega(s, x))$$

is a terminal σ-normal system.

The reader can verify that the generalized automaton bears essentially the same relationship to the automaton (as the latter is formulated by Nelson [13], for example) as the abstract machine does to the sequential machine. It is a characteristic of automata theory that most authors treat the automaton as a terminal system. The same is more or less true of the sequential machine, although Ginsburg's generalization of it, the abstract machine, leads to a non-terminal σ-normal system in general.

TURING MACHINES

The study of σ-normal systems is relevant to the theory of Turing machines; hence, to recursive function theory, the theory of computability [15], and to the theoretical study of computing systems. We shall here demonstrate this by proving that every Turing machine is a generalized automaton, and hence an abstract machine. By this chain of development, it follows that every Turing machine defines a σ-normal system.

An alphabet is a finite set (of symbols). The set W_A of all *words* on a finite alphabet A is the set of all finite sequences on A. More precisely, it is the free semi-group over A [5].

We shall take the set $W_{Q,A}$ of all *well-formed words* on an ordered pair (Q, A) of alphabets to be the class of all elements of the form wqw', where $q \in Q$ and $w, w' \in W_A$. An element e of alphabet A is a *unit* on $W_{Q,A}$ if and only if for all $\alpha \in W_{Q,A}$

(48) $$\alpha e = e\alpha = \alpha.$$

Definition 7. A *Turing machine* is an ordered triple $\langle Q, A, F\rangle$ where:

(i) Q is a non-void alphabet.

(ii) A is a non-void alphabet containing a unit, e, on the set $W_{Q,A}$ of all well-formed words on (Q, A). Q and A are disjoint.

(iii) F is a transformation mapping the set $W_{Q,A}$ into itself such that for all $q \in Q$ and all $a, b \in A$ there exists some $q' \in Q$ and some $b' \in A$ such that either

(49) $$F(aqb) = aq'b',$$

or

$$F(aqb) = q'ab, \tag{50}$$

or

$$F(aqb) = abq', \tag{51}$$

and for all $w, w' \in W_A$, all $a, b \in A$, and every $q \in Q$,

$$F(waqbw') = wF(aqb)w'.^* \tag{52}$$

Let $T = \langle Q, A, F\rangle$ be an arbitrary Turing machine, and let $W_{Q,A}$ be the set of all well-formed words on (Q, A). Let P^+ denote the set of all positive integers. Define a function

$$\omega: W_{Q,A} \times P^+ \to W_{Q,A}, \tag{53}$$

such that for all $\alpha \in W_{Q,A}$

$$\begin{cases} \omega(\alpha, 1) = F(\alpha), \\ \omega(\alpha, n+1) = F(\omega(\alpha, n)) \qquad (n = 1, 2, \ldots). \end{cases} \tag{54}$$

Consider the quintuple $M = \langle W_{Q,A}, P^+, W_{Q,A}, \omega, \omega\rangle$.

THEOREM 5. *M is a generalized automaton.*

Proof. $W_{Q,A}$ is a set. Hence, by the above lemma, it is a trivial effect object, i.e., $W_{Q,A}$ is trivially a semi-group with left-cancellation. Conditions (ii) and (iii) of definition 6 are satisfied trivially. Now, P^+ is obviously a semi-group under ordinary addition of integers. Hence, it remains only to show that the function ω defined in (54) has the property

$$\omega(\alpha, n + m) = \omega(\omega(\alpha, n), m), \tag{55}$$

for all $\alpha \in W_{Q,A}$ and all positive integers n and m. This can be done by a simple induction on m using the definition (54), which is left to the reader. Thus the theorem is proved.

COROLLARY. *Every Turing machine $\langle Q, A, F\rangle$ defines a terminal σ-normal system*

$$\Sigma = \bigcup_{\alpha \in W_{Q,A}} \bigcup_{n \in P^+} (n, \omega(\alpha, n)), \tag{56}$$

where $W_{Q,A}$ is the set of all well-formed words on (Q, A) and P^+ is the set of positive integers.

In developing the theory of Turing machines, the concept of the Turing machine itself is ordinarily supplemented by introduction of the concept of a "computation by a Turing machine." Relative to definition 7, this reads:

*This definition is non-standard, but is in essential agreement with the definitions given by Davis [15] and Ritchie [16].

Definition 8. A *computation* on a well-formed word $\alpha_1 \in W_{Q,A}$ by a Turing machine $T = \langle Q, A, F\rangle$ is a finite sequence $\alpha_1, \alpha_2, \ldots, \alpha_n$ of well-formed words in $W_{Q,A}$ such that

$$(57)\qquad \begin{cases} \alpha_{i+1} = F(\alpha_i) & \text{for } 1 \leqslant i < n, \\ \alpha_n = F(\alpha_n), & \\ \alpha_i \neq F(\alpha_i) & \text{for } i < n^*. \end{cases}$$

The integer n is the *length* of the computation. The well-formed word α_n is called the *resultant* of α_1 with rsepect to T, i.e.,

$$(58)\qquad \alpha_n = \mathrm{Res}_T(\alpha_1).$$

The following proposition is obvious.

COROLLARY. *Given a Turing machine* $T = \langle Q, A, F\rangle$ *and a well-formed word* $\alpha_1 \in W_{Q,A}$, *there exists a computation by* T *on* α_1 *if and only if there exists a positive integer* $n < \infty$ *such that*

$$(59)\qquad \omega(\alpha_1, n+1) = \omega(\alpha_1, n),$$

where ω *is as defined in* (54). *Also,*

$$(60)\qquad \mathrm{Res}_T(\alpha_1) = \omega(\alpha_1, n).$$

In developing the theory of systems on an abstract level, one can expect certain similar concepts existing in different areas to be formally related. Viewed in the given manner, the issue of the existence of a computation on a well-formed word α_1 by a Turing machine T (i.e., the "halting problem" [15]) is very nearly the same as the issue of the existence of "finite stability" starting from a given system state as defined elsewhere, e.g., in the input-output theory of linear differential equation systems [17]. In particular, the condition (59) implies for all $m \geqslant 1$,

$$(61)\qquad \omega(\alpha_1, n+m) = \omega(\alpha_1, n).$$

In the latter field, an element $\omega(\alpha_1, n)$ with the property (61) would be referred to as an "equilibrium position" of the system.

In passing, we note that there exists a clear relationship between the Turing machine and the concept of a normal algorithm [18]. Thereby, the σ-normal system can be proved to be relevant to the study of an important class of formal systems. This relationship is further enhanced through Nelson's treatment [13] of automata theory as a special case of a broader theory of formal systems. Thus, it may be possible to formally identify the field of mathematical linguistics, for example, with the proposed algebraic theory of systems.

*The reader knowledgeable in Turing machine theory will note that, since e is defined to be an identity element (i.e., a unit) on $W_{Q,A}$ and not on W_A, well-formed words of the form wqw', $wqew'$, and $weqw'$ are not in general identical. Hence, an indefinitely long translation of q to the right or left does *not* by definition 8 yield a computation. This is in agreement with the usual convention.

DYNAMICAL SYSTEMS

The proposed algebraic theory of systems is directly relevant to the general input-output theory of systems defined by sets of finite difference equations and by sets of ordinary (and partial) differential equations. In connection with his treatment of the Lyapunov stability theory of systems defined by sets of autonomous (i.e., unforced) ordinary differential equations, Zubov [7] employs the following general definition of a dynamical system.

Definition 9. A *dynamical system* S_D in a metric space D is a one-parameter family of transformations $F_t\colon D \to D$ (into) (where t ranges over the real line R) having the following properties.

(i) For all $p \in D$,

$$F_0(p) = p. \tag{62}$$

(ii) $F_t(p)$ is continuous on $R \times D$ into D in the sense that for every real $\epsilon > 0$ there exist real numbers $\delta_1 > 0$ and $\delta_2 > 0$ such that

$$\rho(F_t(p), F_t{}'(p')) < \epsilon, \tag{63}$$

whenever

$$\rho(p, p') < \delta_1 \quad \text{and} \quad |t - t'| < \delta_2, \tag{64}$$

where ρ is the metric on D.

(iii) For all $p \in D$ and all $t, t' \in R$

$$F_t(F_{t'}(p)) = F_{t+t'}(p). \tag{65}$$

It is easy to see that S_D is a group of transformations. That is (i) S_D is closed under product of transformations (composition), i.e., $F_t\, F_{t'} = F_{t+t'}$, (ii) product of transformations is associative, (iii) F_0 serves as identity on $S_{D'}$, i.e., $F_0\, F_t = F_t\, F_0 = F_t$, and (iv) F_{-t} is the inverse of F_t for all $t \in R$, i.e.,

$$F_{-t}\, F_t = F_t\, F_{-t} = F_0. \tag{66}$$

Let S_D be an arbitrary dynamical system on a metric space D. With each element $p \in D$, associate a function $G_p\colon R \to D$ such that for all $t \in R$,

$$G_p(t) = F_t(p). \tag{67}$$

Let $\mathfrak{F}$ be the set of all functions G_p thus defined. Clearly, by (ii) of definition 9, $G_p(t)$ is a continuous function on $D \times R$ into D. Define Σ to be the union of the functions in $\mathfrak{F}$, i.e.,

$$\Sigma = \bigcup_{G_p \in \mathfrak{F}} G_p. \tag{68}$$

Then, it is immediate that Σ is a relation on $R \times D$. Also, we have the following.

THEOREM 6. *Σ is a terminal σ-normal system.*

Proof. R is obviously a semi-group under addition, i.e., a cause object. D, being a metric space, is a set. Hence, it is a trivial effect object. Choose any $G_p \in \mathfrak{F}$ and any real number t. Then, for all real t',

$$\begin{aligned} G_p(t + t') &= F_{t+t'}(p) && \text{by (67)}, \\ &= F_t(F_{t'}(p)) && \text{by definition 9}, \\ &= F_{t'}(F_t(p)), && \\ &= F_{t'}(G_p(t)) && \text{by (67)}, \\ &= G_{Gp(t)}(t') && \text{by (67)}, \\ &= G_p(t)G_{Gp(t)}(t') && (D \text{ is trivial}), \end{aligned} \tag{69}$$

i.e., for every $G_p \in \mathfrak{F}$ and every real t, there exists a G_p, such that

$$G_p(t + t') = G_p(t)G_{p'}(t')$$

for all real t'. In particular, $p' = G_p(t)$. Therefore, $\mathfrak{F}$ is a complete set of system functions on $R \times D$, where D is a trivial effect object. Hence, Σ is a terminal σ-normal system.

The above definition of a dynamical system can easily be modified to include more general cases of interest without upsetting theorem 6. For example, it suffices to assume that D is a topological space in order to assert the continuity of the transformations F_t. Also, and more importantly, the real line R in definition 9 may be replaced by any (Abelian) subgroup of R, e.g., the set of integers, without overturning theorem 6. With such a modification, the given concept of dynamical system includes certain "discrete" systems, e.g., systems defined by sets of (autonomous) finite-difference equations. Thereby, the concept of the σ-normal system can be made relevant to a class of systems defined in terms of topological spaces and to certain systems with a "discrete-time" character. In fact, given certain minor modifications in the definition of dynamical system, it can be proved that every Turing machine is a dynamical system.

Zubov in [7] discusses general examples of systems defined in function spaces (through the use of partial differential equations) which are dynamical systems according to definition 9. Thus, the relevance of the σ-normal system to such mathematical objects can be established.

An essential generalization of Zubov's definition of a dynamical system is still required, however, in order to include non-autonomous (i.e., forced) systems defined by differential equations and by difference equations. A proposal along these lines is contained in the following. We require as a preliminary a brief mathematical development for which we were unable to find an adequate reference.

CONSTRUCTIVE CAUSE AND EFFECT OBJECTS

Let T be any subset of the positive real numbers R which contains 0 (zero) and is closed under addition. Denote by Λ_T the class of non-empty half-open intervals

$\overline{(0, t)}$ on T, i.e., the class of elements $(0, t) \cap T$ where $t \in T$. Given an arbitrary set K, denote as $\bar{I} = \bar{I}(T, K)$ the class of all functions x mapping elements of Λ_T into K, i.e., define $x \in \bar{I}$ if and only if $x: \lambda \to K$ and $\lambda \in \Lambda_T$.

With every function $x \in \bar{I}$ there is uniquely associated a real number $\mu(x)$ called the *measure* of x such that $\mu(x) = t$ if and only if $x: \overline{[0, t)} \to K$. Clearly, $\mu: \bar{I} \to T$. In fact, as can be readily proved, if $x \in \bar{I}$ and $x: \lambda \to K$, then

$$\mu(x) = \min_{t \in T} (T - \lambda). \tag{70}$$

Define as the product of given functions $x, y \in \bar{I}$ the function xy such that

$$xy(t) = \begin{cases} x(t) & \text{for } 0 \leqslant t < \mu(x), \\ y(t - \mu(x)) & \text{for } \mu(x) \leqslant t < \mu(x) + \mu(y), \end{cases} \tag{71}$$

and xy is undefined otherwise. We shall refer to this operation as (*functional*) *juxtaposition*. Clearly, xy is well-defined for all $x, y \in \bar{I}$. In fact, xy is a function with domain $\overline{[0, \mu(x) + \mu(y))}$ and co-domain K. Further, $\overline{[0, \mu(x) + \mu(y))} \in \Lambda_T$ since T is closed under addition. Therefore, we have proved:

LEMMA. *$\bar{I}$ is a multiplicative set.*

LEMMA. *$\bar{I}$ is an effect (cause) object.*

Proof. The proof consists of showing that juxtaposition on $\bar{I}$ is associative and permits left-cancellation. This is straightforward and is left to the reader.

It is an immediate corollary of this second lemma that any subset $I \subseteq \bar{I}$ which is closed under juxtaposition is an effect (hence, also, a cause) object. Let $\chi(T, K)$ denote the class of all subsets of $\bar{I}(T, K)$ which are effect (cause) objects. Any element $I \in \chi(T, K)$ is called a *constructive effect* (*cause*) *object* on T, K. The set T will be called a *time set*. The reader can verify the fact that the following mathematical objects are effect (cause) objects:

(i) The set W_A of all finite sequences on an alphabet, A. That is, $W_A = \bar{I}(P^+, A)$, where P^+ denotes the set of positive integers (including zero).

(ii) The set Z of all finite sequences of words on a finite alphabet A (i.e., $Z = \bar{I}(P^+, W_A)$).

(iii) The set of all finite sequences of n-tuples on $\{0, 1\}$.

(iv) The set of all finite sequences of n-tuples of reals.

(v) The set B of all bounded sequences of ordered n-tuples of real numbers. that is, $B \in \chi(P^+, R^n)$ where P^+ denotes the set of positive integers (including zero) and R^n is Euclidean n-space [19]. Note that the boundedness of sequences in B is preserved under juxtaposition.

(vi) The set C of all bounded, piecewise continuous functions mapping half-open intervals $[0, t)$ of the real line into R_n. That is, $C \in \chi(R^+, R_n)$. Note that boundedness and piecewise continuity on C are preserved under juxtaposition.

(These last two examples are particularly important in the input-output theory of systems defined by sets of finite-difference equations and by sets of ordinary differential equations, respectively.

FORCED DYNAMICAL SYSTEMS*

Definition 10. A (*complete*) *forced dynamical system* S_T is a pair of mappings $\psi: S \times T \times I \to \Omega$ and $\phi: S \times T \times I \to S$ where

(i) S is an arbitrary set.

(ii) T is a time set.

(iii) I and Ω are, respectively, constructive cause and effect objects, on T, K and T, K'.

(iv) ψ and ϕ are continuous functions with respect to given topologies defined on S, T, I, and Ω.

(v) For all $s \in S$, $t \in T$, and $x, y \in I$, ψ has the property

$$\psi(s, t, xy) = \psi(s, t, x)\psi(\phi(s, t, x), t + \mu(x), y), \tag{72}$$

where the indicated products are functional juxtaposition on I and Ω, respectively, and where $\mu(x)$ is the measure of x.

S is called a set of states, I is a (complete) set of inputs, and Ω is the output set of S_T.

Let S_T be a (complete) forced dynamical system as above. Define a relation

$$\Sigma = \bigcup_{t \in T} \bigcup_{s \in S} \bigcup_{x \in I} (x, \psi(s, t, x)).$$

THEOREM 7. *Σ is a σ-normal system.*

Proof. With every ordered pair $(s, t) \in S \times T$ associate a function $F_{s,t}: I \to \Omega$ such that for all $x \in I$,

$$F_{s,t}(x) = \psi(s, t, x). \tag{74}$$

Let $\mathfrak{F}$ denote the set of all functions thus defined. It is immediate (as the reader can verify) by (v) of definition 10 that $\mathfrak{F}$ is a complete set of system functions on $I \times \Omega$. Further, it is easily proved that

$$\Sigma = \bigcup_{F_{s,t} \in \mathfrak{F}} F_{s,t}. \tag{75}$$

Hence, Σ is a σ-normal system.

*References [8] and [9] have been particularly useful here.

CONCLUSION

The σ-normal system is a loosely-structured characterization of the input-output relation that is naturally associated with a variety of mathematical objects of proven use in the study of systems. These mathematical objects seem invariably to possess a constructive character. It is this that leads one to identify the input-output relation (i.e., that which is constructed) with the abstract system itself. The loose structure of the σ-normal system makes it possible (as has been demonstrated in the foregoing) to formally identify it with a wide variety of mathematical objects in various areas of systems study—from mathematical machine theory to systems in functions spaces. However, the loose structure that is assumed for the σ-normal system is still *non-trivial.*

Because of this, there is reason to believe that a suitable theory of the σ-normal system will likewise be non-trivial with respect to the study of systems—particularly in developing tools to aid the systems investigator conceptually. There is no valid reason now to take the position that, with respect to the "real" problems in systems work, such a theory will be sterile of useful results due to the absence of more sophisticated structure in the objects to be studied. We need only point to the existence in mathematics of a theory of considerable depth and breadth about semi-groups, such as is revealed in reference [5], in order to refute this argument.

Also, there has recently appeared in mathematical machine theory, for example, a quite non-trivial algebraic theory of decomposition [20]. Finally, we expect the proposed theory to be an axiomatic theory, i.e., a theory where additional structure is assumed (when it is called for) for the development of various concepts of interest in applications.

At present, it can be argued that what is really needed in the systems field is a general theory in which assumptions about structure of systems are very closely related to theorems about behaviour. This may counter an unfortunate trend in the study of systems to apply probability theory to models having enormous mathematical structure because these models fail to reflect reality. An alternative is to retreat from the use of such detailed models (and from the corresponding assumptions about the structure of the "real" system) and properly consider the use of more abstract models having less structure.

ACKNOWLEDGMENTS

The author gratefully acknowledges the leadership, aid, and encouragement of M. D. Mesarovič in the conduct of the present research. Conversations with R. B. Banerji, N. Jones, and L. Birta have greatly enhanced our efforts. The research has been partially supported by ONR contract no. 1141(12).

REFERENCES

[1] Mesarovič, M. D., and Eckman, D. P., "On Some Basic Concepts of General Systems Theory," *Proceedings of the Third International Conference on Cybernetics* (Namur, Belgium, 1961).

[2] Mesarovič, M. D., "Toward the Development of a General Systems Theory," *Neue Technik* (Zurich, Aug., 1963).

[3] ——— "Foundations for a General Systems Theory," *Views on General Systems Theory* (New York: John Wiley & Sons, 1964), 1–24.

[4] ——— "General Systems Theory," notes on the class lectures given at the Case Institute of Technology, Cleveland, Ohio, 1963.

[5] Ljapin, E. S., *Semi-Groups* (English translation of the Russian original; Providence, Rhode Island: American Mathematical Society, 1963).

[6] Bellman, R., *Adaptive Control Processes* (Princeton: Princeton University Press, 1961).

[7] Zubov, V. I., *The Methods of A. M. Lyapunov and Their Applications* (AEC-tr-4439, Oct., 1961).

[8] Kalman, R., "Mathematical Description of Linear Dynamical Systems," RIAS Technical Report 62-18 (Nov., 1962).

[9] Zadeh, L. A., "The Concept of State in System Theory," *Views on General Systems Theory* (New York: John Wiley & Sons, 1964), 39–50.

[10] Ginsburg, S., "Abstract Machines: A Generalization of Sequential Machines," *Proc. Symp. Math. Theory of Automata* (New York: Polytech. Inst. of Brooklyn, April, 1962), 125–38.

[11] Banerji, R. B., "The State Space as a Unifying Principle in Systems Theory" (Cleveland, Ohio: Case Institute of Technology, 1965).

[12] Gill, A., *Introduction to the Theory of Finite-State Machines* (New York: McGraw-Hill Book Co., 1962).

[13] Nelson, R. J., "Introduction to Automata" (Cleveland, Ohio: Case Institute of Technology, 1965).

[14] Ginsburg, S., *An Introduction to Mathematical Machine Theory* (Addison-Wesley, 1962).

[15] Davis, M., *Computability and Unsolvability* (New York: McGraw-Hill Book Co., 1958).

[16] Ritchie, R. W., "Classes of Predictably Computable Functions," *Trans. Am. Math. Soc.*, 106 (1963), 139–73.

[17] Dorato, P., "Short-Time Stability in Linear, Time-Varying Systems," *OSR Report No. AFOSR 748* (May 9, 1961).

[18] Markov, A. A., *Theory of Algorithms*, Academy of Sciences of U.S.S.R. (English translation by NSF and U.S. Dept. of Commerce), 42 (Moscow, 1954).

[19] Kolmogorov, A. N., and Fomin, S. V., *Elements of the Theory of Functions and Functional Analysis*, 1 (English translation of the Russian original of 1954) (Rochester, New York: Graylock Press, 1957).

[20] Krone, K. B., and Rhodes, J. L., "Algebraic Theory of Machines," *Proc. Symp. Math. Theory of Automata* (New York: Polytech. Inst. of Brooklyn, April, 1962), 341–78.

[21] Windeknecht, T. G., "Toward a General Theory of Discrete Systems" (Cleveland, Ohio: Case Institute of Technology, Dec., 1964).